Hands-On Mobile D
with .NET Core

Build cross-platform mobile applications with Xamarin, Visual
Studio 2019, and .NET Core 3

Can Bilgin

Packt>

BIRMINGHAM - MUMBAI

Hands-On Mobile Development with .NET Core

Copyright © 2019 Packt Publishing

Commissioning Editor: Kunal Choudhari
Acquisition Editor: Devanshi Doshi
Content Development Editor: Keagan Carneiro
Technical Editor: Leena Patil
Copy Editor: Safis Editing
Project Coordinator: Kinjal Bari
Proofreader: Safis Editing
Indexer: Manju Arasan
Graphics: Alishon Mendonsa
Production Coordinator: Shraddha Falebhai

First published: May 2019

Production reference: 1310519

Published by Packt Publishing Ltd.
Livery Place
35 Livery Street
Birmingham
B3 2PB, UK.

ISBN 978-1-78953-851-9

www.packtpub.com

Mapt

Mapt is an online digital library that gives you full access to over 5,000 books and videos, as well as industry leading tools to help you plan your personal development and advance your career. For more information, please visit our website.

Why subscribe?

- Spend less time learning and more time coding with practical eBooks and Videos from over 4,000 industry professionals

- Improve your learning with Skill Plans built especially for you

- Get a free eBook or video every month

- Mapt is fully searchable

- Copy and paste, print, and bookmark content

Packt.com

Did you know that Packt offers eBook versions of every book published, with PDF and ePub files available? You can upgrade to the eBook version at www.packt.com and as a print book customer, you are entitled to a discount on the eBook copy. Get in touch with us at customercare@packtpub.com for more details.

At www.packt.com, you can also read a collection of free technical articles, sign up for a range of free newsletters, and receive exclusive discounts and offers on Packt books and eBooks.

Contributors

About the author

Can Bilgin currently works with Authority Partners Inc. as a Solution Architect. He has been working in the software industry, primarily with Microsoft technologies, for over a decade and has been recognized as a Microsoft Most Valuable Professional for his technical contributions between 2014 and 2018. In this period, he took key roles in projects for high profile clients using technologies such as BizTalk, SharePoint, Dynamics CRM, Xamarin, WCF, Azure Serverless and other web/cloud technologies. He is passionate about mobile and IoT development using the modern tools available for developers.

He shares his experience on his blog, social media and through speaking engagements in local and international community events.

Dedicated to my beloved wife, Sanja Grebovic Bilgin.

About the reviewer

Ahmed Ilyas is a BENG degree holder from Napier University, Edinburgh, Scotland. He is a major in software development with 19 years of professional experience. After leaving Microsoft, currently focusing on his consultancy company, Sandler Ltd, and Sandler Software LLC, offering the best possible real-world solutions for issues faced by a plethora of industries. Using the Microsoft stack, his venture focuses on suggesting and applying best practices, patterns, and software solutions to meet their clients' needs, targeting long-term stability and compliance in the dynamic software industry. He has been awarded with an MVP in C# by Microsoft on three occasions for providing excellent real-world solutions to problems faced by developers.

> *I would like to thank the author and publisher of this book for giving me the great honor and privilege of reviewing it. I would also like to thank my client base, and especially Microsoft Corporation and my colleagues over there, for enabling me to become a reputable leader as a software developer in the industry, which is my passion.*

Packt is searching for authors like you

If you're interested in becoming an author for Packt, please visit `authors.packtpub.com` and apply today. We have worked with thousands of developers and tech professionals, just like you, to help them share their insight with the global tech community. You can make a general application, apply for a specific hot topic that we are recruiting an author for, or submit your own idea.

Table of Contents

Section 3: Azure Cloud Services

Preface

In this book, you will learn how to design and develop highly attractive, maintainable, and robust mobile applications for multiple platforms, including iOS, Android, and UWP, with the toolset provided by Microsoft using Xamarin, .NET Core, and Azure Cloud Services.

Who this book is for

This book is for mobile developers who wish to develop cross-platform mobile applications. Programming experience with C# is required. Some knowledge and understanding of core elements and of cross-platform application development with .NET is required.

What this book covers

Chapter 1, *Getting Started with .NET Core*, gives you a brief introduction to .NET Core while explaining the different tiers of the .NET Core infrastructure. Languages, runtimes, and extensions that can be used together with .NET Core will be discussed and analyzed.

Chapter 2, *Xamarin, Mono, and .NET Standard*, tries to explain the relationship between .NET Core and Xamarin. You will learn about how the Xamarin source code is executed with MonoTouch on iOS and Mono Runtime on Android.

Chapter 3, *Universal Windows Platform*, discusses the components that allow UWP apps to be portable within the Windows 10 ecosystem and how they are associated with .NET Core.

Chapter 4, *Developing Mobile Applications with Xamarin*, explains Xamarin and Xamarin.Forms development strategies, and we will create a Xamarin.Forms application that we will develop throughout the remainder of the book. We will also discuss the architectural models that might help us along the way.

Chapter 5, *UI Development with Xamarin*, takes a look at certain UI patterns that allow developers and UX designers to create a compromise between the user expectations and product demands in order to create a platform and product with a consistent user experience across platforms.

Chapter 6, *Customizing Xamarin.Forms*, goes through the steps and procedures of customizing Xamarin.Forms without compromising the performance or user experience. Some of the features that will be analyzed include effects, behaviors, extensions, and custom renderers.

Chapter 7, *Azure Services for Mobile Applications*, discusses the fact that there are a number of services that are offered as services (SaaS), platform (PaaS), or infrastructure (IaaS), such as Notification Hub, Cognitive Services, and Azure Functions, that can change the impressions of the users regarding your application with little or no additional development hours. This chapter will give you a quick overview of using some of these services while developing .NET Core applications.

Chapter 8, *Creating a Datastore with Cosmos DB*, explains that Cosmos DB offers a multi-model and multi-API paradigm that allows applications to use multiple data models while storing application data with the most suitable API for the application, such as SQL, JavaScript, Gremlin, and MongoDB. In this chapter, we will create the data store for our application and implement the data access modules.

Chapter 9, *Creating Microservices Azure App Services*, goes through the basics of Azure App Services, and we will create a simple data-oriented backend for our application using ASP.NET Core with authentication provided by Azure Active Directory. Additional implementation will include offline sync and push notifications.

Chapter 10, *Using .NET Core for Azure Serverless*, shows how to incorporate Azure Functions into our infrastructure to process data on different triggers, and integrate Azure Functions with a Logic App that will be used as a processing unit in our setup.

Chapter 11, *Fluid Applications with Asynchronous Patterns*, explains that when developing Xamarin applications and ASP.NET Core applications, both the task's framework and the reactive modules can help distribute the execution threads and create a smooth and uninterrupted execution flow. This chapter will go over some of the patterns associated with these modules and apply them to various sections of the application.

Chapter 12, *Managing Application Data*, explains that, in order to avoid data conflicts and synchronization issues, developers must be diligent regarding the procedures implemented according to the type of data at hand. This chapter will discuss the possible data synchronization and offline storage scenarios using products such as SQLite and Realm, as well as the out-of-the-box offline support provided by Azure App Services.

Chapter 13, *Engaging Users with Notifications and the Graph API*, briefly explains how notifications and the graph API can be used to improve user engagement by taking advantage of push notifications and the graph API. We will create a notification implementation for cross-platform applications using Azure Notification Hub. We will also create so-called activity entries for our application sessions so that we can create a timeline that is accessible on multiple platforms.

Chapter 14, *Introducing Cognitive Services*, adds speech recognition to our application using the speech API. Additionally, we will be creating a computer vision training set to categorize user-uploaded images. After the Xamarin tasks for Android and iOS are completed, we will be creating a simple machine learning example using the client SDK that's available for UWP.

Chapter 15, *Azure DevOps and Visual Studio App Center*, shows how to use Visual Studio Team Service and App Center to set up a complete automated pipeline for Xamarin applications that will connect the source repository to the final store submission.

Chapter 16, *Application Telemetry with Application Insights*, explains how Application Insights is a great candidate for collecting telemetry from Xamarin applications that use an Azure-hosted web service infrastructure because of its intrinsic integration with Azure modules, as well as the continuous export functionality for App Center telemetry.

Chapter 17, *Automated Testing*, discusses how to create unit and coded UI tests, and the architectural patterns that revolve around them. Data-driven unit tests, mocks, and Xamarin UI tests are some of the concepts that will be discussed.

Chapter 18, *Deploying Azure Modules*, demonstrates how to configure the ARM template for the Azure web service implementation, as well as other services (such as Cosmos DB and Notification Hub) that we used previously so that we can create deployments using the Visual Studio Team Services build and release pipeline. Introducing configuration values into the template and preparing it to create staging environments are our primary focus of this chapter.

Chapter 19, *CI/CD with Azure DevOps*, explains how developers can create fully automated templates for builds, testing, and deployments using the toolset provided with Visual Studio Team Services. In this chapter, we will set up the build and release pipeline for Xamarin in line with the Azure deployment pipeline.

To get the most out of this book

The book is primarily aimed at .NET developers with slim to moderate experience with Xamarin and .NET Core. The cloud infrastructure-related sections heavily use various services in the Azure cloud infrastructure. However, familiarity with basic management concepts of Azure portal should be enough for the more advanced topics.

For the code samples, a combination of Windows and macOS development environments are used throughout the book. The ideal setup to utilize the samples would be to use macOS together with a Windows 10 virtual machine. This way, samples from both environments can be used.

The IDE of choice for implementing the code walk-throughs is Visual Studio 2019 on Windows, and Visual Studio for Mac on macOS. Visual Studio Code, which supports both platforms, can be used to create the scripting and Python examples.

Download the example code files

You can download the example code files for this book from your account at www.packt.com. If you purchased this book elsewhere, you can visit www.packt.com/support and register to have the files emailed directly to you.

You can download the code files by following these steps:

1. Log in or register at www.packt.com.
2. Select the **SUPPORT** tab.
3. Click on **Code Downloads & Errata**.
4. Enter the name of the book in the **Search** box and follow the onscreen instructions.

Once the file is downloaded, please make sure that you unzip or extract the folder using the latest version of:

- WinRAR/7-Zip for Windows
- Zipeg/iZip/UnRarX for Mac
- 7-Zip/PeaZip for Linux

The code bundle for the book is also hosted on GitHub at `https://github.com/PacktPublishing/Hands-On-Mobile-Development-with-.NET-Core`. In case there's an update to the code, it will be updated on the existing GitHub repository. Refer GitHub README for more details on prerequisites of services.

We also have other code bundles from our rich catalog of books and videos available at `https://github.com/PacktPublishing/`. Check them out!

Download the color images

We also provide a PDF file that has color images of the screenshots/diagrams used in this book. You can download it here: `https://www.packtpub.com/sites/default/files/downloads/9781789538519_ColorImages.pdf`.

Conventions used

There are a number of text conventions used throughout this book.

`CodeInText`: Indicates code words in text, database table names, folder names, filenames, file extensions, pathnames, dummy URLs, user input, and Twitter handles. Here is an example: "Mount the downloaded `WebStorm-10*.dmg` disk image file as another disk in your system."

A block of code is set as follows:

```
namespace FirstXamarinFormsApplication
{
    public partial class MainPage : ContentPage
    {
        public MainPage()
        {
            InitializeComponent();
            BindingContext = new MainPageViewModel();
        }
    }
}
```

When we wish to draw your attention to a particular part of a code block, the relevant lines or items are set in bold:

```
public App ()
{
    InitializeComponent();
    MainPage = new NavigationPage(new ListItemView());
}
```

Any command-line input or output is written as follows:

```
docker run -p 8000:80 netcore-usersapi
```

Bold: Indicates a new term, an important word, or words that you see onscreen. For example, words in menus or dialog boxes appear in the text like this. Here is an example: "Both the ALM process and the version control options are available under the **Advanced** section of the project settings."

> Warnings or important notes appear like this.

> Tips and tricks appear like this.

Get in touch

Feedback from our readers is always welcome.

General feedback: If you have questions about any aspect of this book, mention the book title in the subject of your message and email us at customercare@packtpub.com.

Errata: Although we have taken every care to ensure the accuracy of our content, mistakes do happen. If you have found a mistake in this book, we would be grateful if you would report this to us. Please visit www.packt.com/submit-errata, selecting your book, clicking on the Errata Submission Form link, and entering the details.

Piracy: If you come across any illegal copies of our works in any form on the Internet, we would be grateful if you would provide us with the location address or website name. Please contact us at `copyright@packt.com` with a link to the material.

If you are interested in becoming an author: If there is a topic that you have expertise in and you are interested in either writing or contributing to a book, please visit `authors.packtpub.com`.

Reviews

Please leave a review. Once you have read and used this book, why not leave a review on the site that you purchased it from? Potential readers can then see and use your unbiased opinion to make purchase decisions, we at Packt can understand what you think about our products, and our authors can see your feedback on their book. Thank you!

For more information about Packt, please visit `packt.com`.

Section 1: .NET Core and Cross-Platform Philosophy

.NET Core, .NET Standard, and Xamarin make up the cross-platform toolset for .NET developers. Using these sets of frameworks, developers are able to create applications for various platforms, including but not limited to, iOS, Android, and Universal Windows Platform, as well as IoT devices and Tizen. Understanding how these runtime and framework references work hand in hand and core concepts of cross-platform application will help you to better craft mobile projects.

The following chapters will be covered in this section:

- Chapter 1, *Getting Started with .NET Core*
- Chapter 2, *Xamarin, Mono, and .NET Standard*
- Chapter 3, *Universal Windows Platform*

Getting Started with .NET Core

1

.NET Core (previously known as .NET vNext) is the general umbrella term used for Microsoft's cross-platform toolset that aims to solve the shortcomings of centralized/machine-wide frameworks (NET framework) by creating a portable, platform agnostic, modular runtime and framework. This decentralized and modular development platform allows developers to create applications for multiple platforms using the common .NET base class libraries (NET standard), as well as various runtimes and application models, depending on the target platforms. This chapter will give you a brief introduction to .NET Core while explaining different tiers of .NET Core infrastructure. Languages and runtimes, as well as extensions that can be used together with .NET Core will be discussed and analyzed.

The combination of.NET Core, .NET Standard, and Xamarin is the key to cross platform projects, and opens many doors that were previously for Windows-only developers. Creating web applications that can run on Linux machines and containers, and the implementation of mobile applications that target iOS, Android, **Universal Windows Platform (UWP)**, and Tizen are just a couple of examples to emphasize the capabilities of this cross-platform approach.

In this chapter, the following sections will guide you through getting started with .NET Core:

- Cross-platform development
- Introduction to .NET Core
- .NET foundation
- Developing with .NET Core

Cross-platform development

The term cross-platform application development refers to the process of creating a software application that can run on multiple operating systems. In this book, we will not try to answer the question of *why*, but *how* – more specifically, will try to create a cross-platform application using the toolset provided by Microsoft and .NET Core.

Before we start talking about .NET Core, let's take a look at the process of developing an application for multiple platforms. Faced with the cross-platform requirement, the product team can choose multiple paths that will lead the developers through different application life cycles.

Throughout this book, we will have hypothetical user stories defined for various scenarios. We will start with an overall user story that underlines the importance of .NET Core:

> *"I, as a product owner, would like to have the client ShopAcross application running on iOS, Android mobile platforms, as well as Windows, Linux, and macOS desktop runtimes, so that I can increase my reach and user base."*

In order to meet these demands, we can choose to implement the application in several different ways:

- Fully native applications
- Hybrid applications
- Cross platform

Developing fully native applications

Following this path would create probably the most performant application, with increased accessibility to the platform APIs for the developers. However, the development team for this type of development would require specific know-how and skill sets so that the same application can be created on multiple platforms, also increasing the developer hours that need to be invested in the application.

Considering the platform set we mentioned previously, we would potentially need to develop the client application in Cocoa and CocoaTouch (macOS and iOS), Java (Android), .NET (Windows), or C++ (Linux), and finally build a web service infrastructure in another language of our choice. In other words, this approach is, in fact, implementing a multi-platform application rather than a cross-platform one.

Hybrid applications

Native hosted web applications (also known as hybrid applications) are another popular choice for (especially mobile) developers. In this architecture, a responsive web application would be hosted on a thin native harness on the target platform. The native web container would also be responsible for providing access to the web runtime on native platform APIs. These hybrid applications wouldn't even need to be packaged as application packages, but as **Progressive Web Apps** (**PWAs**) so that users can access them right from their web browsers. While the development resources are even more efficiently used than the native cross-platform framework approach, this type of application is generally prone to performance issues.

In reference to the business requirements at hand, we would probably develop a web service layer and a small **Single Page Application** (**SPA**), part of which is packaged as a hybrid application. The other parts can be hosted as a web application.

Native cross-platform frameworks

Development platforms such as React Native, Xamarin, and .NET Core provide the much-required abstraction for the target platforms, so that development can be done using one framework and development language for multiple runtimes. In other words, the developers are still using the APIs provided by the native platform (for example, Android or iOS SDK), but the development is executed in a single language and framework. This approach not only decreases the development resources, but also saves you from the burden of managing multiple source repositories for multiple platforms. This way, the same source is used to create multiple application heads.

For instance, using .NET Core, the development team can implement all target platforms using the same development suite and language, thus creating multiple client applications for each target platform, as well as the web service infrastructure.

In a cross-platform implementation, architecturally speaking, the application is made up of three distinct tiers:

- Application model (implementation layer for the consumer application)
- Framework (the toolset available for developers)
- Platform abstraction (the harness or runtime to host the application)

In this context, we, in essence, are in pursuit of creating a platform-agnostic application layer that will be catered for on a platform abstraction layer. The platform abstraction layer, whether it's on the native web host or the native cross-platform framework, is responsible for providing the bridge between the single application implementation and the polymorphic runtime component.

.NET Core and Mono provide the runtime, while the .NET Standard provides the framework abstraction, which means that cross-platform applications can be implemented and distributed on multiple platforms. Using Xamarin with the .NET Standard framework on mobile applications and .NET Core on the web infrastructure, sophisticated cloud-supported native mobile applications can be created.

Introduction to .NET Core

In order to understand the roots of and motivation for let us start with a quote:

> *"Software producers who maximize their product's potential for useful combination with other software, while at the same time minimizing any restrictions upon its further re-combination, will be the survivors within a software industry that is in the process of reorganizing itself around the network exchange of commodity data."*

> *-- David Stutz – General Program Manager for Shared Source Common Language Infrastructure, Microsoft, 2004.*

.NET Core dates back as early as 2001 when **Shared Source Common Language Infrastructure (SSCLI)** was shared sourced (not for commercial use) under the code name **Rotor**. This was the ECMA 335, that is, the **Common Language Infrastructure (CLI)** standard implementation. Rotor could be built on FreeBSD (version 4.7 or newer) and macOS X 10.2. It was designed in such a way that a thin **Platform Abstraction Layer (PAL)** was the only thing that was needed to port the CLI to a different platform. This release constitutes the initial steps to migrate .NET to a cross-platform infrastructure.

2001 was also the year the Mono project was born as an open source project that ports parts of .NET to the Linux platform as a development platform infrastructure. In 2004, the initial version of Mono was released for Linux, which would lead to ports on other platforms such as iOS (MonoTouch) and Android (MonoDroid), and would eventually be merged into the .NET ecosystem under the Xamarin name.

One of the driving factors for this approach was the fact that the .NET framework was designed and distributed as a system-wide monolithic framework. Applications that are dependent on only a small portion of the framework required the complete framework to be installed on the target operating system. It did not support application-only scenarios where different applications can be run on different versions without having to install a system-wide upgrade. However, more importantly, applications that were developed with .NET were implicitly bound to Windows because of the tight coupling between the .NET framework and Windows API components. .NET Core was born out of these incentives and opened up the doors of various platforms for .NET developers.

Semantically speaking, .NET Core describes the complete infrastructure for the whole set of cross-development tools that rely on a common language infrastructure and multiple runtimes, including .NET Core runtime, .NET, also known as Big CLR, Mono runtime, and Xamarin:

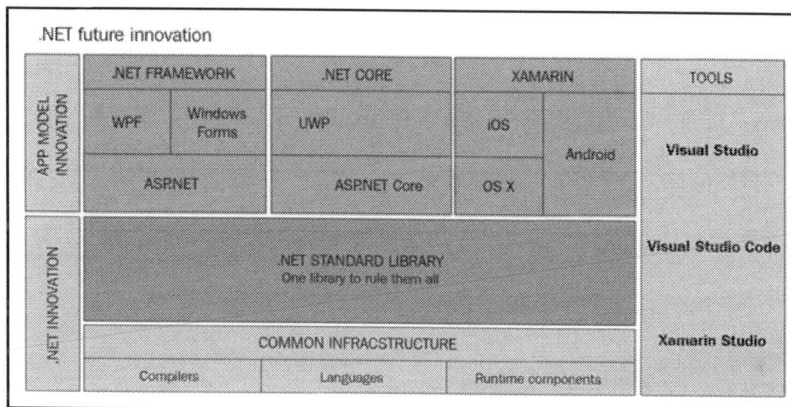

Adapted from: Soumyasch [CC BY-SA 3.0 (http://creativecommons.org/licenses/by-sa/3.0/)]

In this setup, the .NET Core CLI is made up of the base class library implementation, which defines the standards that need to be provided by the supported runtimes. The base class library is responsible for providing the PAL, which is provided by the hosting runtime under the name of the Adaption Layer. This infrastructure is supported by compiler services such as Roslyn and **Mono Compiler (MCS)**, as well as **Just-In-Time (JIT)** and **Ahead-of-Time (AOT)** compilers such RyuJIT (.NET Core), mTouch, and LLVM (for Xamarin.iOS) in order to produce and execute the application binaries for the target platform.

Overall, .NET Core is a rapidly growing ecosystem with various dynamic frameworks, runtimes, and tools. Most of these components can be found on GitHub as open source projects under the supervision of the .NET Foundation.

.NET Foundation

The .NET Foundation is an independent organization that supports open source development and collaboration within the .NET ecosystem. The .NET foundation supports the development of active projects within the ecosystem by evangelizing the technologies through organizing/sponsoring meetups and by active involvement in community-driven projects.

The .NET Foundation portfolio grew especially large due to the projects that were brought in by the acquisition of Xamarin.

Notable .NET Foundation projects

Some of the most notable projects that are generally used in modern .NET applications, as well as cross-platform mobile applications, are as follows:

- .NET Core
- ASP.NET Core
- Roslyn
- Reactive Extensions
- Entity Framework
- Identity Server
- ML.NET
- Xamarin and Xamarin.Forms
- xUnit.NET

.NET Core

The .NET Core project is composed of the .NET framework implementation and the common language runtime for .NET Core (CoreFX and CoreCLR). Additionally .NET Core tools such as the .NET Core command-line interface can also be found as separate repositories. The community is free to make contributions, as well as submit issue reports.

ASP.NET Core

ASP.NET Core is the cross-platform implementation of ASP.NET. As a platform agnostic web development framework, applications created with it can be hosted on multiple platforms, as well as on Windows using classic .NET. ASP.NET MVC, Web API, web pages, and SignalR are some of the repositories under the ASP.NET Core project.

Roslyn

Complete implementation of Roslyn (.NET compiler platform for C# and Visual Basic) can be found on GitHub as part of the .NET Foundation group. Roslyn is the implementation of the compiler as a service paradigm and has various extensibility points, including customizable code analyzers.

Reactive Extensions for .NET

Reactive Extensions for .NET is a library that provides developers with event-based asynchronous observable sequences and LINQ style query operators. Extensions can be used in .NET applications using the system's reactive namespaces and its children.

Entity Framework

The Entity Framework is the recommended data access technology for modern .NET applications. The newest version of the Entity Framework was built from scratch using .NET, Core so that it can be used in cross-platform applications, from ASP.NET Core applications to device-specific scenarios such Xamarin and UWP.

IdentityServer

OpenID Connect and the OAuth 2.0 Framework for Katana and ASP.NET Core are the components of identity server project. They provides tools that developers can use to enable authentication as a service, **Single Sign-on (SSO)**, and federation gateways in their applications.

ML.NET

This project allows developers to include cognitive functions and AI-related implementations in their applications with .NET. The same open-source is used by the Hello feature on Windows 10. Developers can use this framework to integrate custom machine learning features into their applications without any prior knowledge about neural networks, artificial intelligence, or machine learning. This library also allows integration with other machine learning libraries such as TensorFlow, ONNX, and Infer.NET.

Xamarin and Xamarin.Forms

Miguel de Icaza, part of the board of directors of the .NET Foundation, publicly announced that Xamarin, `Xamarin.SDK`, and `Xamarin.Forms`, as well as the Mono runtime ports for iOS and Android, are to be part of the .NET foundation and open sourced in Evolve 2016. Even though these projects are not listed on the Foundation site, they can be found on GitHub under the common MIT license.

xUnit.net

This is a free, open source, community-based unit testing tool for the .NET Framework. This testing framework is used in many of the aforementioned repositories and projects as the main testing tool. It has strong integration with newer versions of Visual Studio and Team Foundation Server.

Developing with .NET Core

.NET Core applications can be developed with Visual Studio on the Windows platform and Visual Studio for Mac (inheriting from Xamarin Studio) on macOS. Visual Studio Code (an open source project) and Rider (JetBrain's development IDE) provide support for both of these platforms, as well as Unix-based systems. While these platforms provide the desired user-friendly development UI, technically, .NET core applications can be written with a simple text editor and compiled using the .NET Core command-line toolset:

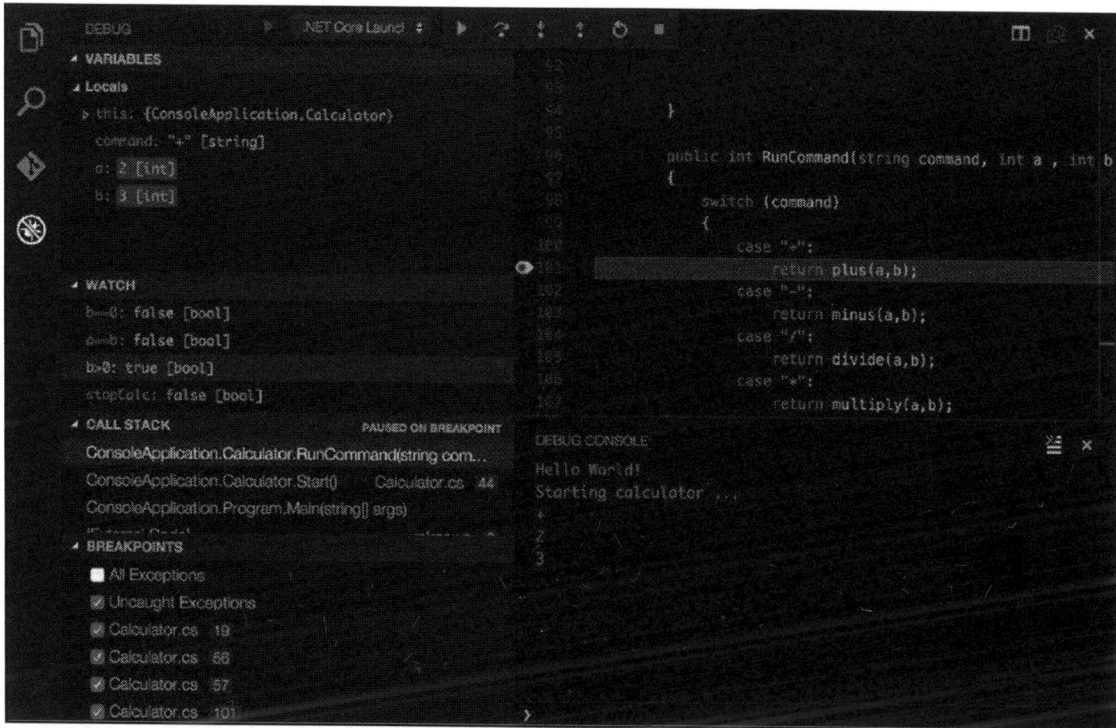

As we mentioned previously, the only intrinsic runtime in the .NET Core CLI is the .NET Core runtime, which is primarily used for creating console applications with access to the complete base class library.

Without further ado, let's create our first cross-platform application with the CLI tools and see how it behaves on multiple target platforms.

Creating a runtime agnostic application

To begin with, we will create our console application on macOS that has .NET Core installed:

```
$ mkdir demo && cd $_
$ dotnet --version
2.1.301
$ dotnet new console
The template "Console Application" was created successfully.

Processing post-creation actions...
Running 'dotnet restore' on /demo/demo.csproj...
 Restoring packages for /demo/demo.csproj...
 Generating MSBuild file /demo/obj/demo.csproj.nuget.g.props.
 Generating MSBuild file /demo/obj/demo.csproj.nuget.g.targets.
 Restore completed in 236.91 ms for /demo/demo.csproj.

Restore succeeded.
```

> **TIP**
> In this example, we have used the console template, but there are many other templates available out of the box, such as class library, unit test project, asp.net core, as well as more specific templates, such as razor page or MVC ViewStart.

Now, the *helloworld* console application should have been created in the folder that you specified in the first step.

In order to restore the NuGet packages associated with any project, you can use the `dotnet restore` command in a command line or Terminal window, depending on your operating system.

> Generally, you don't need to use the restore command, as the compilation already does this for you. In the case of template creation, the last step actually restores the NuGet packages.

Now that our application project is ready (after editing the `program.cs` file), we can build and run the console application:

```
Cans-MBP:demo can.bilgin$ dotnet new console
The template "Console Application" was created successfully.

Processing post-creation actions...
Running 'dotnet restore' on /Volumes/Data/book/demo/demo.csproj...
  Restoring packages for /Volumes/Data/book/demo/demo.csproj...
  Generating MSBuild file /Volumes/Data/book/demo/obj/demo.csproj.nuget.g.props.
  Generating MSBuild file /Volumes/Data/book/demo/obj/demo.csproj.nuget.g.targets.
  Restore completed in 236.91 ms for /Volumes/Data/book/demo/demo.csproj.

Restore succeeded.

Cans-MBP:demo can.bilgin$ pico program.cs
Cans-MBP:demo can.bilgin$ dotnet build
Microsoft (R) Build Engine version 15.7.179.6572 for .NET Core
Copyright (C) Microsoft Corporation. All rights reserved.

  Restore completed in 56.6 ms for /Volumes/Data/book/demo/demo.csproj.
  demo -> /Volumes/Data/book/demo/bin/Debug/netcoreapp2.1/demo.dll

Build succeeded.
    0 Warning(s)
    0 Error(s)

Time Elapsed 00:00:03.85
Cans-MBP:demo can.bilgin$ dotnet run
Hello .NET Core
Cans-MBP:demo can.bilgin$
```

Here, we used the `run` command to compile and run our application in the current platform (macOS). If you were to navigate to the `build` folder, you would notice that, instead of an executable, the CLI actually created a **Dynamic Link Library** (DLL) file. The reason for this is that, since no other compilation option was defined, the application was created as a framework-dependent application. We can try running the application with the `dotnet` command, which is called the driver:

```
$ cd bin/Debug/netcoreapp2.1/
$ ls
demo.deps.json demo.pdb demo.runtimeconfig.json
demo.dll demo.runtimeconfig.dev.json
$ dotnet demo.dll
Hello .NET Core
```

Here, it is important to note that we used the description **framework-dependent** (in this case, the NETCore.App 2.1 runtime). If we were discussing the .NET framework prior to .NET Core, this would strictly refer to the Windows platform. In this context, however, it refers to an application that is only dependent on the framework itself while being platform-agnostic. In order to test our application on Windows, we can copy the bin folder to a Windows machine with the target framework installed and try running our application:

```
Command Prompt                                          —    □    ✕

\U:\demo\bin>cd Debug\netcoreapp2.1

\U:\demo\bin\Debug\netcoreapp2.1>dotnet demo.dll
lHello .NET Core

\U:\demo\bin\Debug\netcoreapp2.1>
```

> **TIP**
>
> In order to verify that the required framework is installed on the target machine, you can use the `dotnet --info` or `dotnet --list-sdks` commands, which will list the installed runtimes on the target machine.

In order to test the runtime independence of the created `demo.dll` file, we can try running it with the mono runtime. On macOS, you can try the following command to execute our application:

```
$ cd bin/Debug/netcoreapp2.1/
$ mono demo.dll
Hello .NET Core
```

Defining a runtime and self-contained deployment

In the previous example, we created a console application that is operating system-agnostic. However, it had a dependency on the `NETCore.App` runtime. What if we want to deploy this application to a target system that doesn't have .NET Core runtime and/or SDK installed?

When the .NET Core applications need to be published, you can include the dependencies from the .NET Core framework and create a so-called **self-contained** package. However, by going down this path, you would need to define the target platform (operating system and CPU architecture) using a **Runtime Identifier** (**RID**) so that the .NET CLI can download the required dependencies and include them in your package.

The runtime can be defined either as part of the project file or as a parameter during `publish` execution:

```
  GNU nano 2.0.6                        File: demo.csproj

<Project Sdk="Microsoft.NET.Sdk">

  <PropertyGroup>
    <OutputType>Exe</OutputType>
    <TargetFramework>netcoreapp2.1</TargetFramework>
    <RuntimeIdentifier>win10-x64</RuntimeIdentifier>
  </PropertyGroup>

</Project>




^G Get Help    ^O WriteOut    ^R Read File   ^Y Prev Page   ^K Cut Text    ^C Cur Pos
^X Exit        ^J Justify     ^W Where Is    ^V Next Page   ^U UnCut Text  ^T To Spell
```

Here, we have edited the project file to target Windows 10 with the x64 architecture. Now, if we were to publish the application (note that the publishing process is going to take place on macOS), it would create an executable for the defined target platform:

```
$ nano demo.csproj
$ dotnet publish
Microsoft (R) Build Engine version 15.7.179.6572 for .NET Core
Copyright (C) Microsoft Corporation. All rights reserved.

Restoring packages for /demo/demo.csproj...
  Installing runtime.win-x64.Microsoft.NETCore.DotNetAppHost 2.1.1.
  Installing runtime.win-x64.Microsoft.NETCore.DotNetHostResolver 2.1.1.
  Installing runtime.win-x64.Microsoft.NETCore.DotNetHostPolicy 2.1.1.
  Installing runtime.win-x64.Microsoft.NETCore.App 2.1.1.
  Generating MSBuild file /demo/obj/demo.csproj.nuget.g.props.
  Generating MSBuild file /demo/obj/demo.csproj.nuget.g.targets.
```

```
Restore completed in 18.81 sec for /demo/demo.csproj.
demo -> /demo/bin/Debug/netcoreapp2.1/win10-x64/demo.dll
demo -> /demo/bin/Debug/netcoreapp2.1/win10-x64/publish/
```

The publish folder, in this case, would include all the necessary packages from the .NET Core runtime and framework targeting the Windows 10 runtime:

Name	Date modified	Type	Size
api-ms-win-crt-string-l1-1-0.dll	6/6/18 1:58 PM	DLL File	25 KB
api-ms-win-crt-time-l1-1-0.dll	6/6/18 1:58 PM	DLL File	21 KB
api-ms-win-crt-utility-l1-1-0.dll	6/6/18 1:58 PM	DLL File	19 KB
clrcompression.dll	6/6/18 1:58 PM	DLL File	713 KB
clretwrc.dll	6/6/18 1:58 PM	DLL File	235 KB
clrjit.dll	6/6/18 1:58 PM	DLL File	1,423 KB
coreclr.dll	6/6/18 1:58 PM	DLL File	5,824 KB
dbgshim.dll	6/6/18 1:58 PM	DLL File	152 KB
demo.deps.json	7/15/18 6:24 AM	JSON File	40 KB
demo.dll	7/15/18 6:24 AM	DLL File	4 KB
demo.exe	7/15/18 6:09 AM	Application	135 KB
demo.pdb	7/15/18 6:24 AM	Program Debug Data...	1 KB
demo.runtimeconfig.json	7/15/18 6:24 AM	JSON File	1 KB
hostfxr.dll	6/6/18 2:04 PM	DLL File	390 KB
hostpolicy.dll	6/6/18 2:04 PM	DLL File	572 KB
Microsoft.CSharp.dll	6/6/18 1:58 PM	DLL File	755 KB
Microsoft.DiaSymReader.Native.amd64.dll	6/6/18 1:58 PM	DLL File	1,236 KB
Microsoft.VisualBasic.dll	6/6/18 1:58 PM	DLL File	455 KB
Microsoft.Win32.Primitives.dll	6/6/18 1:58 PM	DLL File	22 KB

Notice that, once the deployment target platform is defined, an executable file is created and there is no more need for the driver. In fact, the executable's sole purpose here is to act as the access point (host) to the dynamic class library that is created by .NET Core.

Some of the most notable runtimes include Windows 7 to Windows 10 on three different architectures (x86, x64, and arm), multiple macOS versions, and various distributions and versions of Linux, including OpenSuse, Fedora, Debian, Ubuntu, RedHat, Tizen, and so on.

Defining a framework

In the previous examples, we have been using netcoreapp2.1 as the target framework. While, for the self-contained deployment for this console application, this proves to be sufficient, if we were preparing a Xamarin application or a UWP. Net Standard, we would have been better off using target platform frameworks such as Xamarin.iOS.

The target platform framework can be changed using the `<TargetFrameworks>` project property. We would have to use the moniker assigned to the desired framework:

Target framework	Latest stable version	Moniker	.NET Standard
.NET Standard	2.0	netstandard2.0	N/A
.NET Core	2.1	netcoreapp2.1	2.0
.NET Framework	4.7.2	net472	2.0

Summary

The .NET ecosystem is growing at an exponential velocity with the new, open-source oriented approach being adopted by Microsoft. Various runtimes and frameworks are part of community-driven projects that cover bigger portions of the original .NET framework which was, ironically, destined to be part of Windows itself.

Using the .NET Core infrastructure and the provided runtimes, developers can, nowadays, develop applications for mobile platforms such as iOS, Android, and UWP, as well as micro runtimes such as Windows IoT Core, Raspbian, and Tizen. Setting device-specific runtimes aside, Azure and web development can also be accomplished using .NET Core.

In the remainder of this book, we will be implementing a Xamarin.Forms application using the Mono runtime and creating a web infrastructure composed of Serverless (Logic Apps and Functions) components, as well as ASP.NET Core using the .NET Core infrastructure. We will also take a look at additional projects that are closely related to the .NET ecosystem, such as cognitive services and machine learning, and how they can be used to enhance the user experience.

Xamarin, Mono, and .NET Standard

2

Xamarin is the app model implementation for the .NET Core infrastructure. As part of the cross-platform infrastructure, Xamarin uses the Mono runtime, which in return acts as the adaption layer for the .NET Standard base class library/libraries. By means of the abstraction provided by Mono runtime (MonoTouch and MonoDroid), Xamarin can target mobile platforms such as iOS and Android. This chapter will try to venture the relationship between .NET Core and Xamarin. You will learn about how the Xamarin source code is executed with MonoTouch on iOS and Mono runtime on Android.

The following sections will help you create your first Xamarin application:

- Introduction to Xamarin
- Creating your first Xamarin application
- Xamarin on Android – Mono Droid
- Xamarin on iOS – Mono Touch
- Xamarin.Forms
- Using .NET Standard with Xamarin
- Extending the reach

Introduction to Xamarin

Where native cross-platform development is concerned, especially on mobile platforms, Xamarin is one the main technologies that's used. Xamarin as a platform can be identified as the legacy of the Mono project, which was an open source project that was led by some of the key people that later on established the Xamarin group. Mono was initially a **Common Language Infrastructure (CLI)** implementation of .NET for Linux that allowed developers to create Linux applications using the .NET (2.0) framework modules. Later on, Mono's runtime and compiler implementation was ported to other platforms until Xamarin took its place within the Microsoft .NET Core ecosystem. The Xamarin suite is one of the flagships of .NET Core and the key technologies for cross-platform development.

Creating your first Xamarin application

As a developer planning to create native mobile applications using Xamarin, you have several options for setting up your development environment. In terms of development, both macOS and Windows can be utilized, using either Visual Studio or Rider IDEs.

As a .NET developer, if you are looking for a familiar environment and IDE, the best option would be to use Visual Studio on Windows.

In order to use the Xamarin-related templates and available SDKs, the first step would be to install the required components using the Visual Studio installer:

Chapters

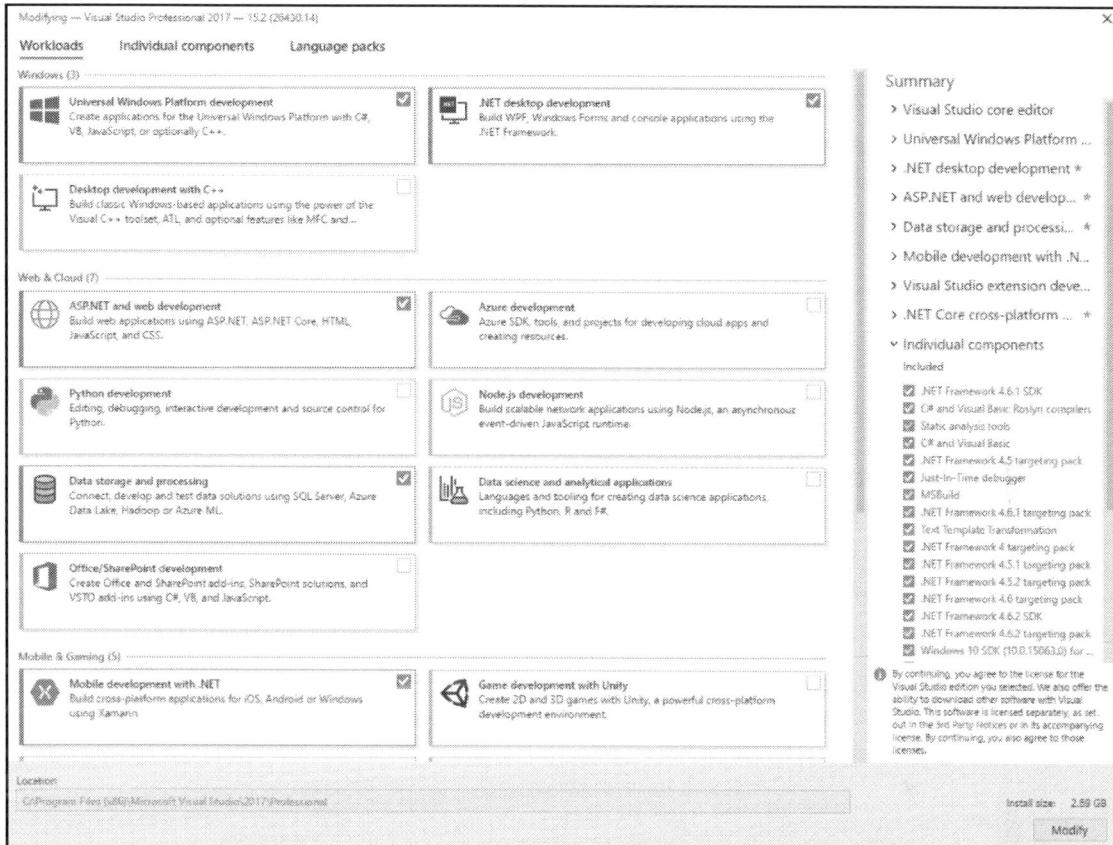

When you install the **Mobile Development with .NET** component, the required SDKs (for Android and iOS) are automatically installed so that the developers don't need to do any additional prerequisite installation.

Once the setup is complete, various project templates become available under the **Cross Platform App** section, as well as platform-specific sections, namely **Android** and **iOS**. The multi-project template for the cross-platform Xamarin app is to guide you through the project creation process using Xamarin.Forms, while the available **Android App** and **iOS App** templates create application projects using the classic Xamarin infrastructure:

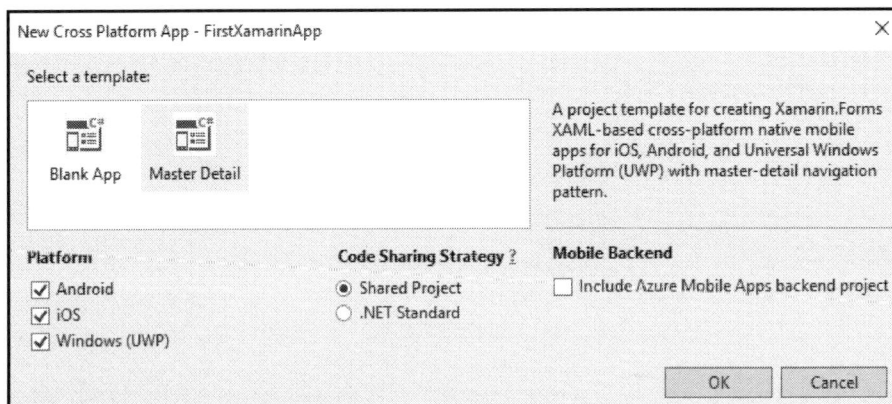

Using this template, Visual Studio will create a common project (shared or .NET Standard) and a project for each selected platform (selected platforms out of iOS, Android, and UWP). For this example, we will be using the **Shared Project** code sharing strategy and selecting iOS and Android as target platforms.

> It is important to note that if you are developing on a Windows machine, a macOS build service (a macOS device with Xamarin.iOS and Xcode installed) is required to be able to compile and use the simulator with the iOS project.

> If you, in the first compilation of the iOS project, receive an error pointing to missing Xcode components or frameworks, you need to make sure that the Xcode IDE is run at least once manually so that you can agree on the terms and conditions. This allows Xcode can complete the setup by installing additional components.

In this solution, you will have platform-specific projects, along with the basic boilerplate code and a shared project that contains the `Main.xaml` file, which is a simple XAML view. While the platform-specific projects are used to host the views that are created using the declarative XAML pages, the `MainActivity.cs` file on an Android project and the `Main.cs` file on an iOS project, are used to initialize the Xamarin.Forms UI framework and render the views.

In the `Main.xaml` file, the code that is used to create a label and center it both vertically and horizontally would look similar to the following:

```
<ContentPage xmlns="http://xamarin.com/schemas/2014/forms"
xmlns:x="http://schemas.microsoft.com/winfx/2009/xaml"
xmlns:local="clr-namespace:FirstXamarinFormsApplication"
x:Class="FirstXamarinFormsApplication.MainPage">
    <StackLayout>
        <!-- Place new controls here -->
        <Label Text="Welcome to Xamarin.Forms!"
            HorizontalOptions="Center"
            VerticalOptions="CenterAndExpand" />
    </StackLayout>
</ContentPage>
```

This XAML view tree is rendered on the target platforms using the designated renderers. It uses the page, layout, and view hierarchy:

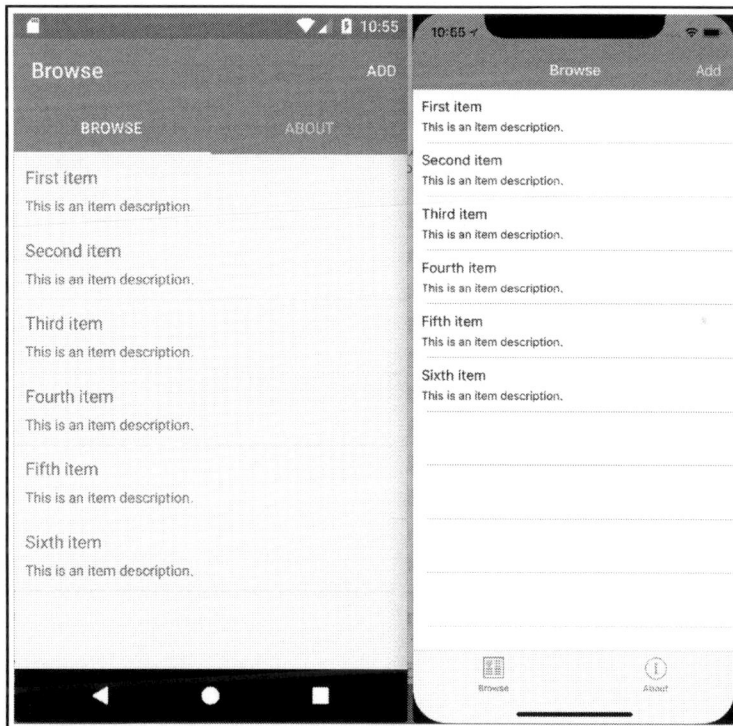

Xamarin on iOS – Mono Touch

The project that we created in the previous section used the Xamarin.Forms UI rendering. While this can be the most efficient way to implement a cross-platform application, in some cases, you might need to implement a very platform-specific application (this includes many platform APIs and specialized UI components). In these type of situations, you can resort to creating a classic Xamarin.iOS and/or Xamarin.Android application.

In a classic Xamarin application, views are created using native SDK components and toolsets. In the case of Xamarin.iOS, developers can create the application views either using an iOS namespace and UI elements using code or using XIBs or storyboards with backing controllers.

In order to recreate the implementation that was done on Xamarin.Forms for Xamarin.iOS, you can use the iOS application template listed under the iOS section.

This template will create a simple Xamarin.iOS application with a single view, an associated storyboard, and a controller for the main view that was created.

With either Visual Studio on Windows or macOS, you can make use of the designer, create the label control, and align it to be centered both horizontally and vertically with constraints referring to the containing view. In order to add the label, you can drag and drop the label control from the toolbox. Once the toolbox is added, constraints can be added using the designer. The iOS designer has two available modes, one of which is the constraint mode. In this mode, you can hold the center pin and drag it to the superview vertical center. You can repeat this process for the horizontal center:

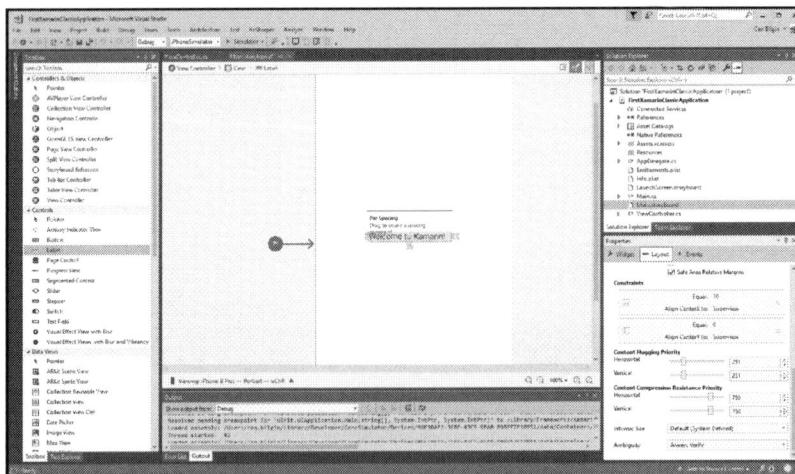

Now, if we compile and run the application, you will get the same view that was rendered by Xamarin.Forms in the previous section:

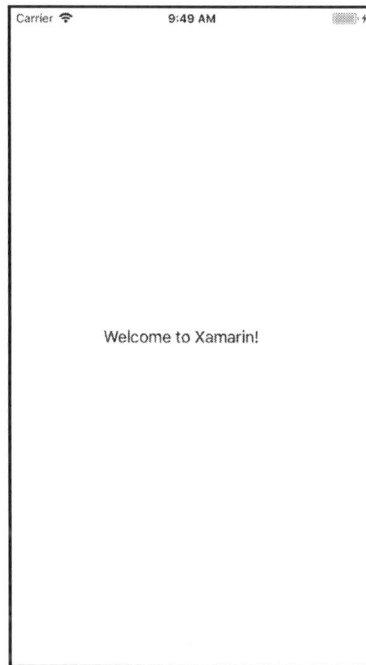

During the compilation process, the project that we created with C# and .NET (standard) modules is first compiled into a **Microsoft Intermediate Language** (**MSIL**), just like any other .NET Core project, and is then compiled into native code with AOT compilation. At this point, one of the most crucial components is the monotouch runtime, which acts as the adaption layer that sits on top of the iOS kernel, allowing the .NET Standard libraries access to the system-level functions. During compilation, just like the application code, the monotouch runtime libraries, together with the .NET Standard packages, are linked and trans(com)piled into native code.

> The AOT compilation is only a requirement when the compiled package is being deployed to a real device because of the code generation restrictions on iOS. For other platforms or when running the application on an iOS simulator, a JIT compiler is used to compile MSIL into native code – not at compile-time, but at runtime.

The following screenshot outlines the transcompilation process of APT and LLVM for Xamarin.iOS applications:

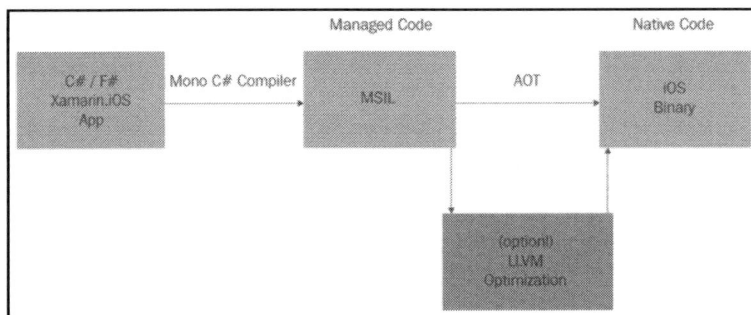

Xamarin on Android – Mono Droid

Following the same methodology, we can recreate the Xamarin.Forms view we created using Xamarin.Android using a native project template. In order to do this, we can reuse the existing Xamarin classic project that we used for iOS and add an Android application project instead:

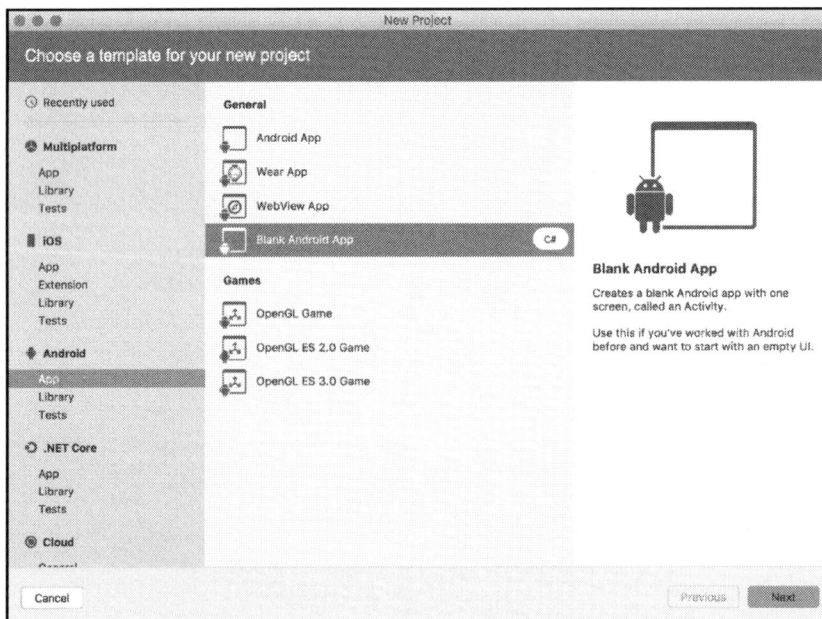

This will create a standard boilerplate application project for Xamarin.Android with a single view and associated layout file. If you open the created `Main.axml` file, the designer view will be loaded, which can be used to create our welcome view:

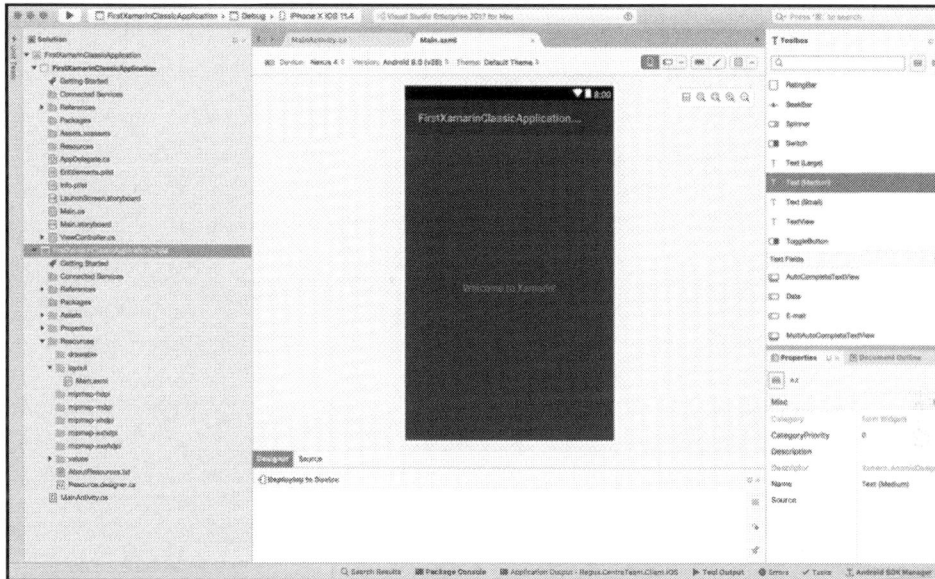

When handling the Android XML layout files, developers are given the option to either use the designer or the source view. By using the designer view to create the welcome view, you would have to drag and drop the text view control and adjust the alignment, layout, and gravity properties for the label.

Using the source view, you can also paste the following layout declaration to see what the application looks like when run on the Android platform:

```xml
<?xml version="1.0" encoding="utf-8"?>
<LinearLayout xmlns:android="http://schemas.android.com/apk/res/android"
    android:orientation="vertical"
    android:layout_width="match_parent"
    android:layout_height="match_parent"
    android:minWidth="25px"
    android:minHeight="25px">
    <TextView
        android:textAppearance="?android:attr/textAppearanceMedium"
        android:layout_width="match_parent"
        android:layout_height="match_parent"
        android:id="@+id/textView1"
        android:text="Welcome to Xamarin!"
```

```
        android:gravity="center_vertical"
        android:textAlignment="center" />
</LinearLayout>
```

When we finally compile and run our first Xamarin.Android application, you will see the welcome view that was created on Xamarin.Forms and Xamarin.iOS.

The Xamarin.Android platform functions a little more like .NET Core. Unlike Xamarin.iOS, there are no restrictions on code generation, so the Mono Droid runtime execution is done using the JIT compiler, which is responsible for providing the IL packages that are part of the application package. The Mono Droid runtime exists in the application package in the form of native code that replaces the .NET Core runtime:

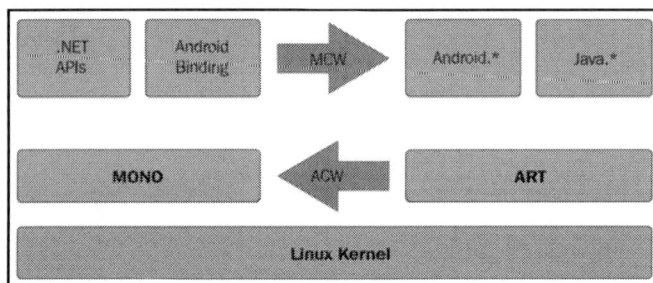

For Xamarin.Forms applications, the same compilation and runtime procedures, such as AOT and JIT, apply, depending on the targeted platform.

Xamarin.Forms

In the previous examples, you can easily see that Xamarin.Forms greatly simplifies the process of creating UI mobile applications on two complete different platforms using the same declarative view tree, even though the native approaches on these platforms are, in fact, almost completely different.

From a UI renderer perspective, Xamarin.Forms provides the native rendering with two different ways of using the same toolset at compile-time (compiled XAMLs) and at runtime (runtime rendering). In both scenarios, page-layout-view hierarchies that are declared in XAML layouts are rendered using renderers. Renderers can be described as the implementations of the view abstractions on target platforms. For instance, the renderer for the label element on iOS is responsible for translating label control (as well as its layout attributes) into a UILabel control.

Nevertheless, Xamarin.Forms can't just be categorized as a UI framework, since it provides various modules out of the box which are essential to most mobile application projects, such as dependency services and messenger services. Being among the main patterns for creating SOLID applications, these components provide the tools to create abstractions on platform-specific implementations, thus unifying the cross-platform architecture in order to create application logic that spans across multiple platforms.

Additionally, the **Data Binding** concept, which is the heart and soul of any **Model-View-ViewModel** (**MVVM**) implementation, can be directly introduced in the XAML level, saving the developers from having to create their own data synchronization logic.

For instance, if we were to expand the implementation from the previous example by creating a ViewModel for our main view, we could make use of our shared project, which was created as part of the multi-project template.

The first step here would be to create the `MainPageViewModel` class under the shared project that contains the view that was created previously. Inside this class, we will create a single (get-only) property, which we will use in our view:

```
namespace FirstXamarinFormsApplication
{
    public class MainPageViewModel
    {
        public MainPageViewModel()
        {
        }

        public string Platform
        {
            get

            {
#if __IOS__
                return "iOS";
#elif __ANDROID__
                return "Android";
#endif
            }
        }
    }
}
```

Now, let's assign this view model as our binding context on the main view:

```
namespace FirstXamarinFormsApplication
{
```

```
public partial class MainPage : ContentPage
{
    public MainPage()
    {
        InitializeComponent();

        BindingContext = new MainPageViewModel();
    }
}
}
```

Then, we will update our XAML to use the bound view model data:

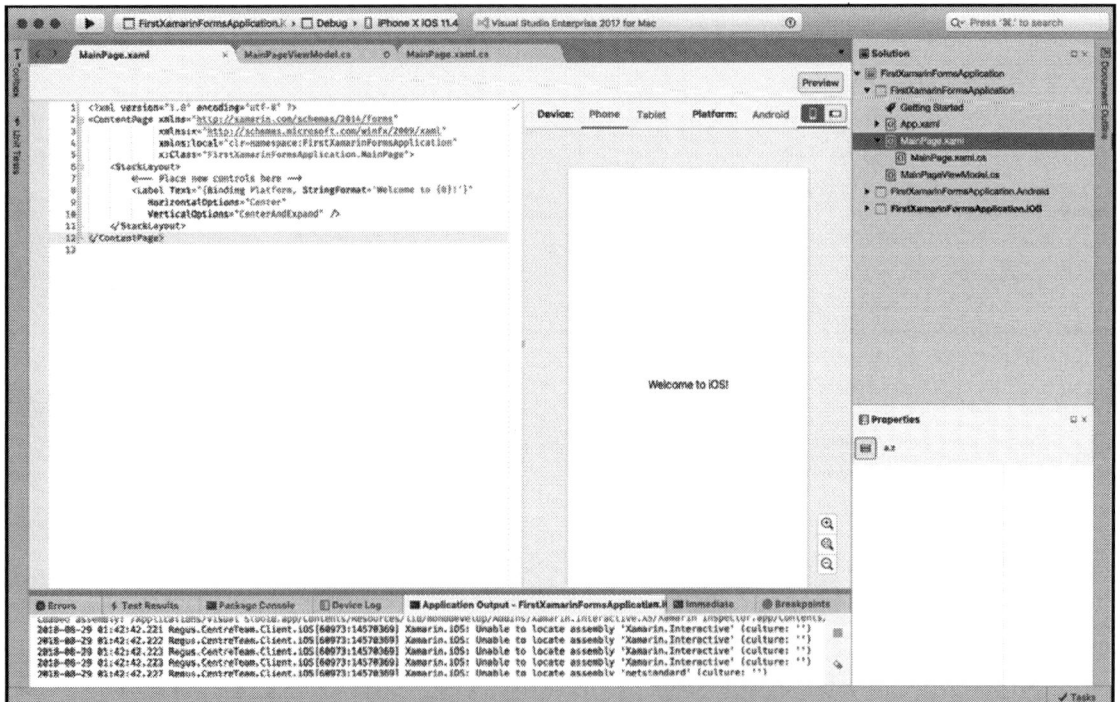

In this example, our binding of the label view looks similar to the following:

```
<Label Text="{Binding Platform, StringFormat='Welcome to {0}!'}"
        HorizontalOptions="Center"
        VerticalOptions="CenterAndExpand" />
```

Using data bindings can substantially decrease the complexity of your applications, creating highly robust and maintainable mobile applications.

Using .NET Standard with Xamarin

Even though the Xamarin and/or .NET Core target platforms (Platform APIs) are treated as if they have the same setup, capabilities, and functionalities as a platform-agnostic framework (.NET Standard), each of these target platforms are different from each other. The adaption layer (implementation of .NET Standard) allows us, developers, to treat these platforms in the same way.

Before the unification and standardization of .NET modules, together with shared projects, cross-platform compatibility was maintained by common denominators of implemented functionality on target platforms. In other words, the available APIs on each selected platform made up a profile that determined the subset of functionality that could be used for these platforms. These platform-agnostic projects that were used to implement the application logic were then packaged into so-called **Portable Class Libraries** (**PCLs**). PCLs were an essential part of cross-platform projects, since they could create and share application code that would be executed on multiple platforms:

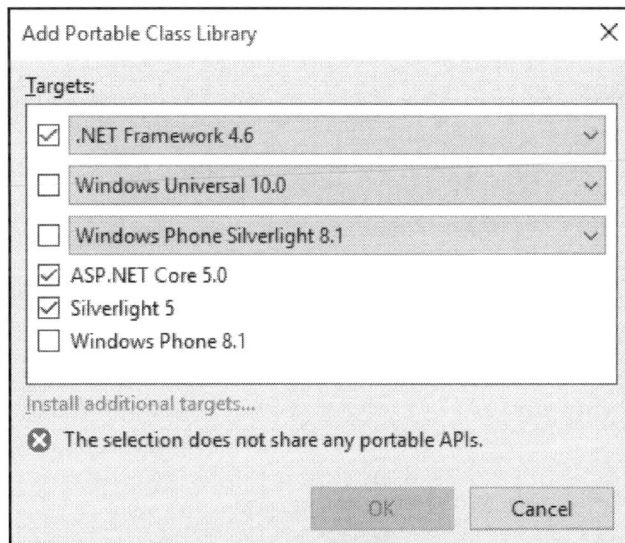

Nowadays, since .NET API implementations on various platforms have all converged into (almost) the same subset, a standard set of .NET APIs were defined as the common implementation ground for cross-platform implementation – .NET Standard. As a simple analogy, .NET Standard can be considered the interface that's used to access the platform APIs which are implemented by target platform runtimes.

Using .NET Standard, we can replace the shared projects in our previous examples for both Xamarin.Forms and Xamarin classic applications. This would allow us to create a testable set of platform-agnostic interfaces which can be tested as a standalone library, unlike the shared projects, which are simply made up of a bundle of linked source code files. While shared source code files are directly compiled into the target platform projects, .NET Standard libraries are simply assigned as reference assemblies.

In order to replace the shared project with a NET Standard library in the Xamarin.Forms application, we would have to do the following:

1. Create a **.NET Standard Library** project.
2. Add the Xamarin.Forms NuGet package with the same version as the platform projects.
3. Copy the `App.xaml` (and `App.xaml.cs`) file and added views to the Standard library.
4. Reference the Standard library instead of the shared one:

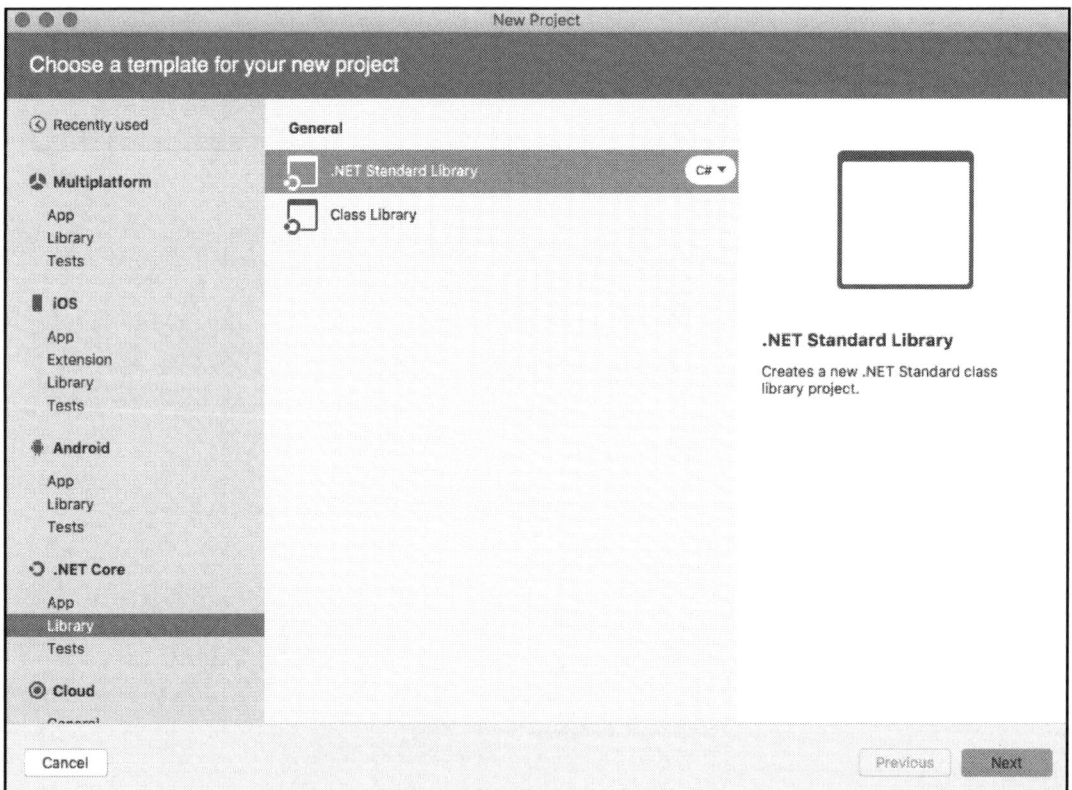

Once these steps are completed, our Xamarin.Forms application would be using .NET Standard to share code between the platforms. However, this would be problematic for the conditional compilation symbols we introduced in our ViewModel, which return the platform string according to the application we are compiling (iOS or Android):

```
public class MainPageViewModel
{
    public MainPageViewModel()
    {
    }

    public string Platform
    {
        get 'MainPageViewModel.Platform.get': not all code paths return a value
        {
#if __IOS__
            return "iOS";
#elif __ANDROID__
            return "Android";
#endif
        }
    }
}
```

What's happening here is that iOS and Android compilation symbols are only applicable to the platform-specific projects, so, unless you manually introduce these into platform-specific build configurations, they will not be available in .NET Standard projects. In order to remedy this, we can use a simple conditional to check the current runtime. (Alternatively, we can introduce an interface that will be used to inject the platform-specific implementations.):

```
public string Platform
{
    get

    {
        if (Device.RuntimePlatform.Equals(Device.Android))
        {
            return "Android";
        }
        else
        {
            return "iOS";
        }
    }
}
```

Extending the reach

Finally, since we are talking about Xamarin, it is important to mention that Xamarin and/or Xamarin.Forms do not bind the developers onto Android and iOS phone or tablet devices. By using Xamarin and Xamarin.Forms, developers can target devices, varying from simple wearables such as smart watches, to IoT devices and home appliances.

When developing applications for iOS or Android-based appliances, exactly the same toolset can be used, while more specialized platforms (such as Tizen) can constitute a target platform, given that the .NET Standard implementation exists natively:

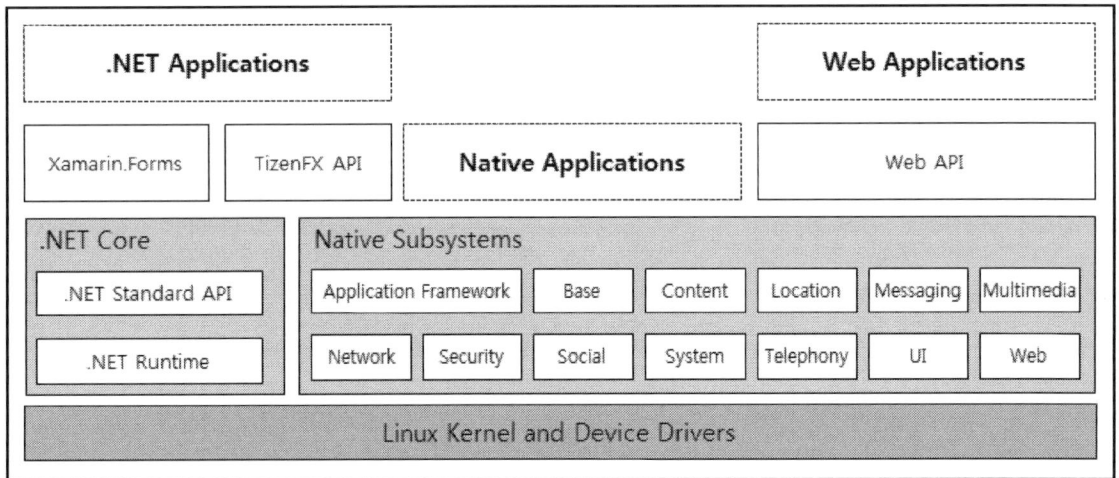

https://developer.tizen.org/development/training/overview#type / [CC BY-SA 3.0 (https://creativecommons.org/licenses/by-sa/3.0)]

Tizen implementation is also a nice example of .NET Core being used by Xamarin.Forms and the Linux kernel.

Summary

In this chapter, we learned about Xamarin, one of the main supported runtimes of .NET Standard, and how to use it to create mobile applications for multiple platforms. UWP, being the most mature member of the cross platform .NET initiative, can provide developers with a completely separate market for development.

In the next chapter, we will have a deeper look into UWP and how it can contribute to .NET developers who are executing cross-platform development projects.

Universal Windows Platform

3

Universal Windows Platform (**UWP**) is a common API layer that allows developers to create applications for various platforms, from desktop PCs to niche devices such as HoloLens. In comparison with the Xamarin setup, UWP applications are a little more involved with .NET framework and runtime components. UWP makes use of two completely different sets of .NET Framework: .NET Native and .NET Core. Here, .NET Core acts as the BCL library, while .NET Native is part of the application model. This chapter will discuss the components that allow UWP apps to be portable within the Windows 10 ecosystem and how they are associated with .NET Core.

The following sections will help you create your first Xamarin application:

- Universal Windows Platform
- Creating UWP applications
- XAML Standard
- .NET Standard and .NET Native
- Platform extensions

Universal Windows Platform

Windows, prior to the release of Windows 8 (and Windows runtime), exposed a flat set of Windows APIs and COM extensions, allowing developers to access system-level functions. .NET modules that rely on these functions, included the invokes to this API layer in order to make use of the operating system-level functionality.

Windows Runtime (**WinRT**) provided a more accessible and managed development interface that is available for a wide range of development languages. WinRT can be used in common .NET languages (C# and VB), as well as C++ and JavaScript.

Using the common ground WinRT created, UWP provided the much needed convergence of multiple platforms within the Microsoft ecosystem. Developers were able to create applications using the same SDK for various devices, which were esoteric targets. Using the UWP development tools, applications with shared modules and user interfaces can target desktop devices, game consoles, and augmented reality devices, as well as mobile and IoT implementations:

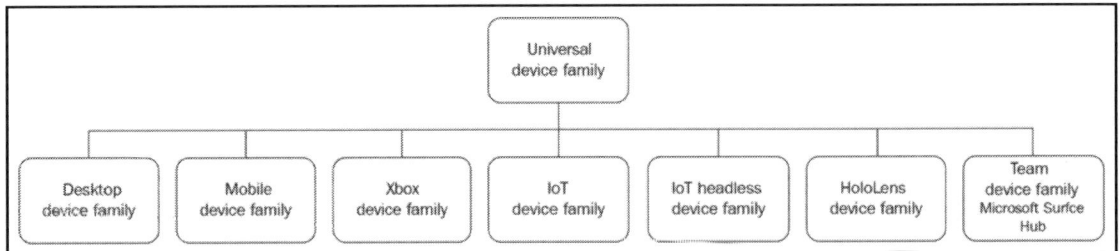

Each device family allows a subset of APIs that are available within the UWP, and it is up to the developer to decide the platform he or she wants to implement. Additionally, each platform brings in extension APIs that are only available for that platform. These differences between the device families are handled with platform extensions that can be included in your projects at compile time, as well as possible device family checks that can be executed at runtime.

UWP should not only be evaluated as a set of development tools, but truly an application platform. As a platform, it imposes certain security policies on how the applications should be handled by the runtime environment. More specifically, the application sandbox model, which is a common concept on other mobile platforms, is also imposed by UWP. Even desktop applications written for UWP should abide by the installation and execution policies to standardize the installation process for the users and protect the runtime by compartmentalizing the applications. Finally, as a result of this platform standardization, the common application store can be used for multiple platforms.

Creating UWP applications

In a cross-platform and .NET Core context, UWP relies on the .NET Framework itself. However, the .NET Framework does implement .NET Standard and, as a result, the portable modules of cross-platform applications can be consumed by UWP applications. In other words, similar to the Xamarin implementation, shared (possibly platform-agnostic) application code can be extracted from UWP applications to leave only the native UI implementation as a UWP-specific module. In return, UWP projects can be included as part of any mobile development endeavour involving .NET Standard and/or Xamarin.

When implementing the native UI, developers have two inherently similar options; depending on the existing project architecture in a Xamarin project, they can create the UWP UI using the native XAML approach (that is, create the user interface within the platform-specific project and share only the business logic) or using Xamarin.Forms and reserving the platform-specific project only for platform dependencies.

Using our previous Xamarin and Xamarin.Forms applications, we can add the UWP project so that we can deploy our application to Windows 10 devices:

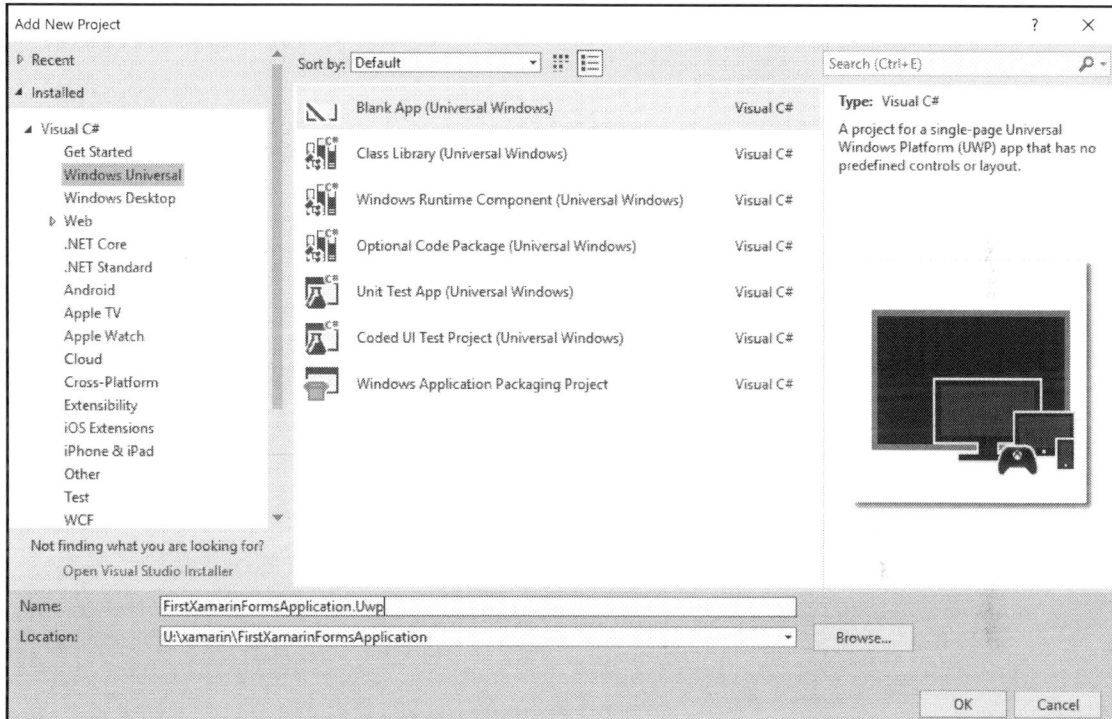

Once the project is created, we can now reference the shared (or .NET standard) platform-agnostic project and reuse the business logic. For Xamarin.Forms, we can choose to include the forms project and bootstrap the Xamarin.Forms application. Bootstrapping the Xamarin.Forms application is as simple as installing the Xamarin.Forms NuGet package for UWP and loading the Xamarin.Forms application that was created previously.

First, we install the Xamarin.Forms package and make sure that all the target platform projects have the same version installed:

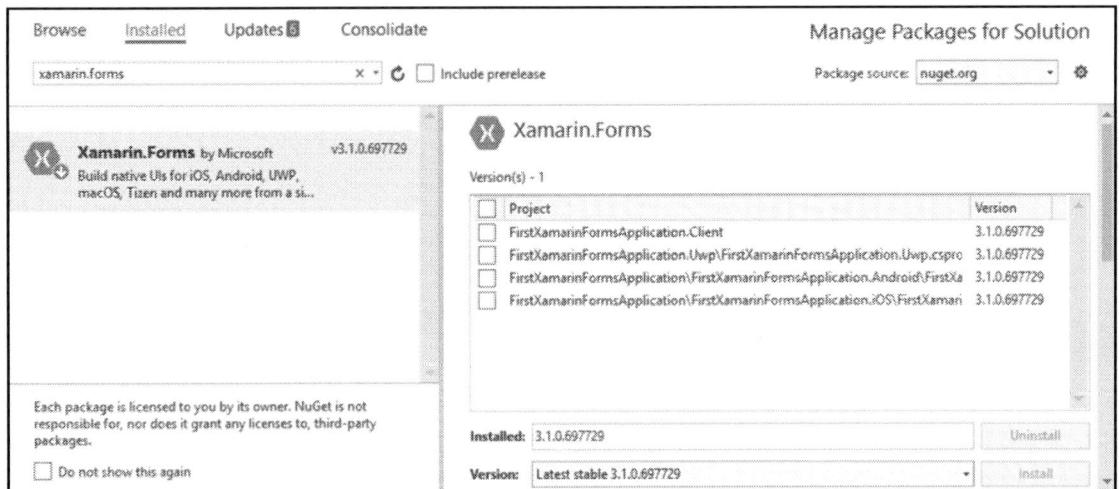

Once the forms package is installed, we can now modify the `MainPage.xaml` file, as well as `MainPage.xaml.cs`. First, we will convert the MainPage view into a forms page:

```
<forms:WindowsPage
    x:Class="FirstXamarinFormsApplication.Uwp.MainPage"
    xmlns="http://schemas.microsoft.com/winfx/2006/xaml/presentation"
    xmlns:x="http://schemas.microsoft.com/winfx/2006/xaml"
    xmlns:d="http://schemas.microsoft.com/expression/blend/2008"
    xmlns:forms="using:Xamarin.Forms.Platform.UWP"
    xmlns:local="using:FirstXamarinFormsApplication.Uwp"
    xmlns:mc="http://schemas.openxmlformats.org/markup-compatibility/2006"
    Background="{ThemeResource ApplicationPageBackgroundThemeBrush}"
mc:Ignorable="d">
    <Grid Background="{ThemeResource ApplicationPageBackgroundThemeBrush}"
/>
</forms:WindowsPage>
```

Then, we will load the Xamarin.Forms application:

```
public sealed partial class MainPage
{
    public MainPage()
    {
        this.InitializeComponent();
        LoadApplication(new FirstXamarinFormsApplication.App());
    }
}
```

Running the application now may result in an exception, stating that Xamarin.Forms should have been initialized. The initialization can be included in the OnLaunched event override method that can be found in the App.xaml.cs file.

Finally, we need to modify our view model so that it returns the correct information about the current runtime:

```
public string Platform
{
    get
    {
        if (Device.RuntimePlatform.Equals(Device.Android))
        {
            return "Android";
        }
        else if (Device.RuntimePlatform.Equals(Device.iOS))
        {
            return "iOS";
        }
        else if (Device.RuntimePlatform.Equals(Device.UWP))
        {
            return "Universal Windows Platform";
        }
        else
        {
            return "Unknown";
        }
    }
}
```

Now, running the application would display the same UI as on the previous platforms, but as a UWP application:

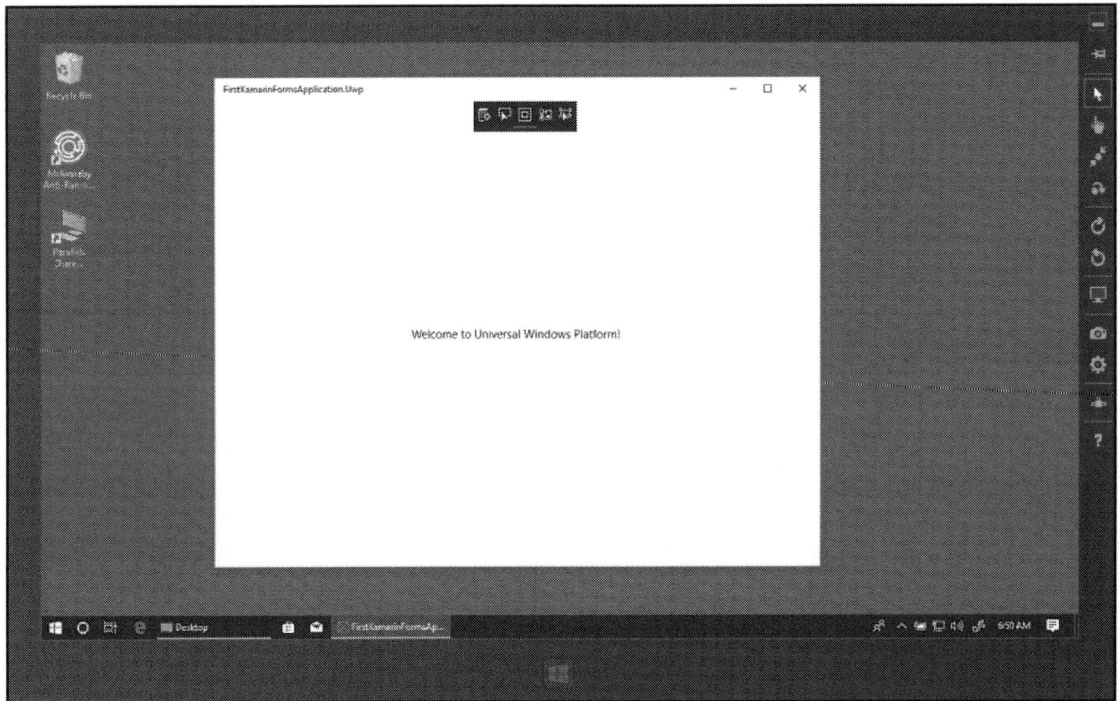

As you can see, with a minimal amount of platform code, we were able to include the Windows platform as one of the targets for our application. The same could have been done either using the Xamarin classic approach and only using the view-model, as well as by creating a native UI using the UWP toolset.

XAML Standard

Both Xamarin.Forms and UWP can utilize declarative UI pages with the **EXtensible Application Markup Language** (**XAML**). It was initially introduced as part of Windows Presentation Foundation and has been extensively used in .NET applications, starting with .NET 3.0.

While both development platforms offer similar UI elements, they use a slightly different set of controls and layouts, which might cause UI inconsistencies while you are creating cross-platform applications that target iOS/Android (with Xamarin.Forms) and UWP.

Let's take a look at the layout structures:

UWP	Xamarin.Forms	Notes
StackPanel	StackLayout	Left-to-right or top-to-bottom infinite stacking
Grid	Grid	Tabular format (rows and columns)
Canvas	AbsoluteLayout	Pixel/coordinate positioning
WrapPanel	FlexLayout	Wrapping stack
RelativePanel	RelativeLayout	Relative rule-based positioning
UniformGrid	n/a	Provides a tabular grid of uniform size
ScrollViewer	ScrollView	Provides scrolling container for content

Similar to layouts are controls. By looking at the controls that are used on these platforms, you can easily spot the subtle differences:

UWP	Xamarin.Forms	Notes
RichTextBox	Editor	Editor does not support rich text
TextBlock	Label	N/A
TextBox	Entry	N/A
ToggleSwitch	Switch	N/A
RadioButton	n/a	Switch is used in most scenarios/or custom controls
Slider	Slider	Provides a tabular grid of uniform size

If we were to recreate the MainPage XAML on UWP, the difference between the two platforms would be apparent.

On Xamarin.Forms, we have the following XAML structure:

```
<StackLayout>
    <Label Text="{Binding Platform, StringFormat='Welcome to {0}!'}"
           HorizontalOptions="Center"
           VerticalOptions="CenterAndExpand" />
</StackLayout>
```

This would have to be translated to the following (using a new page that we created in the UWP application – `MainPageAlternative.xaml`):

```
<StackPanel VerticalAlignment="Center">
    <TextBlock
        HorizontalAlignment="Center"
        VerticalAlignment="Center"
        Text="{Binding Platform,
               Converter={StaticResource FormatConverter},
               ConverterParameter='Welcome to {0}!'}" />
</StackPanel>
```

Notice that `StackLayout` was translated into `StackPanel`, while `TextBlock` was translated into `TextBlock`. Additionally, the `HorizontalOptions` and `VerticalOptions` attributes were changed to the `HorizontalAlignment` and `VerticalAlignment` attributes. Another big difference is the fact that the `StringFormat` property of the Binding markup extension is not supported on UWP. In order to support the string formatting, we would need to create our own value converter.

> **TIP**
>
> The Window Community Toolkit is a collection of helper functions and custom controls that simplify a developer's tasks in regards to creating UWP apps. The toolkit already contains commonly used converters, one of which is called `StringFormatConverter`.

Now, if we set the `DataContext` of `MainPageAlternative` to `MainPageViewModel` and assign this page as the first navigation page instead of the MainPage that was used to harness Xamarin.Forms, the resulting view would be the same:

```
rootFrame.Navigate(typeof(MainPageAlternative), e.Arguments);
```

In order to diminish the subtle differences between the two platforms, a new standardization initiative was introduced that aims to create a XAML standard, which can then be used on multiple development platforms (Xamarin.Forms and UWP).

Our current control mappings are as follows:

Xamarin.Forms	XAML Standard
Frame	Border
Picker	ComboBox
ActivityIndicator	ProgressRing
StackLayout	StackPanel
Label	TextBlock
Entry	TextBox
Switch	ToggleSwitch
ContentView	UserControl

At the moment, in order to use the XAML standard, you can introduce the `Xamarin.Forms.Alias` preview module for Xamarin.Forms and use the XAML Standard references while creating your views. After this introduction, our MainPage (we created a copy called `MainPageAlternative`) in the Xamarin.Forms project would look like this:

```
<?xml version="1.0" encoding="utf-8" ?>
<ContentPage
    x:Class="FirstXamarinFormsApplication.MainPageAlternative"
    xmlns="http://xamarin.com/schemas/2014/forms"
    xmlns:x="http://schemas.microsoft.com/winfx/2009/xaml"
    xmlns:alias="clr-
namespace:Xamarin.Forms.Alias;assembly=Xamarin.Forms.Alias"
    xmlns:local="clr-namespace:FirstXamarinFormsApplication">
    <alias:StackPanel>
        <!--  Place new controls here  -->
        <alias:TextBlock
            HorizontalOptions="Center"
            Text="{Binding Platform, StringFormat='Welcome to {0}!'}"
            VerticalOptions="CenterAndExpand" />
    </alias:StackPanel>
</ContentPage>
```

Following this initial phase of implementation and alignment, UWP and Xamarin.Forms are expected to continue to converge on both control and attribute levels.

.NET Standard and .NET Native

As we saw previously, UWP using the .NET Framework utilizes .NET Standard. .NET Standard is used as the common BCL (base class library), while the **Core Common Language Runtime** (**Core CLR**) is responsible for executing the modules that are implemented by .NET Standard. Besides .NET Core and .NET Standard, another .NET concept is invaluable for Universal Windows Applications: .NET Native.

.NET Native provides a set of tools that are responsible for generating native code from .NET applications for UWP, bypassing the intermediate language. Using the .NET Native toolchain, .NET Standard class libraries, as well as the common language runtime infrastructure modules such as garbage collection, are linked to smaller, dynamic link libraries (similar to the Xamarin build process for iOS and Android).

In order to enable the native compilation, you need to enable the .NET Native tool chain for the current configuration (for example, Release x64), which will be used for preparing the appx package in the application, as well as the appx bundle that will need to be created for the supported architectures (ARM, x86, and x64):

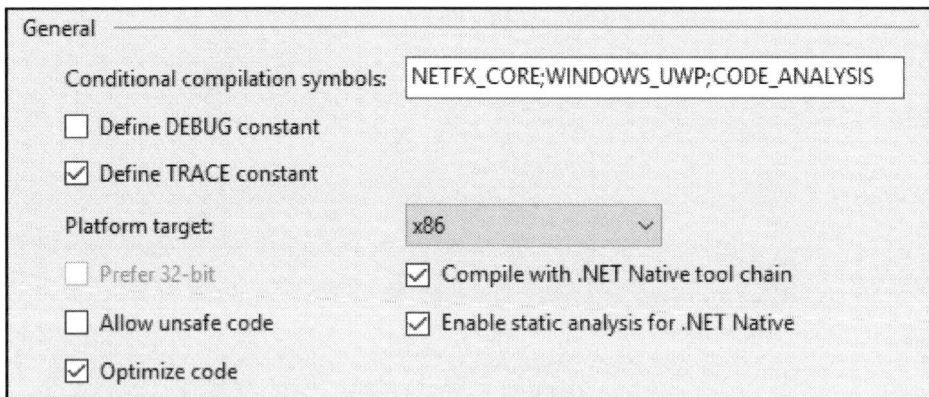

General

Conditional compilation symbols:	NETFX_CORE;WINDOWS_UWP;CODE_ANALYSIS

☐ Define DEBUG constant

☑ Define TRACE constant

Platform target: x86

☐ Prefer 32-bit   ☑ Compile with .NET Native tool chain

☐ Allow unsafe code   ☑ Enable static analysis for .NET Native

☑ Optimize code

During the linking process, several actions are executed:

- Certain code paths that utilize reflection and metadata are replaced with static native code
- Eliminates all metadata (where applicable)
- Links out the unused third-party libraries, as well the .NET Framework class libraries
- Replaces the full common language runtime with a refactored version that primarily includes the garbage collector

As a result, similar to .NET Core runtime applications (with a specified platform target), UWP applications can be cleaned up from direct dependencies to the .NET Framework by means of converting these dependencies into local references for the application itself. This, in turn, reflects on the application as performance and portability enhancements.

Platform extensions

As we mentioned previously, UWP supports a wide range of devices. Each of these devices executes its own implementation of .NET Standard and the UWP app model.

Nevertheless, the surface area of this complete API layer might not always apply to the target platform. The UWP app model contains certain APIs that are specific to only a subset of these devices. These types of API modules are in fact left as placeholder methods in the core UWP SDK, while the actual implementation is included in extension modules that can be referenced in your UWP applications:

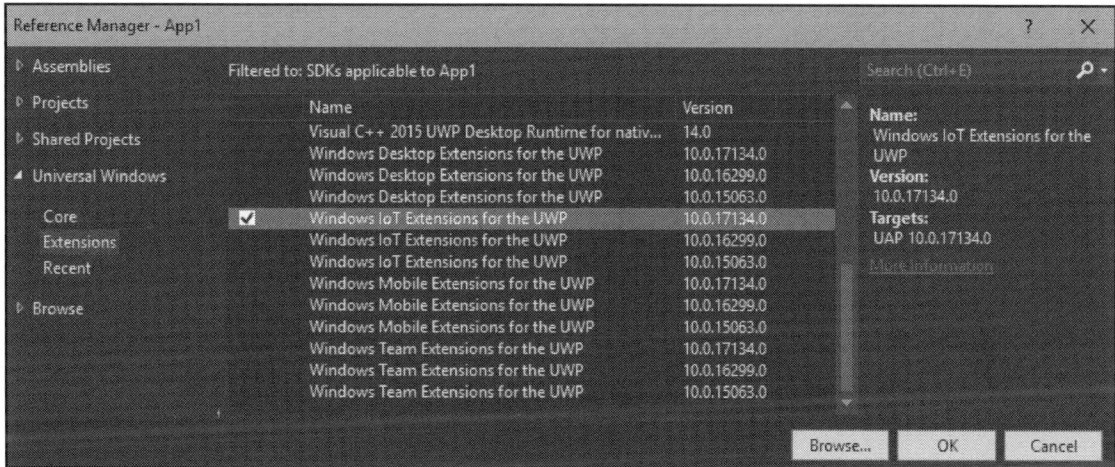

Without adding the specific SDK, the developers are confined to only universal APIs. Without adding the extension modules, it is highly likely that certain platform-specific methods would throw `NotImplementedException` or similar, since the actual implementation of these methods only exists in the platform extensions libraries.

After including the target platform extension, the developers are also responsible for executing runtime checks for the methods and events to see whether these APIs are supported in the current device runtime. Developers can make use of various `ApiInformation` methods, such as `IsTypePresent`, `IsEventPresent`, `IsMethodPresent`, and `IsPropertyPresent`.

For instance, in order to check whether the current device supports the `CameraPressed` event (it might be present on a mobile device, though unlikely to be supported on a desktop PC), we would need to resort to `IsEventPresent`:

```
bool isHardwareButtons_CameraPressedAPIPresent =
Windows.Foundation.Metadata.ApiInformation.IsEventPresent
("Windows.Phone.UI.Input.HardwareButtons", "CameraPressed");
```

This runtime check can also be executed at the contract level to see whether a group of events, or other class members that are used to execute a certain action, are supported or not:

```
bool isWindows_Devices_Scanners_ScannerDeviceContract_1_0Present =
Windows.Foundation.Metadata.ApiInformation.IsApiContractPresent
("Windows.Devices.Scanners.ScannerDeviceContract", 1, 0);
```

This way, developers can avoid situations where the application behavior and compliance with requirements is neither predetermined nor predictable.

Summary

As you can see, UWP is one of the most sophisticated members of the .NET Core family. While being quite similar to the Xamarin.Forms architecture in nature and utilizing a similar compilation/execution process, it differs in certain fundamental aspects. However, it can easily be included in, and executed with, Xamarin and Xamarin.Forms projects without increasing development timeline or maintenance costs, with the added benefit of you being able to deliver your applications to various UWPs.

Over the next few chapters, we will concentrate on Xamarin and Xamarin.Forms by creating our own Xamarin.Forms application. We will reference UWP where necessary.

2
Section 2: Xamarin and Xamarin.Forms

Xamarin and the contained Mono runtime variants for iOS and Android, as well as .NET Core for UWP, implement the .NET Standard to create the operating system abstraction for creating native mobile applications for target platforms using the familiar Microsoft stack. Xamarin.Forms, on the other hand, is a UI framework that allows developers to use common declarative user interfaces for the target platforms.

The following chapters will be covered in this section:

- Chapter 4, *Developing Mobile Applications with Xamarin*
- Chapter 5, *UI Development with Xamarin*
- Chapter 6, *Customizing Xamarin.Forms*

Developing Mobile Applications with Xamarin

4

When you're dealing with cross-platform development with Xamarin, it is important to understand that the application source cannot be completely cross-platform. Platform-agnostic modules of a Xamarin application vary, depending on the application's content, as well as the development approach that's used. Xamarin classic and Xamarin.Forms are two different approaches to creating native applications for (mainly) iOS and Android platforms. While Xamarin classic uses a more native approach, literally migrating the native platform implementation strategy to the .NET ecosystem, Xamarin.Forms delivers an additional abstraction layer for the native UI implementation.

In this chapter, we will learn about Xamarin and Xamarin.Forms development strategies and create a Xamarin.Forms application that we will develop throughout the remainder of this book. We will also discuss architectural models that might help us along the way.

The following sections will guide you through implementing a cross-platform native mobile application using the Xamarin framework and toolset:

- Xamarin versus Xamarin.Forms
- Xamarin application anatomy
- Selecting the presentation architecture
- Useful architectural patterns

Xamarin versus Xamarin.Forms

Xamarin, as a runtime and framework, provides developers with all the necessary tools to create cross-platform applications. In this quest, one of the key goals is to create a code base with a minimal amount of resources and time; another is to decrease the maintenance costs of the project. This is where Xamarin.Forms comes into the picture.

As we explained previously, by using the Xamarin classic approach, developers can create native applications. In this approach, we aren't really worried about creating a cross-platform application since we are creating an application, for all the target platforms using the same development tools and language. The shared components between the target platforms would, in this case, be limited to the business logic (that is, view-models) and the data access layer (models). However, in a modern mobile application, the actual business logic would have been migrated to a service-oriented implementation. The application is generally responsible for executing simple service calls through a gateway API facade on the bundle of downstream microservices, as shown in the following diagram:

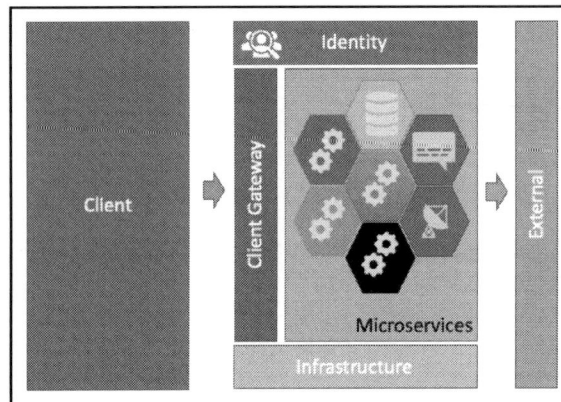

In an infrastructure setup similar to this, where each client can benefit from a tailored mobile API gateway, the platform implementations diverge from each other, mostly because of the separate UI layer. Moreover, shared business logic and the data access layer cannot really increase the amount of shared code.

As a general rule of thumb, Xamarin classic applications are advised to be used with applications with key features that are dependent on the platform they are running on (peripheral APIs, intrinsic UI components, performance requirements, and so on). However, for a general purpose mobile application with a cloud-base service backend, it might be a better option to use Xamarin.Forms.

The Xamarin.Forms framework aims to standardize the UI implementation process while preserving the nativity of the application on multiple platforms. In essence, applications that are created with Xamarin.Forms are rendered with native UI elements from the target platform. As a matter of fact, once compiled and linked, a Xamarin application is not any different than a Xamarin.Forms application for any given target platform.

Xamarin application anatomy

When developing a Xamarin.Forms application, the essentials of the application includes the target platform projects – which act as a harness to initialize the Xamarin.Forms framework and application, as well as the native rendering or API implementations – and a platform-agnostic project that contains the Xamarin.Forms views, as well as the abstractions, so that the custom components can be implemented on platform-specific projects.

As the project grows in size, developers will need to create a separate project that would only contain the view-model and platform-agnostic services implementation. In this case, the project would become the main target of the unit testing process, since this layer does not depend on the UI elements or platform services directly. Additionally, a separate project can be used to share **data transfer object** (DTO) models between the services layer and the client applications. In a setup like this, the overall architectural layout will look similar to the following:

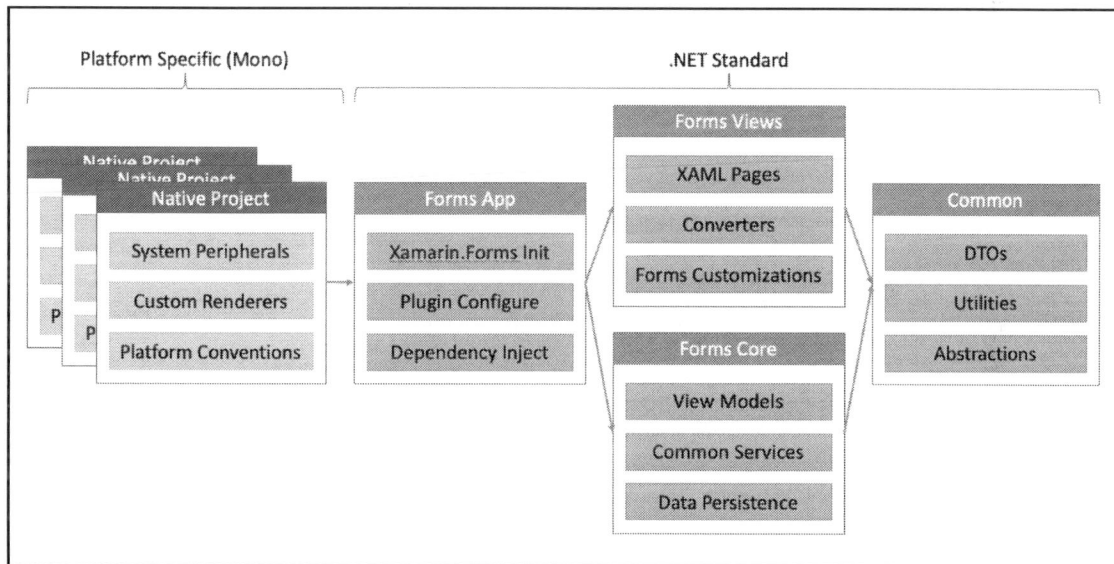

In some implementations where platform-specific APIs need to be tested, platform-specific unit tests are used, which are executed on the target platform rather than the development platform itself.

Selecting the presentation architecture

When developing a cross-platform mobile application, it is perhaps one of the most crucial decisions to select the presentation architecture. The view and the business logic implementation should be factoring in the architectural concepts that the selected pattern entails.

Model-View-Controller (MVC) implementation

Both iOS and Android platforms are inherently designed to be used with a derivative of the **Model-View-Controller** (MVC) pattern. If we were dealing with a native application, it would have been the most logical path to use MVC for iOS and **Model-View-Presenter** (**MVP**) or a slightly derived version, the **Model-View-Adapter** (**MVA**), pattern, for Android:

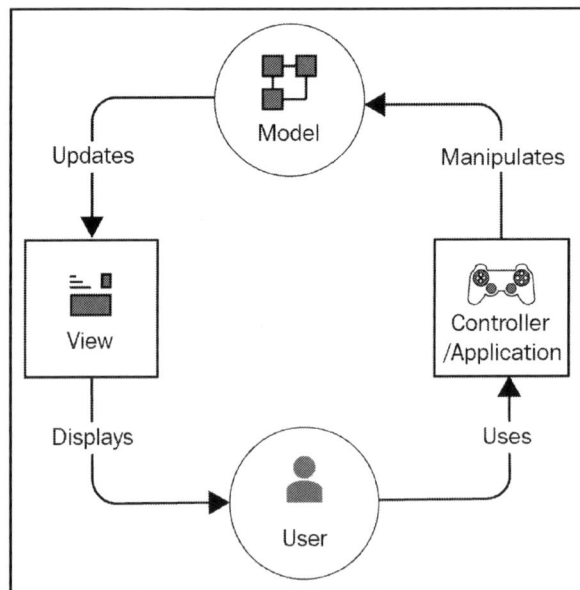

The MVC pattern was born as a reaction to the single responsibility principle. In this pattern, the **View** (UI implementation) component is responsible for presenting the data that's received from the **Model** (service layer) and delegating the user input to the **Controller** so that the data changes can be propagated to the **Model**.

While it is being used widely with web applications, generally, a derived version is used for mobile and desktop applications. Derivatives of this pattern include MVA and MVP.

In an MVA architecture (or Mediating Controller), the Adapter acts as a mediator between the **View** and **Model**, and is responsible for defining the strategy for one or more view components, as well as acting as the observer for these UI components:

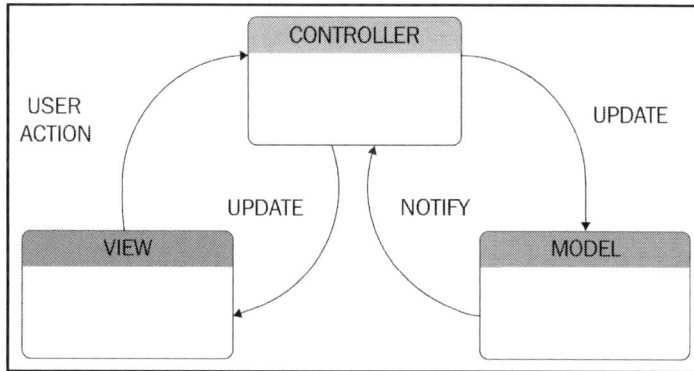

In an MVC implementation, while using both classic and mediator patterns, the **Controller** becomes the heart and soul of the application. It needs to be aware of the **Model**, as well as the **View** (which is tightly coupled), since it implements a strategy for the view events (user input) that are acting as both the strategy implementer and observer.

Let's demonstrate this pattern while implementing a login view for our application:

1. First, we need to create the view. Create a content page with a XAML design component:

```
<ContentPage xmlns="http://xamarin.com/schemas/2014/forms"
xmlns:x="http://schemas.microsoft.com/winfx/2009/xaml"
x:Class="FirstXamarinFormsApplication.LoginView">
    <ContentPage.ToolbarItems>
        <ToolbarItem x:Name="signUpButton" Text="Sign Up" />
    </ContentPage.ToolbarItems>
    <ContentPage.Content>
        <StackLayout VerticalOptions="CenterAndExpand"
        Padding="20">
            <Label Text="Username" />
            <Entry x:Name="usernameEntry" Placeholder="username"/>
            <Label Text="Password" />
            <Entry x:Name="passwordEntry" IsPassword="true"
            Placeholder="password" />
            <Button x:Name="loginButton" Text="Login" />
            <Label x:Name="messageLabel" />
        </StackLayout>
    </ContentPage.Content>
</ContentPage>
```

2. Next, create the two entries (that is, username and password), two buttons (that is, for login and signup), and finally a label field, which we will use to display the result of the login function.

3. In order to support the implementation of a controller for this view that will handle the field validations, as well as the login and signup actions, expose the entry and button components to the associated controller:

```
public partial class LoginView : ContentPage
{
    // private LoginViewController _controller;

    public LoginView()
    {
        InitializeComponent();

        // _controller = new LoginViewController(this);
    }

    internal Entry UserName { get { return this.usernameEntry; }}

    internal Entry Password { get { return this.passwordEntry; }}

    internal Label Result { get { return this.messageLabel; }}

    internal Button Login { get { return this.loginButton; }}

    internal ToolbarItem SignUp { get { return this.signUpButton;
}}
}
```

4. Now, create the controller that will define the strategies for various events that we will need to handle:

```
public class LoginViewController
{
    private LoginView _loginView;

    public LoginViewController(LoginView view)
    {
        _loginView = view;

        _loginView.Login.Clicked += Login_Clicked;

        _loginView.SignUp.Clicked += SignUp_Clicked;

        _loginView.UserName.TextChanged += UserName_TextChanged;
    }

    void Login_Clicked(object sender, EventArgs e)
    {
        // TODO: Login
        _loginView.Result.Text = "Successfully Logged In!";
    }

    void SignUp_Clicked(object sender, EventArgs e)
    {
        // TODO: Navigate to SignUp
    }

    void UserName_TextChanged(object sender,
    TextChangedEventArgs e)
    {
        // TODO: Validate
    }
}
```

As you can see, even though we can further refactor our code so that we can insert abstraction layer(s) between the controller and the view, there is a strong coupling between the two. In order to remedy this bound, we can resort to a **Model–View–ViewModel (MVVM)** setup.

Model-View-ViewModel (MVVM) implementation

In response to the tight-coupling problem, another derivative of MVC was born with the release of the **Windows Presentation Foundation** (**WPF**). The idea that was coined by Microsoft was the concept of outlets being exposed by a controller (or ViewModel, in this case) and the outlets being coupled by the view elements. The concept of these outlets and their coupling is called the binding:

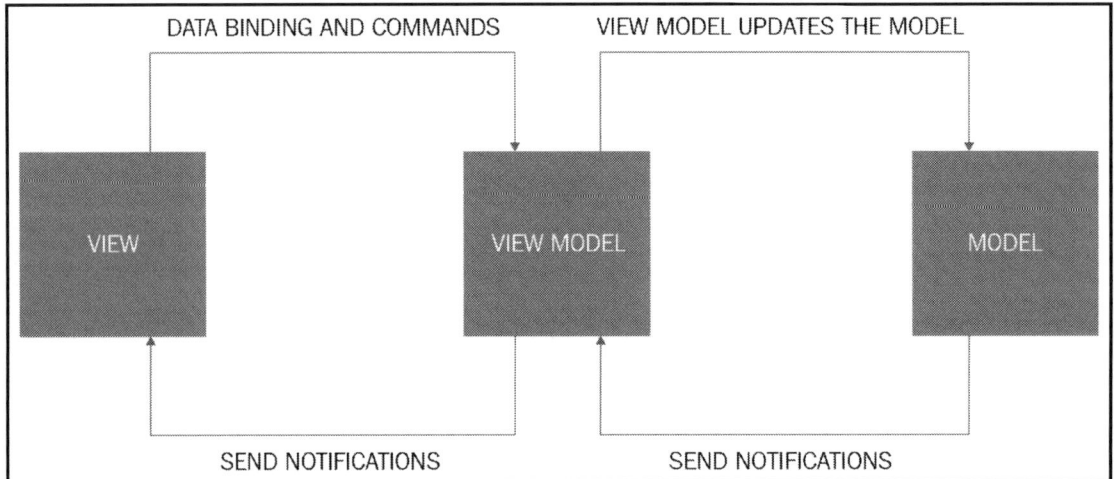

Source: https://commons.wikimedia.org/wiki/File:MVVMPattern.png

Using bindings, we can decrease the amount of knowledge of the view-model about the inner workings of the view elements and let the application runtime handle the synchronization on the outlets on View and ViewModel fronts.

In addition to bindings, the concept of command becomes invaluable so that we can delegate the user actions to the ViewModel. A command is a stateful single execution unit that represents a function and the related data to execute this function.

Using our previous example, we can create a view-model to demonstrate the benefits of using MVVM:

1. Let's start by creating a class that represents the user interaction points on our view (that is, two string fields for username and password, and two functions for login and signup):

```
public class LoginViewModel
{
    private string _userName;
```

```csharp
private string _password;

private string _result;

public LoginViewModel()
{

}

public string UserName
{
    get
    {
        return _userName;
    }
    set
    {
        if (_userName != value)
        {
            _userName = value;
        }
    }
}

public string Password
{
    get
    {
        return _password;
    }
    set
    {
        if (_password != value)
        {
            _password = value;
        }
    }
}

public string Result
{
    get
    {
        return _result;
    }
    set
    {
        if (_result != value)
```

```
            {
                _result = value;
            }
        }
    }

    public void Login()
    {
        //TODO: Login
        Result = "Successfully Logged In!";
    }

    public void Submit()
    {
        // TODO:
    }
}
```

2. At this stage of implementation, we can bind the `Entry` fields from our view to the view-model. In order to assign the view-model to the view, use the `BindingContext` of our view:

```
public partial class LoginView : ContentPage
{
    public LoginView()
    {
        InitializeComponent();
        BindingContext = new LoginViewModel();
    }
}
```

3. We can now set up the bindings for the `Entry` fields:

```
<Label Text="Username" />
 <Entry x:Name="usernameEntry" Placeholder="username"
Text="{Binding UserName}" />
 <Label Text="Password" />
 <Entry x:Name="passwordEntry" IsPassword="true"
Placeholder="password" Text="{Binding Password}" />
 <Button x:Name="loginButton" Text="Login" />
 <Label x:Name="messageLabel" Text="{Binding Result}" />
```

When executing this sample, you will notice that the values for the entries with unidirectional data flow (that is, the `UserName` and `Password` fields are only propagated from the View to the view-model) are behaving as expected; the values that are entered in the associated fields are pushed to the properties, as expected.

The view to view-model binding context setup can also be done in XAML as well. `<ContentPage.BindingContext>` can be used to set the binding context to the view-model, which is initialized using the correct `clr` namespace (for example, `<local:LoginViewModel />`). In order for this to work as expected, the view-model class needs to have a parameterless constructor.

In order to increase the binding's performance and decrease the resources that are used for a certain binding, it is important to define the direction for the binding. There are various **BindingMode** available, as follows:

- **OneWay**: This is used when the `ViewModel` updates a value. It should be reflected on the view.
- **OneWayToSource**: This is used when the view changes a value. The value change should be pushed to the view-model.
- **TwoWay**: The data flow is bi-directional.
- **OneTime**: The synchronization of data occurs only once when the binding context is bound, and data is propagated from the view-model to the view.

With this information at hand, the username and password fields should be using the OneWayToSource binding, whereas the message label should use a OneWay binding mode, since the result is only updated by the view-model.

The next step would be to set up the commands for the functions to be executed (that is, login and signup). Semantically, a command is composed of a method (with its enclosed data and/or arguments) and a state (whether it can be executed or not). This structure is described by the `ICommand` interface:

```
public interface ICommand
{
    void Execute(object arg);
    bool CanExecute(object arg);
    event EventHandler CanExecuteChanged;
}
```

In Xamarin.Forms, there are two implementations of this interface: `Command` and `Command<T>`. Using either of these classes, command bindings can be accomplished. For instance, in order to expose the `Login` method as a command, follow these steps:

1. First, declare our `Command` property:

```
private Command _loginCommand;

public ICommand LoginCommand { get { return _loginCommand; } }
```

2. In order to initialize `_loginCommand`, use the constructor:

```
public LoginViewModel()
{
    _loginCommand = new Command(Login, Validate);
}
```

Note that we used two actions to initialize the command. The first action is the actual method execution, while the second function is a method that returns a Boolean indicating whether the method can be executed.

3. The `Validate` method's implementation could look like this:

```
public bool Validate()
{
    return !string.IsNullOrEmpty(UserName) &&
!string.IsNullOrEmpty(Password);
}
```

4. Finally, in order to complete the implementation, send the `CanExecuteChanged` event whenever the `UserName` or `Password` fields are changed:

```
public string UserName
{
    get
    {
        return _userName;
    }
    set
    {
        if (_userName != value)
        {
            _userName = value;
            _loginCommand.ChangeCanExecute();
        }
    }
}

public string Password
{
    get
    {
        return _password;
    }
    set
    {
        if (_password != value)
```

```
            {
                _password = value;
                _loginCommand.ChangeCanExecute();
            }
        }
    }
```

5. Now, if we were to run the application, you would see how the disabled and enabled states of the command are reflected on the UI.

   The same setup can be used with methods that require an input argument using the `Command<T>` class:

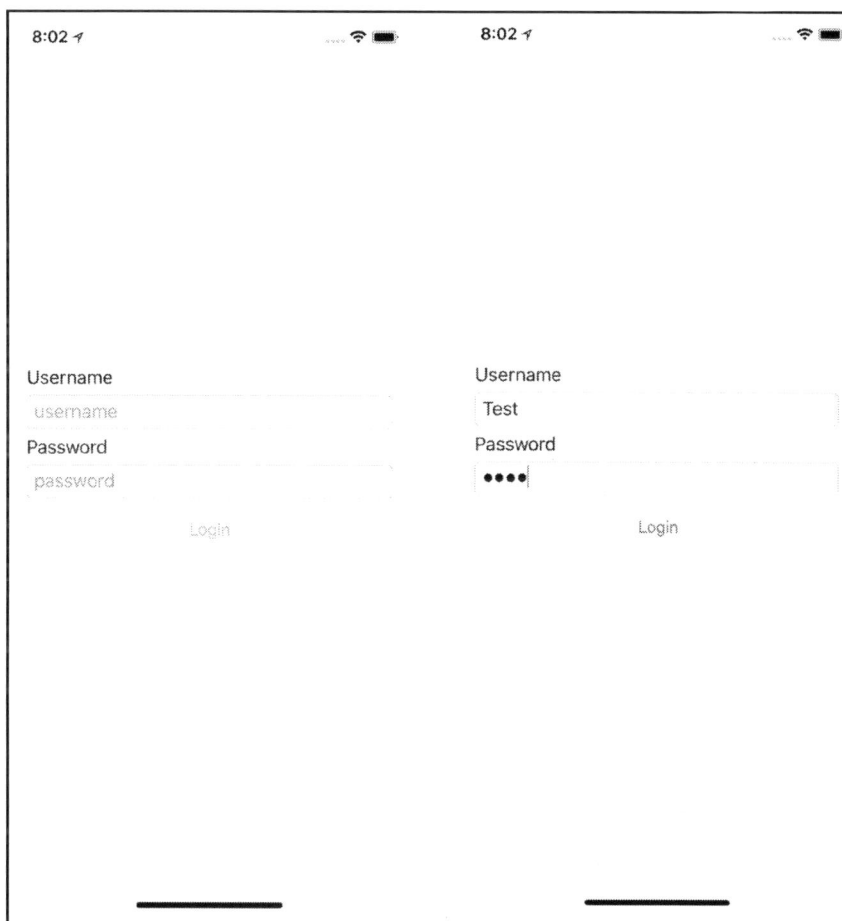

6. Once the command setup is complete, we only have the result message binding, which is still not working as expected. At this point, tapping the login button will update the view-model data, and yet the user interface will not reflect this data change. The reason for this is the fact that this field should be bound with a OneWay binding (changes in the source should be reflected on the target) and the main requirement for this is that the source (view-model) should be implementing the INotifyPropertyChanged interface. INotifyPropertyChanged is the essential mechanism for propagating the changes on the binding context to the view elements:

```
/// <summary>Notifies clients that a property value has
changed.</summary>
public interface INotifyPropertyChanged
{
    /// <summary>Occurs when a property value changes.</summary>
    event PropertyChangedEventHandler PropertyChanged;
}
```

A simple implementation would require the invocation of the PropertyChanged event with the property that is currently being changed.

> **TIP**
>
> If the change of a property is affecting multiple data points (for example, assigning a list data source changes the item count property), then the view-model is responsible for firing the same event for all the properties that the UI needs to invalidate.

7. Finally, by including the event trigger on the setter of the Result property, we should able to see the outcome of the Login command:

```
public string Result
{
    get
    {
        return _result;
    }
    set
    {
        if (_result != value)
        {
            _result = value;
            PropertyChanged?.Invoke(this, new
            PropertyChangedEventArgs(nameof(Result))); ;
        }
    }
}
```

This finalizes the view and view-model setup for the login page. In this example, we have created a setup where the view is responsible for creating the view-model; however, by using an implementation of **Inversion of Control (IoC)**, such as dependency injection or the service locator pattern, the view can be dismissed of this duty.

Useful architectural patterns

Xamarin.Forms as a framework contains modules that help developers implement well-known architectural patterns so that they can create maintainable and robust applications.

Inversion of Control

IoC is a design principle in which the responsibility of selecting concrete implementations for dependencies of a class is delegated to an external component or source. This way, the classes are decoupled from their dependencies so that they can be replaced/updated without much hassle.

The most common implementation of this principle is using the service locator pattern, where a container is created to store the concrete implementations, often registered via an appropriate abstraction, as shown in the following diagram:

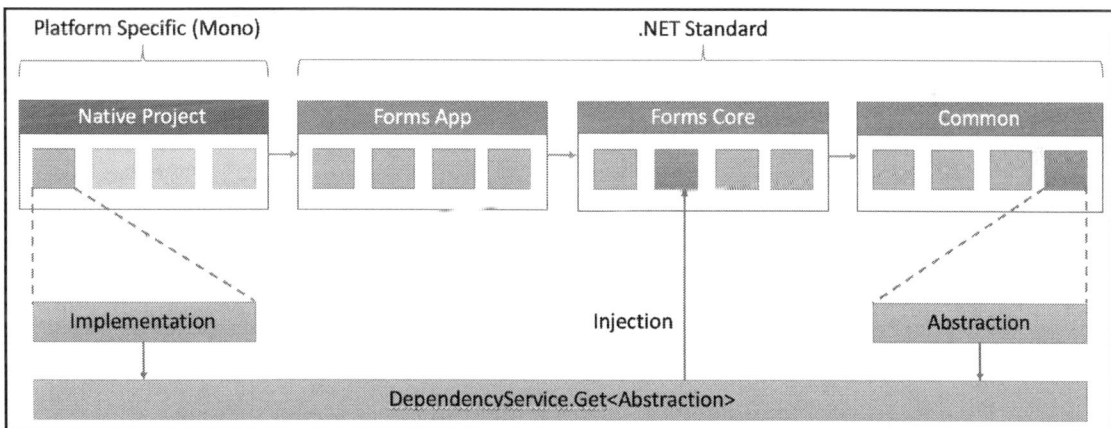

Xamarin.Forms offers **DependencyService**, which can be especially helpful when you're creating platform-specific implementations for platform-agnostic requirements. We must also remember that it should only be used in Xamarin.Forms platform projects; otherwise, we would be creating an unnecessary dependency on Xamarin.Forms libraries.

Event aggregator

An event aggregator (also known as publisher/subscriber or Pub-Sub) is a messaging pattern where senders of messages/events, called publishers, do not program the messages to be sent directly to a single/specific receiver, called subscribers, but, instead, categorize published messages into classes without knowledge of which subscribers, if any, there may be. Messages are then funneled through a so-called aggregator and delivered to the subscribers. This approach provides a complete decoupling between the publishers and subscribers while maintaining an effective messaging channel.

Event aggregator implementation within the Xamarin.Forms framework is done via MessagingCenter. MessagingCenter exposes a simple API that is composed of two methods for subscribers (that is, subscribe and unsubscribe) and one method for publishers (send).

We can demonstrate the application of the event aggregator pattern by creating a simple event sink for service communication or authentication issues in our application. In this event sink (which will be our subscriber), we can subscribe to error messages that are received from various view-models (given that they all implement the same base type) and alert the user with a friendly dialog:

```
MessagingCenter.Subscribe<BaseViewModel> (this, "ServiceError", (sender,
arg) =>
    {
        // TODO: Handler the error
    });
```

We would have the following in the event of an unhandled exception on our view-model:

```
public void Login()
{
    try
    {
        //TODO: Login
        Result = "Successfully Logged In!";
    }
    catch(Exception ex)
    {
        MessagingCenter.Send(this, "ServiceError", ex.Message);
    }
}
```

Decorator

The decorator pattern is a design pattern that allows behavior to be added to an individual object dynamically, without affecting the behavior of other objects from the same class. Xamarin.Forms makes use of this pattern to use platform-agnostic visual elements (the views used in XAML) and attaches renderers to these elements that define the way they are rendered (creating native platform-specific controls) on target platforms. The composition of Xamarin.Forms elements does not in any way change the behavior of the renderers and vice versa, allowing the developers to create custom renderers and attach them to views without affecting other visual elements. The following diagram shows the abstraction of renderer class's interaction with the decorator pattern:

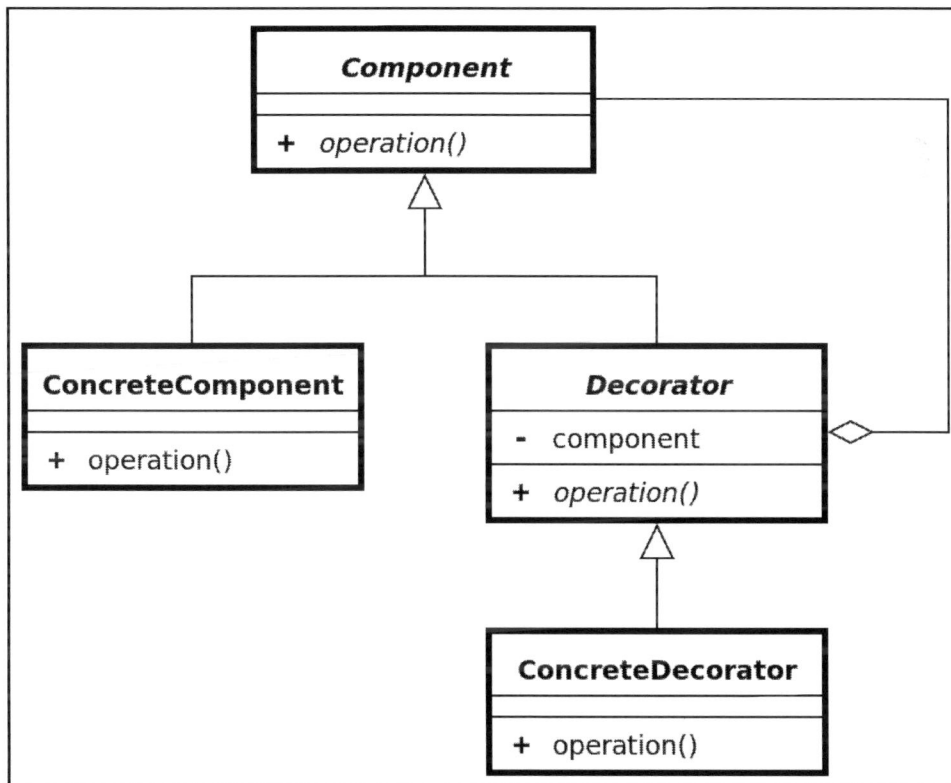

A similar approach is used to create so-called effects, which are simple behavioral modifiers that are attached to existing visual elements, as well as their native counterparts.

Summary

In this chapter, we had a deep dive into the architectural aspects of implementing a Xamarin application, and set up the foundation for a MVVM application. We implemented the login view using both the MVC and MVVM patterns to demonstrate the architectural differences between the two. We have also briefly browsed through several other patterns that we might need in order to implement Xamarin applications.

In the next chapter, we will implement the initial views of our application with standard Xamarin.Forms components. In the remainder of this book, we will try to implement the components of the application that were discussed in this chapter.

UI Development with Xamarin 5

Material Design, which is the most prominent UI pattern for Android applications; Apple's human interface guidelines; and finally, UWP's Fluid UI language, can make it overwhelming for UX designers and developers to decide on an application design. Factors to consider include, but are not limited to, user expectations of the target platform and branding-related product owners requirements regardless of the platform.

In this chapter, we will take a look at certain UI patterns that allow developers and UX designers to create a compromise between user expectations and product demands to create a product with a consistent UX across all platforms. The following topics will walk you through creating the skeleton of our sample application:

- Application layout
- Implementing navigation structure
- Using Xamarin.Forms and native controls
- Creating data-driven views

Application layout

For designers, as well as developers, probably one of the most exciting phases of the application life cycle is the design phase. In this phase, there are multiple factors that need to be carefully considered, avoiding any rash decisions. An application's design, in simple terms, should satisfy the following:

- The consumers' expectations
- The platform imperatives
- Development costs

Consumer expectations

The feature-set of an application should really correlate with customers' expectations. Layout options and navigation hierarchy should serve the purpose of the application. According to the requirements, an application can be designed as a single-page application or with a complex hierarchy of navigation pages; the content can be text-only or rich media elements can be used; and context actions can provide access to user actions or the interaction can be laid over multiple application pages.

On the view level, in general terms, an application view contains three different types of elements: content, navigation, and actions. It is the developers' and designers' responsibility to create the optimal blend of these elements. In most modern applications, these elements can take on multiple functionalities, where the content elements become user actions as well as navigation elements.

For instance, if we had a list of items defined with an image, a simple title, and a description (that is, a simple two-column, two-row template) and we were to implement actions related to use content items, then the design, the flow, and the behavior of the elements would really depend on the type of those actions.

> As a user, I would like to see a list of items and interact with them so that I can execute certain actions on those items.

Interaction models for the list view and the complimenting pages can differ greatly:

- This list could be a common item list view used for detail navigation, where the item detail page exhibits the actions available for an item. In this case, we are assuming that the user needs to see the details of an item before they can execute the necessary action on an item (for example, if the items are hard to distinguish just using the listing):

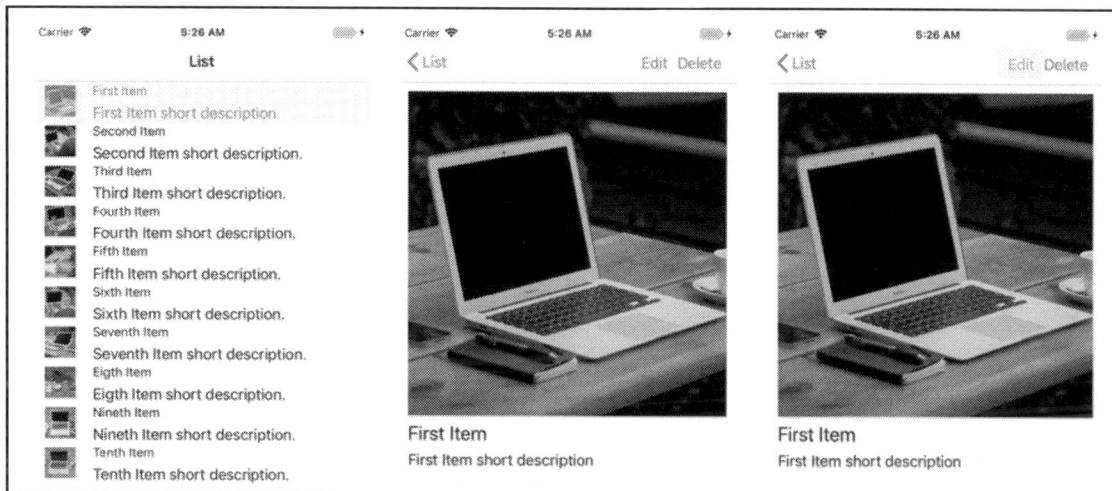

In this case, the content items in the list view are behaving as navigation items, and on the details screen, the user is able to execute the actions that are related to those specific items. However, for a simple action, the user would need to change the view and would lose the context of the list.

- If an action can be executed on the list view, we do not need to take the user to a secondary view, and can allow them to execute the actions by directly interacting with the list:

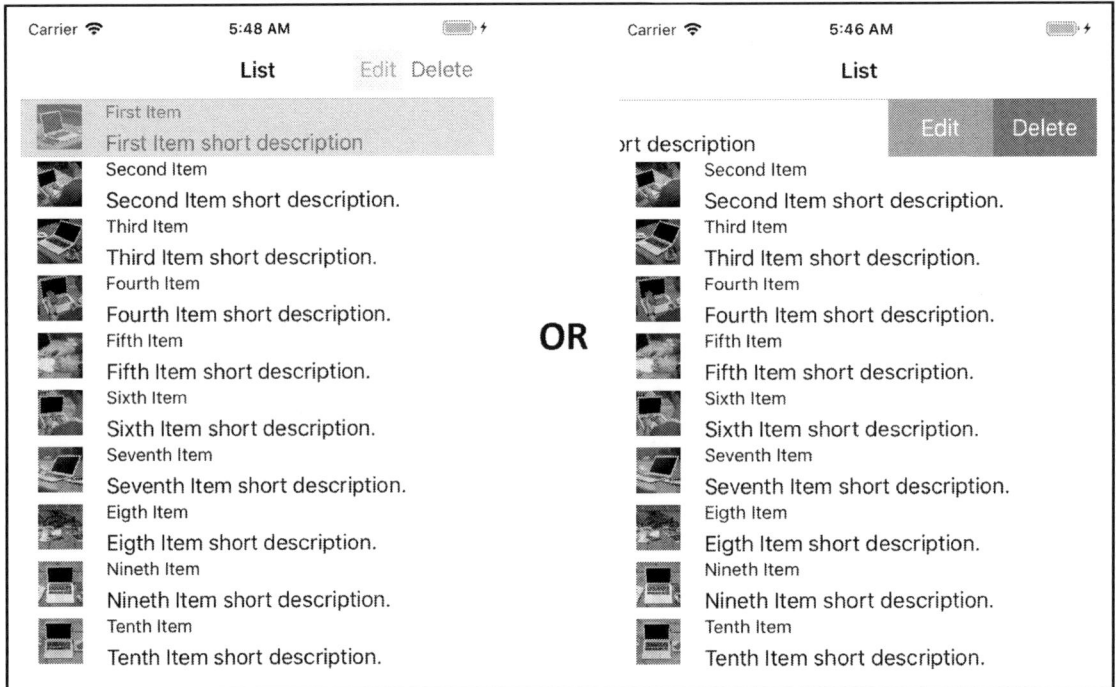

In this implementation, the list view acts as the single interaction context and actions are executed on items directly. In other words, content elements are used as action elements instead of using them for navigation.

- Finally, if there are actions available for execution on multiple content elements, for the economical use of design space, content items themselves could be used with additional styling (for example, an overlay of a checkmark on the image element) to replace the possible use of checkboxes or radio buttons. This would further improve the user experience, since we would, again, be allowing the user to interact with the content itself rather than the user input elements.

Additionally, to decrease the amount of unnecessary control elements and embellishments, the iOS and Windows platforms in particular emphasize the use of calligraphy while creating content elements. Using font variations, the visual priority of certain content elements can be adjusted to provide the correct information. For instance, in the previous examples, the title of the element was created using a smaller font, emphasizing the description.

Platform imperatives

Dealing with cross-platform mobile applications, developers need to create applications that will satisfy multiple design surfaces, as well as guidelines for multiple operating systems and idioms. Platform imperatives refer to the guidelines that developers and designers need to find a compromise between.

An **idiom** is how the form factor is defined in Xamarin.Forms applications. It can be used to create specific views for various phone form factors, as well as tablet, desktop, TV, and even for design surfaces for wearables such as Tizen watch.

When dealing with different idioms, target device capabilities, design surfaces, and input methods should be taken into consideration.

In order to make best use of the space available for web applications on desktop and mobile devices, developers often use responsive design techniques that can also be applied to Xamarin applications:

- **Fluid layout**: In a fluid layout, items are stacked in a horizontal list and can take as many rows as required to list them, depending on the horizontal space available:

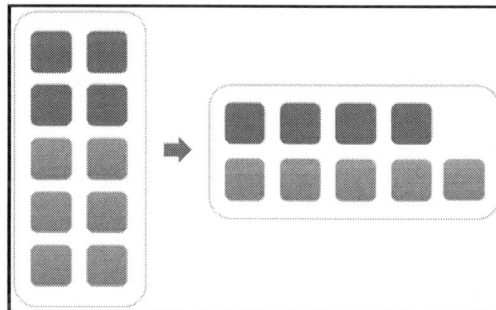

- **Orientation change**: Items listed in a horizontal list on a device with a wider screen can be stacked vertically on devices with smaller screens with a greater height relative to the width.
- **Restructure**: Elements' general layout can completely be restructured according to the available space. For instance, a view can use three segments in a horizontal setup in landscape mode, whereas for the portrait mode, two of these segments can be merged into one:

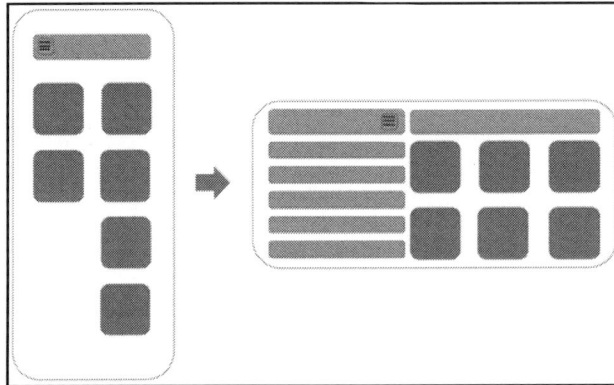

- **Resize**: Rich media content elements, as well as text content, can be resized to make the best use of the design space available:

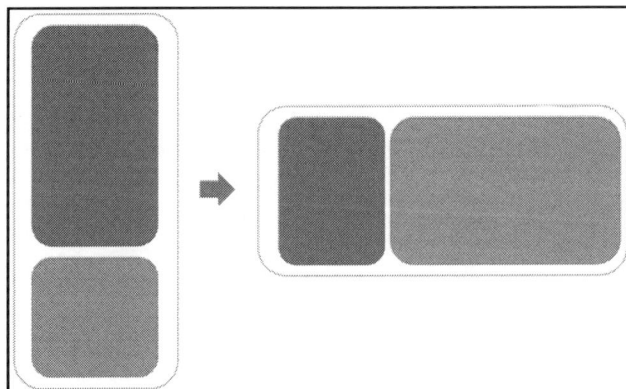

Additionally, as previously mentioned, device capabilities can play a big role in how an application should react to user input. For instance, the hardware back button is a nice example of this design consideration. If we were designing a mobile application that targeted the Android, iOS, and UWP platforms, we would need to remember that only Android offers a hardware or software back button. This capability, or lack of it on other platforms, makes it crucial to include a back navigation element on a second-tier application view. Similarly, if we were designing an application to be used on mobile devices (let's say, on iOS and Android), but at the same time the application should run on TVs with Tizen or Android operating systems, the input method and how the user navigates through the screens would become a crucial design factor.

Development cost

Finally, technical feasibility is another important aspect of objectively analyzing the design requirements of an application. In some cases, the development costs of creating a custom control to mimic a web application outweighs the business or platform value added to the native counterpart of the same application.

Each mobile platform that Xamarin and Xamarin.Forms target offers a different user experience and a different set of controls. Xamarin.Forms create an abstraction on top of this set of native controls so that the same abstraction is rendered using native views on a specific platform. In this context, trying to introduce new design elements or customize controls that are inherently different in appearance and behave like each other can have costly repercussions.

For instance, if the web counterpart of the application uses a checkbox for a certain preference, the mobile view to use in this case would be a toggle switch. Insisting on a checkbox would mean additional development hours, as well as an undesirable user experience on the target platform. Similarly, using checkboxes for (multi) selection rather than highlighting the selected content can lead to UX degradation for the specific mobile platform and platform users:

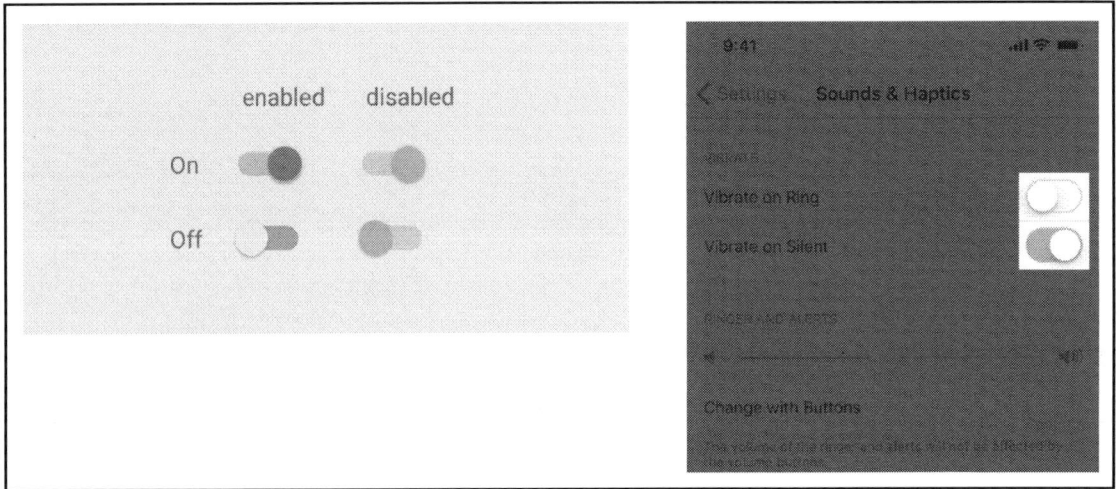

Implementing navigation structure

One of the main decisions to make before starting development is to decide on the navigation hierarchy of your application. Generally, this decision should have been taken care of during the UX design phase.

According to the requirements and target audience of your application, the navigation hierarchy can be designed in different ways. Some of these navigation strategies can be summarized as follows:

- Single-page view
- Simple navigation
- Multi-page views
- Master/detail view

Single-page view

In a single-page view, as the name suggests, a single view is used for the content and possible user interaction, and actions are either executed on this view or on action sheets. Depending on the design requirements, this view can be implemented using either `ContentPage` or `TemplatedPage`:

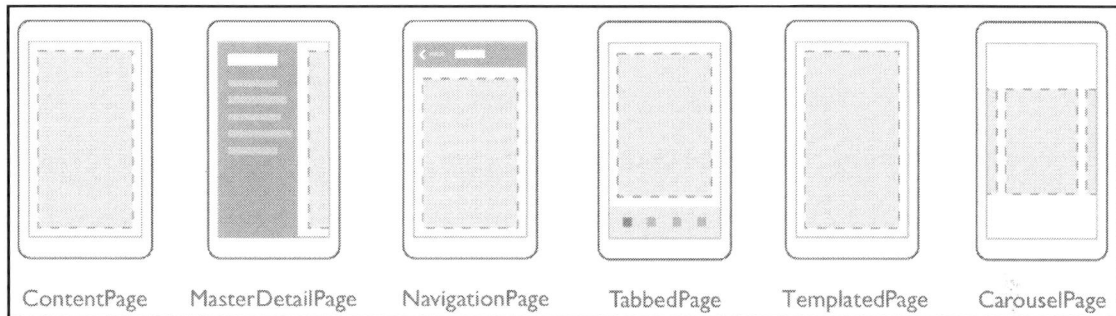

ContentPage   MasterDetailPage   NavigationPage   TabbedPage   TemplatedPage   CarouselPage

`ContentPage` is among the most commonly used page definitions. Using this page structure, developers are free to include any layout and view elements within the content definition of a content page.

In order to create the item list view that was previously demonstrated, we start by creating our content page:

```xml
<?xml version="1.0" encoding="UTF-8"?>
<ContentPage
    Title="List"
    xmlns="http://xamarin.com/schemas/2014/forms"
    xmlns:x="http://schemas.microsoft.com/winfx/2009/xaml"
    x:Class="FirstXamarinFormsApplication.Client.ListItemView">
    <ContentPage.ToolbarItems>
        <!-- Removed for brevity -->
    </ContentPage.ToolbarItems>
    <ContentPage.Content>
        <!-- Removed for brevity -->
    </ContentPage.Content>
</ContentPage>
```

Here, the content containers that are used are the `Content` and `Toolbar` items, to create a list view of items and the toolbar action buttons respectively.

ContentPage is a derivative of TemplatedPage, which is another page type that can be used with Xamarin.Forms applications. TemplatedPage allows developers to create a base style for TemplatePage (that is, ContentPage) so that certain global level customizations can be applied to these pages.

For instance, if we were to expand our previous implementation with a footer, we would need to define a style for this page (in App.xaml):

```
<Application.Resources>
    <ResourceDictionary>
        <ControlTemplate x:Key="PageTemplate">
            <Grid>
                <Grid.RowDefinitions>
                    <RowDefinition />
                    <RowDefinition Height="25" />
                </Grid.RowDefinitions>
                <Grid.ColumnDefinitions>
                    <ColumnDefinition />
                </Grid.ColumnDefinitions>
                <ContentPresenter Grid.Row="0" />
                <BoxView Grid.Row="1" Color="Navy" />
                <Label
                        Grid.Row="1"
                        Margin="10,0,0,0"
                        Text="(c) Hands-On Cross Platform 2018"
                        TextColor="White"
                        VerticalOptions="Center" />
            </Grid>
        </ControlTemplate>
    </ResourceDictionary>
</Application.Resources>
```

In this template, notice that ContentPresenter is used as the placeholder for the ContentPage that is to be used. We would apply this template in the ListItemView (and ItemView) pages with the following code:

```
<ContentPage
    Title="List"
    ControlTemplate="{StaticResource PageTemplate}"
    xmlns="http://xamarin.com/schemas/2014/forms"
    xmlns:x="http://schemas.microsoft.com/winfx/2009/xaml"
    x:Class="FirstXamarinFormsApplication.Client.ListItemView">
```

This would result in the footer appearing on both pages:

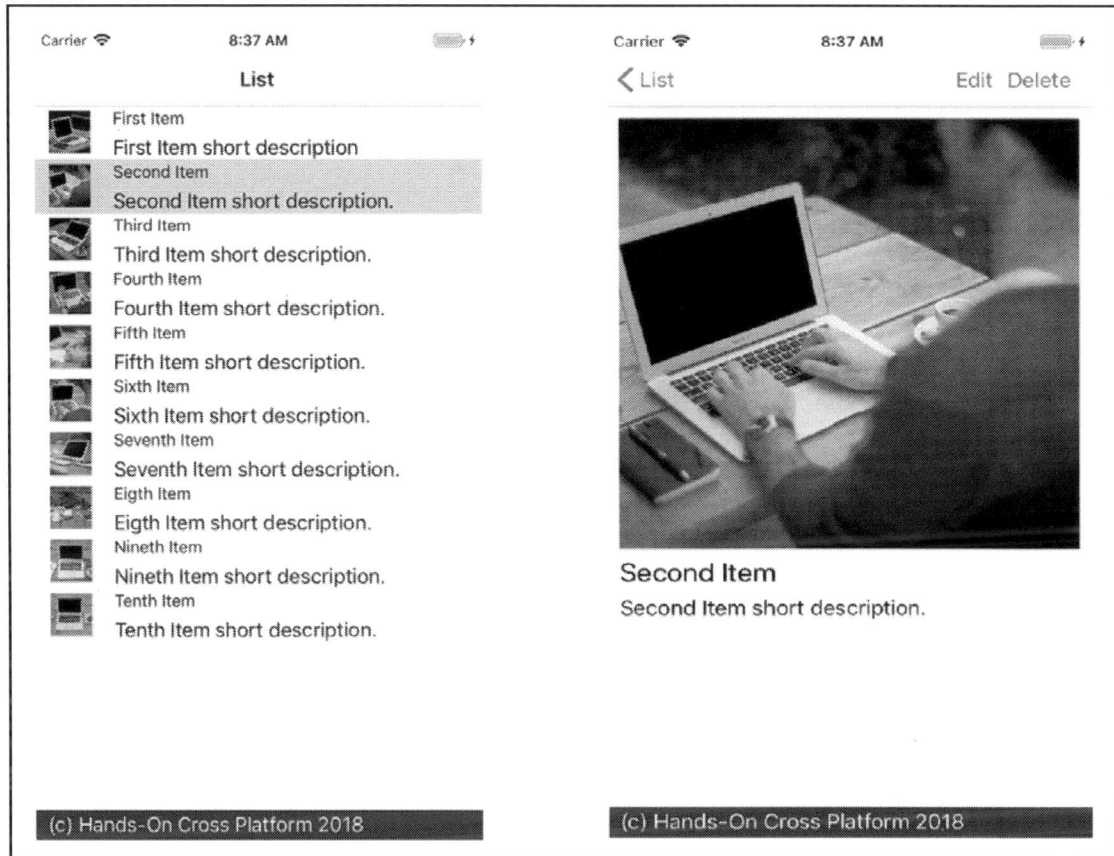

Simple navigation

Within the Xamarin ecosystem, each platform has its own intrinsic navigation stack and applications are built around those stacks. Developers are responsible for maintaining these stacks in order to create the desired UX flow for users.

In order to navigate between pages, Xamarin.Forms exposes a `Navigation` service, which can be used together with the `NavigationPage` abstract page implementation. In other words, `NavigationPage` cannot be categorized as a page type to provide content for users, however, it is a crucial component to maintain the navigation stack, as well as the navigation bar, within Xamarin.Forms applications.

In our sample, we are navigating from `ListItemView` to `ItemView`. After this navigation, on iOS, you will notice that the title of the first page is inserted as back-button text in the navigation bar. Additionally, since the `Title` property is used for the `ListItemView` (that is, `List`), the text is additionally displayed in the navigation bar.

The creation of this navigation infrastructure is achieved by creating a `NavigationPage` and passing the desired page as the root of the navigation stack (looking at `App.xaml.cs`):

```
public App ()
{
    InitializeComponent();

    MainPage = new NavigationPage(new ListItemView());
}
```

The navigation from one page to the next is handled within the `ItemTapped` event handler for `ListView`:

```
private void Handle_ItemTapped(object sender,
Xamarin.Forms.ItemTappedEventArgs e)
{
    var itemView = new ItemView();
    itemView.BindingContext = e.Item;
    Navigation.PushAsync(itemView);
}
```

Prior to Xamarin.Forms 3.2, the only way to customize what and how the navigation bar was displayed was using some form of native customization (for example, a custom renderer for `NavigationPage`). Nevertheless, you can now add custom elements to the navigation bar using the `TitleView` dependency property of a navigation page.

Using the `ListItemView` page for this illustration, we can add the following XAML section to our `ContentPage`:

```
<NavigationPage.TitleView>
    <StackLayout Orientation="Horizontal" VerticalOptions="Center"
Spacing="10">
        <Image Source="Xamarin.png"/>
        <Label
            Text="Custom Title View"
            FontSize="16"
            TextColor="Black"
            VerticalTextAlignment="Center" />
    </StackLayout>
</NavigationPage.TitleView>
```

The resulting view will have the defined `StackLayout` instead of the `List` title that was previously displayed:

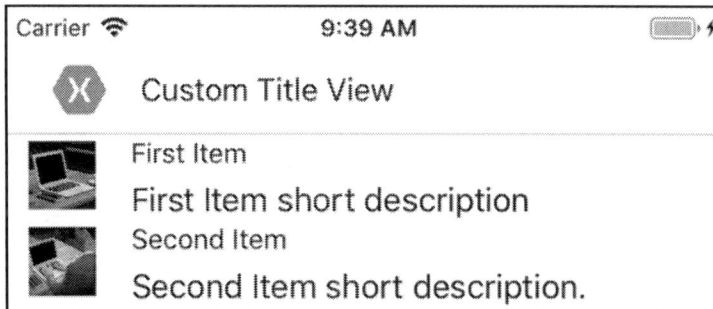

Multi-page views

`CarouselPage` and `TabbedPage` are two Xamarin.Forms page implementations that derive from the `MultiPage` abstraction. These pages can each host multiple pages with unique navigation between them.

To illustrate the usage of `MultiPage` implementations, we can utilize our previously implemented pages:

```xml
<?xml version="1.0" encoding="UTF-8"?>
<CarouselPage
    xmlns="http://xamarin.com/schemas/2014/forms"
    xmlns:x="http://schemas.microsoft.com/winfx/2009/xaml"
    xmlns:local="using:FirstXamarinFormsApplication.Client"
    x:Class="FirstXamarinFormsApplication.Client.ItemsCarousel">
    <CarouselPage.Children>
        <local:ListItemView BindingContext="{Binding .}"
Icon="Xamarin.png"/>
        <local:ItemView BindingContext="{Binding Items[0]}" Title="First"
Icon="Xamarin.png"/>
        <local:ItemView BindingContext="{Binding Items[1]}" Title="Second"
Icon="Xamarin.png"/>
        <local:ItemView BindingContext="{Binding Items[2]}" Title="Third"
Icon="Xamarin.png"/>
    </CarouselPage.Children>
</CarouselPage>
```

In a similar fashion, we can create our tabbed page using the list and item details view pages:

```xml
<?xml version="1.0" encoding="UTF-8"?>
<TabbedPage
    xmlns="http://xamarin.com/schemas/2014/forms"
    xmlns:local="using:FirstXamarinFormsApplication.Client"
    xmlns:x="http://schemas.microsoft.com/winfx/2009/xaml"
    x:Class="FirstXamarinFormsApplication.Client.ItemsTabbed">
    <TabbedPage.Children>
        <local:ListItemView BindingContext="{Binding .}"
Icon="Xamarin.png"/>
        <local:ItemView BindingContext="{Binding Items[0]}" Title="First"
Icon="Xamarin.png"/>
        <local:ItemView BindingContext="{Binding Items[1]}" Title="Second"
Icon="Xamarin.png"/>
        <local:ItemView BindingContext="{Binding Items[2]}" Title="Third"
Icon="Xamarin.png"/>
    </TabbedPage.Children>
</TabbedPage>
```

The resulting pages would, in fact, host all the children in their respective layout and navigation methods (tabs versus swipe gestures):

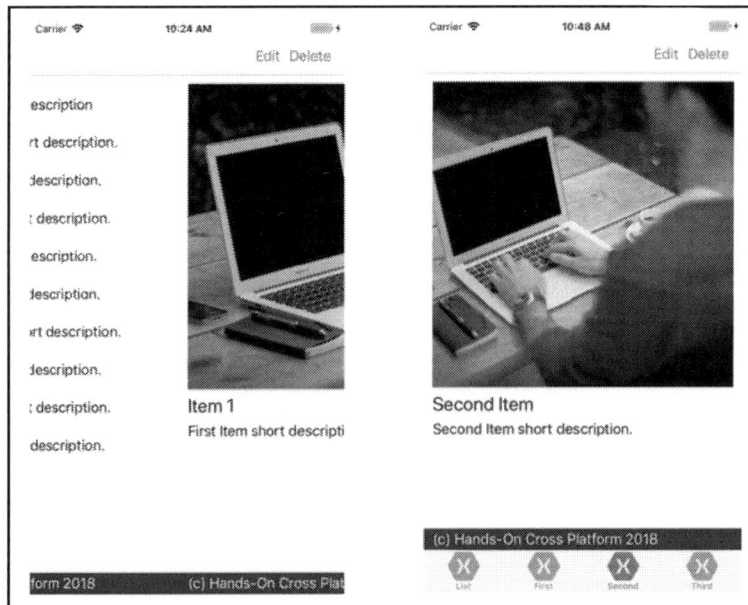

It is important to note that, on iOS, the title and icon properties of children are used to create tabbed navigation items. In order for icons to display properly, the image that is used should be *30x30* for normal resolution, *60x60* for high resolution, and *90x90* for iPhone 6 resolution. On Android, the title is used to create tab items.

In particular, `TabbedPage` is one of the fundamental controls used in iOS applications at the top of the navigation hierarchy. `TabbedPage` implementation can be extended by creating a navigation stack for each of the tabs separately. This way, navigating between tabs preserves the navigation stack for each tab independently, with support for navigating back and forth:

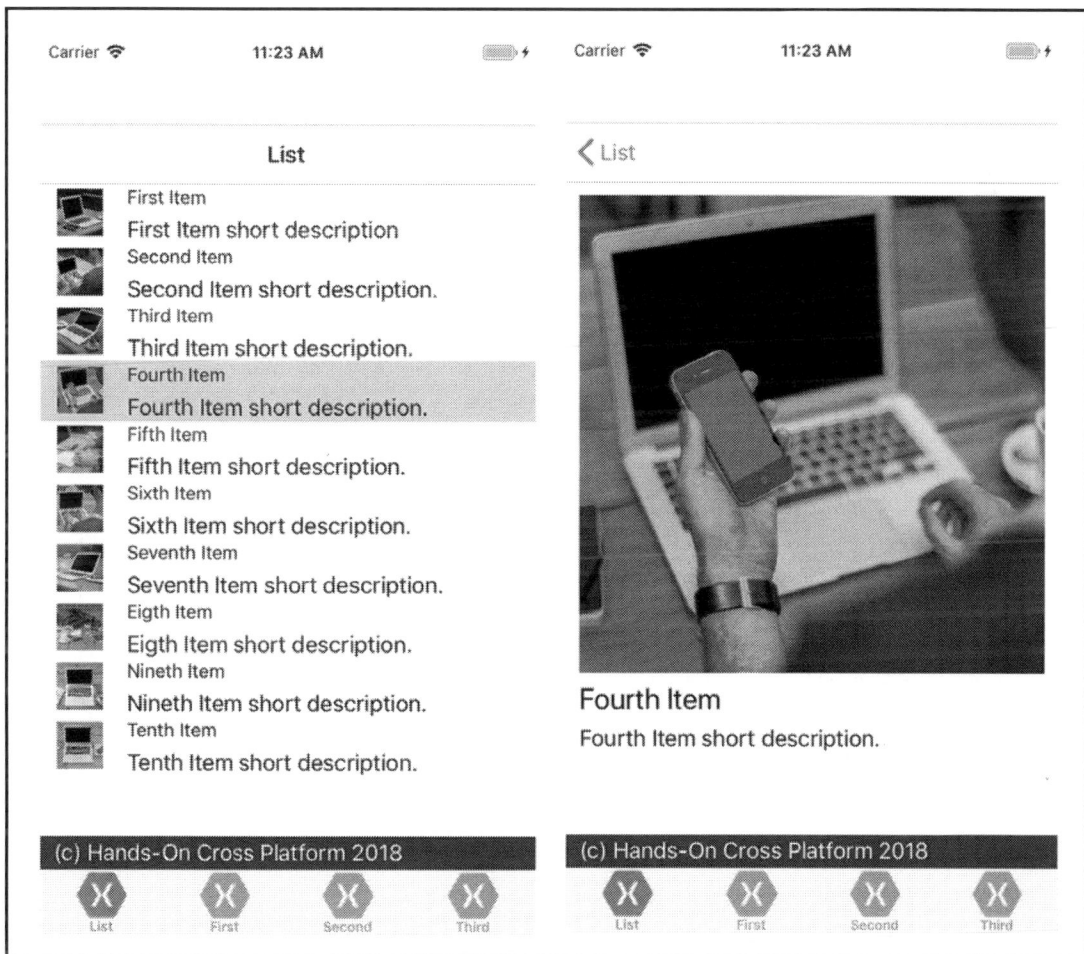

Master/detail view

On Android and UWP, the prominent navigation pattern and the associated page type is master/detail, using a so-called navigation drawer. In this pattern, jumping (across the different tiers of the hierarchy) or cross-navigating (within the same tier) across the navigation structure is maintained with `ContentPage`, which is known as the master page. User interaction with the master page, which is displayed in the navigation drawer, is propagated to the `Detail` view. In this setup, the navigation stack exists for the detail view, whilst the master view is static.

In order to replicate the tab structure in the previous example, we can create `MasterDetailPage`, which will host the list of menu items. `MasterDetailPage` will consist of the `Master` content page and the `Detail` page, which will host `NavigationPage` to create the navigation stack.

The `Master` page could look as follows:

```
<MasterDetailPage.Master>
    <ContentPage Title="Main" Padding="0,60,0,0" Icon="slideout.png">
        <StackLayout>
            <ListView
                x:Name="listView"
                ItemsSource="{Binding .}"
                SeparatorVisibility="None">
                <ListView.ItemTemplate>
                    <DataTemplate>
                        <ViewCell>
                            <Grid Padding="5,10">
                                <Grid.ColumnDefinitions>
                                    <ColumnDefinition Width="30"/>
                                    <ColumnDefinition Width="*" />
                                </Grid.ColumnDefinitions>
                                <Image Source="{Binding Icon}" />
                                <Label Grid.Column="1" Text="{Binding
                                Title}" />
                            </Grid>
                        </ViewCell>
                    </DataTemplate>
                </ListView.ItemTemplate>
            </ListView>
        </StackLayout>
    </ContentPage>
</MasterDetailPage.Master>
```

Notice that the `Master` page simply creates a `ListView` with the menu item entries. It is also important to note that the so-called hamburger menu icon needs to be added as the `Icon` property for the `Master` page (see `slideout.png`), otherwise the title of the master page is used instead of the menu icon.

The `details` page assignment would then look as follows:

```
<MasterDetailPage.Detail>
    <NavigationPage Title="List">
        <x:Arguments>
            <local:ListItemView />
        </x:Arguments>
    </NavigationPage>
</MasterDetailPage.Detail>
```

Now, running the application would create the navigation drawer and the contained `Master` page:

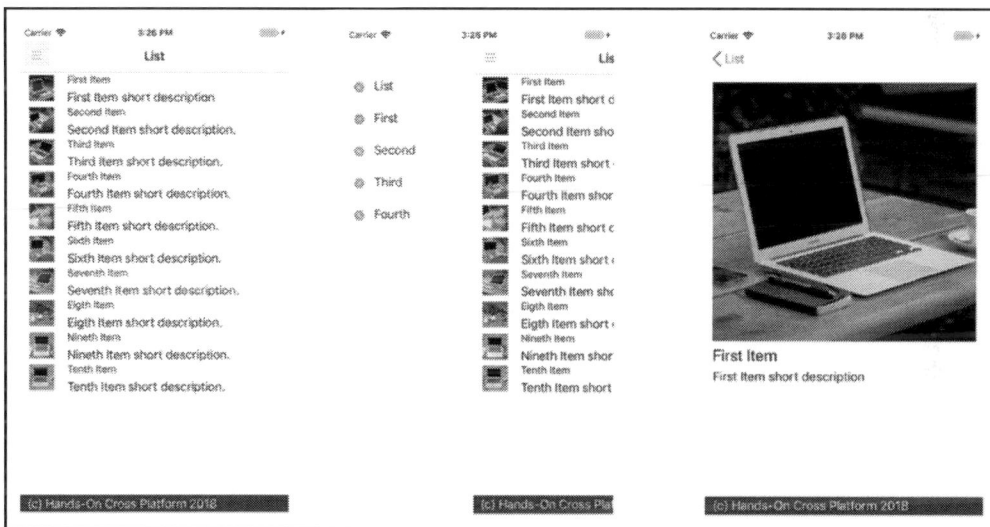

To complete the implementation, we also need to handle the `ItemTapped` event from the master list:

```
private void Handle_ItemTapped(object sender,
Xamarin.Forms.ItemTappedEventArgs e)
{
    if(e.Item is NavigationItem item)
    {
        Page detailPage = null;
```

```
        // Removed for brevity - Initialization of detailPage
        this.Detail = new NavigationPage(detailPage);

        this.IsPresented = false;
    }
}
```

Now, the implementation is complete. Every time a menu item is used, the navigation category is changed and a new navigation stack is created; however, within a navigation category, the navigation stack is intact. Also, note that the `IsPresented` property of the `MasterDetailPage` is set to `false` to dismiss the master page immediately after a new detail view is created.

Using Xamarin.Forms and native controls

Now that we are more familiar with the different page types and navigation patterns, we can move on to creating the actual UI for our pages. Creating a UX that is flexible enough for Xamarin target platforms can be dreadfully complicated; especially if the stakeholders involved are not familiar with the aforementioned UX design factors. Nevertheless, Xamarin.Forms offers various layouts and views that help developers to find the optimal solution for a project's needs.

> In Xamarin.Forms, the visual tree is composed of three layers: pages, layouts, and views. Layouts are used as containers for views, which are the user controls to create pages, which are the main interactive surfaces for users.

Let's take a closer look at the UI components.

Layouts

Layouts are container elements used to allocate user controls across a design surface. In order to satisfy platform imperatives, layouts can be used to align, stack, and position view elements. The different types of layouts are as follows:

- `StackLayout`: This is one of the most overused layout structures in Xamarin.Forms. It is used to stack various view and other layout elements with prescribed requirements. These requirements are defined through various dependency or instance properties, such as alignment options and dimension requests.

For instance, on the `ItemView` page, we use the `StackLayout` to combine the `Image` of the given item with the title and description:

```
<StackLayout Padding="10" Orientation="Vertical">
    <Image Source="{Binding Image}"
HorizontalOptions="FillAndExpand"/>
    <Label Text="{Binding Title}" FontSize="Large" />
    <Label Text="{Binding Description}" />
</StackLayout>
```

In this setup, the important declarations are the `Orientation`, which defines that the stacking should occur vertically. `HorizontalOptions` is defined for the `Image` element, which allows the `Image` to expand both horizontally and vertically, depending on the available space. `StackLayout` can be employed to create orientation-change-responsive behavior.

- `FlexLayout`: This can be used to create fluid and flexible arrangements of view elements that can adapt to the available surface. `FlexLayout` has many available directives that developers can use to define alignment directions. In order to demonstrate just a few of these, let's assume`ItemView` requires an implementation of a horizontal layout, where certain features are listed in a floating stack that can wrap into as many rows as required:

```
<StackLayout Padding="10" Orientation="Vertical" Spacing="10">
    <Label Text="{Binding Title}" FontSize="Large" />
    <Image Source="{Binding Image}"
HorizontalOptions="FillAndExpand" />
    <FlexLayout Direction="Row" Wrap="Wrap">
        <Label Text="Feature 1" Margin="4"
VerticalTextAlignment="Center" BackgroundColor="Gray" />
        <Label Text="Feat. 2" Margin="4"
VerticalTextAlignment="Center" BackgroundColor="Lime"/>
        <!-- Additional Labels -->
    </FlexLayout>
    <Label Text="{Binding Description}" />
</StackLayout>
```

This would create a design structure similar to the one described in the fluid layout responsive UI pattern:

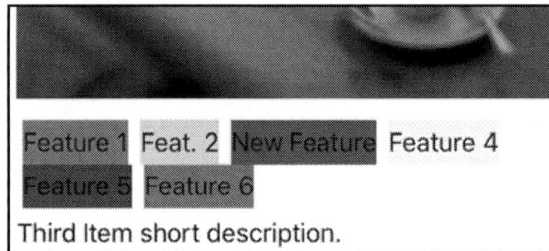

- `Grid`: If it is not desired for the views in a layout to expand and trigger layout cycles—in other words, if a certain page requires a more top-down layout structure (that is, with the parent element determining the layout)—then `Grid` would be the most suitable control. Using the `Grid` layout, controls can be laid out in accordance with column and row definitions, which can be adjusted to respond to control size changes or the overall size of the `Grid`.

While creating the control template for our page, we used a `Grid` to create a rigid structure to place the footer with an absolute height value, while allowing the rest of the screen to be covered by the content presenter:

```
<ControlTemplate x:Key="PageTemplate">
    <Grid>
        <Grid.RowDefinitions>
            <RowDefinition />
            <RowDefinition Height="25" />
        </Grid.RowDefinitions>
        <Grid.ColumnDefinitions>
            <ColumnDefinition />
        </Grid.ColumnDefinitions>
        <ContentPresenter Grid.Row="0" />
        <BoxView Grid.Row="1" Color="Navy" />
        <Label Grid.Row="1" Margin="10,0,0,0" Text="(c) Hands-On
        Cross Platform 2018" TextColor="White"
        VerticalOptions="Center" />
    </Grid>
</ControlTemplate>
```

Note that we used a margin value for the label. To avoid using the margin, we could have created a column definition with a fixed value and, according to the desired outcome, set that margin column to apply to the content presenter as well:

```
<Grid>
    <Grid.RowDefinitions>
        <RowDefinition />
        <RowDefinition Height="25" />
    </Grid.RowDefinitions>
    <Grid.ColumnDefinitions>
        <ColumnDefinition Width="10 />
        <ColumnDefinition />
        <ColumnDefinition Width="10" />
    </Grid.ColumnDefinitions>
    <ContentPresenter Grid.Column="1" Grid.Row="0" />
    <BoxView Grid.Row="1" Grid.ColumnSpan="3" Color="Navy" />
    <Label Grid.Row="1" Grid.Column="1" Text="(c) Hands-On Cross
    Platform 2018" TextColor="White" VerticalOptions="Center" />
</Grid>
```

With this setup, the `BoxView` will expand on three columns, while the footer text and the actual content will be isolated to the second column, Column-1, with Column-0 and Column-2 acting as the margins.

`Grid` can also be used only to structure a certain segment of a view. For instance, if we were to add a specifications section in our `ItemView` page, it would look similar to the following:

```
<StackLayout Padding="10" Orientation="Vertical" Spacing="10">
    <!-- Removed for Brevity -->
    <Label Text="{Binding Description}" />
    <Label Text="Specifications" Font="Bold" />
    <Grid>
        <Grid.RowDefinitions>
            <RowDefinition Height="Auto"/>
            <RowDefinition Height="Auto"/>
            <RowDefinition Height="Auto"/>
            <RowDefinition Height="Auto"/>
            <RowDefinition Height="Auto"/>
        </Grid.RowDefinitions>
        <Grid.ColumnDefinitions>
            <ColumnDefinition Width="3*" />
            <ColumnDefinition Width="5*" />
        </Grid.ColumnDefinitions>
        <Label Text="Specification 1"
                Grid.Column="0" Grid.Row="0"/>
        <Label Text="Value for Specification"
```

```
                    Grid.Column="1" Grid.Row="0" TextColor="Gray"/>
        <Label Text="Another Spec."
                    Grid.Column="0" Grid.Row="1" />
        <Label Text="Value for Specification that is a little
         longer"
                    Grid.Column="1" Grid.Row="1" TextColor="Gray"/>
        <!-- Additional Specs go here -->
     </Grid>
</StackLayout>
```

Notice the columns are set to use 3/8th and 5/8th of the screen to result in the optimal use of the space available. This would create a view similar to the following:

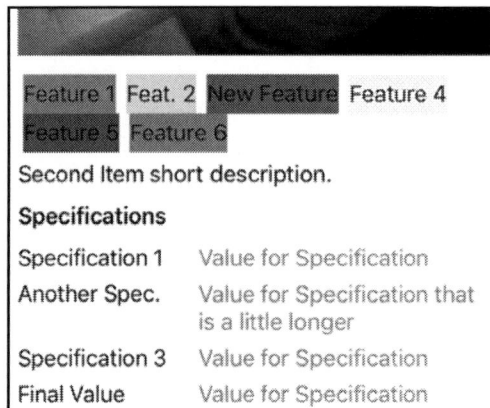

After adding this last element to the screen, you might notice that the screen space is exhausted vertically, so the final grid element might overflow out of the view port, depending on the screen size.

- ScrollView: In order to allow the scrolling of the screen so that the whole content is visible to the user, we can introduce ScrollView. ScrollView is another prominent layout element, which acts as a scrollable container for the contained view elements.

In order to enable the scrolling of the screen so that all the specifications are visible, we can simply wrap the main layout in ItemView.xaml in a ScrollView:

```
<ContentPage.Content>
    <ScrollView>
        <StackLayout Padding="10" Orientation="Vertical"
         Spacing="10">
```

```
            <!-- Removed for brevity -->
        </StackLayout>
    </ScrollView>
</ContentPage.Content>
```

An additional use of `ScrollView` comes into the picture when `Entry` fields are involved. When the user taps on an `Entry` field, the behavior on a mobile device is that the keyboard slides up from the bottom of the screen, creating a vertical offset and decreasing the design space. In the view that the `Entry` is contained in, the keyboard might overlap with the `Entry` that is currently in focus. This would create an undesirable user experience. In order to remedy this behavior, the form content should be placed in `ScrollView` so that the appearance of the keyboard does not push the `Entry` in question out of the bounds of the screen.

- `AbsoluteLayout` and `RelativeLayout`: These are the other layout options that we have not covered so far. Both of these layouts, generally speaking, treat the view almost like a canvas and allow items to be placed on top of each other, using either the current screen (in the case of `AbsoluteLayout`) or the other controls (in the case of `RelativeLayout`) as a reference for positioning.

For instance, if we were to place a **floating action button (FAB)**, from Material Design on our `ListItemView`, we could easily achieve that using an absolute layout, by placing the button in the bottom-right corner of the screen (that is, position proportional) and adding a margin on our FAB:

```
<AbsoluteLayout>
    <ListView
        ItemsSource="{Binding Items}"
        ItemTapped="Handle_ItemTapped"
        SeparatorVisibility="None" >
        <ListView.ItemTemplate>
            <DataTemplate>
                <!-- Removed for brevity -->
            </DataTemplate>
        </ListView.ItemTemplate>
    </ListView>
  <Image
Source="AddIcon.png"
HeightRequest="60"
WidthRequest="60"
AbsoluteLayout.LayoutFlags="PositionProportional"
AbsoluteLayout.LayoutBounds="1.0,1.0"
Margin="10"/>
</AbsoluteLayout>
```

This would create a view where the FAB (that is, the image used instead of a FAB) is displayed over the list view items:

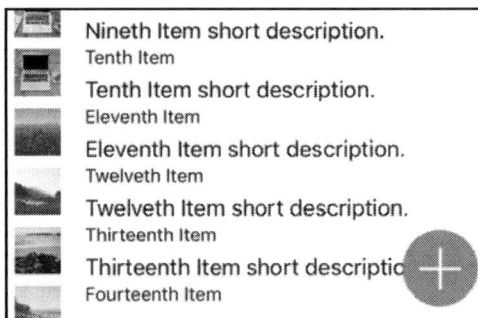

Additionally, `RelativeLayout`, in similar fashion, allows developers to create proportional calculations between elements, as well as for the view itself.

Xamarin.Forms view elements

The main view elements used in our application up until now have been `Label` and `Image` (while creating the list and details views). Additionally, on the login screen, we used the `Entry` and `Button` views. The main difference between these two sets of controls is the fact that, whilst `Label` and `Image` are used to display (generally) read-only content, `Entry` and `Button` are elements used for user input.

If we take a closer look at `Label`, there are various properties used to create a customized display of text content to accentuate the calligraphy/typography (referring to the platform imperatives of iOS and UWP) in our design. Developers can not only customize the look and feel of text content, but also create rich text content using `Span` elements. Spans are analogous to `Run` elements in WPF and web elements that share the same name (that is, `Span`). In later versions of Xamarin.Forms, `Span` are able to recognize gestures, enabling developers to create interactive regions within a single block of text content. In order to utilize Spans, the `FormattedText` attribute of the label can be used.

In order to further customize (and perhaps apply branding to) an application, custom fonts can also be introduced. In order to include a custom font, each platform requires a different step to be executed.

As a first step, the developer needs to have access to the TFF file for the font, and this file needs to be copied to the platform-specific projects. On iOS, the file(s) need to be set as `BundleResource`, and on Android as `AndroidAsset`. On iOS only, custom fonts should be declared as part of the fonts provided by the application entry in the `Info.plist` file:

Bundle version	String 1.0
▼ Fonts provided by application	Array (5 items)
	String Ubuntu-Bold.ttf
	String Ubuntu-Italic.ttf
	String Ubuntu-Light.ttf
	String Ubuntu-Regular.ttf
	String Ubuntu-Medium.ttf
Add new entry	

At this point, the custom font already used can be added with the `FontFamily` attribute to the target label; however, the declarations for the font family differ for Android and iOS:

```
<Label Text="{Binding Description}">
    <Label.FontFamily>
        <OnPlatform x:TypeArguments="x:String">
            <On Platform="iOS" Value="Ubuntu-Light" />
            <On Platform="Android" Value="Ubuntu-Light.ttf#Ubuntu-
            Light" />
            <On Platform="UWP" Value="Assets/Fonts/Ubuntu-
            Light.ttf#Ubuntu-Light" />
        </OnPlatform>
    </Label.FontFamily>
</Label>
```

In order to make it easier to use the font or even apply it to all the labels in the application, the `App.xaml` file can be used to add it to the application's resources:

```
<Application.Resources>
    <ResourceDictionary>
        <!-- Removed for brevity -->
        <OnPlatform x:Key="UbuntuBold" x:TypeArguments="x:String">
            <On Platform="iOS">Ubuntu-Bold</On>
            <On Platform="Android">Ubuntu-Bold.ttf#Ubuntu-Bold</On>
        </OnPlatform>
        <OnPlatform x:Key="UbuntuItalic" x:TypeArguments="x:String">
            <On Platform="iOS">Ubuntu-Italic</On>
            <On Platform="Android">Ubuntu-Italic.ttf#Ubuntu-
            Italic</On>
```

```
        </OnPlatform>

        <!-- Additional Fonts and Styles -->
      </ResourceDictionary>
  </Application.Resources>
```

Now we can define either implicit or explicit styles for certain targets:

```
<Style x:Key="BoldLabelStyle" TargetType="Label">
    <Setter Property="FontFamily" Value="{StaticResource UbuntuBold}" />
</Style>
<!-- Or an implicit style for all labels -->
<!--
<Style TargetType="Label">
    <Setter Property="FontFamily" Value="{StaticResource UbuntuRegular}"
/>
 </Style>
-->
```

> **TIP**
> This can be taken one step further to include a font that includes glyphs (for example, FontAwesome) to use labels as menu icons. A simple implementation would be to create a custom control that derives from `Label` and set up a global implicit style that targets this custom control.

The interactive counterparts of `Label` are `Entry` and `Editor`, which both derive from the `InputView` abstraction. These controls can be placed in user forms to handle single-line or multi-line text input, respectively. In order to improve the user experience, both of these controls expose the `Keyboard` property, which can be used to set the appropriate type of software keyboard for user entries (for example, `Chat`, `Default`, `Email`, `Numeric`, `Telephone`, and so on).

The rest of the user input controls are more scenario-specific, such as `BoxView`, `Slider`, `Map`, and `WebView`.

It is also important to mention that there are three additional user input controls, namely, `Picker`, `DatePicker`, and `TimePicker`. The pickers represent the combination of the data field that is displayed on the form and the picker dialog used once the data field comes into focus.

If the customization of these controls does not satisfy the UX requirements, Xamarin.Forms allows developers to reference and use native controls.

Native components

In some cases, developers need to resort to using native user controls; especially when a certain control only exists for a certain platform (that is, no Xamarin.Forms abstraction exists for that specific UI element). In these types of situations, Xamarin enables users to declare native views within Xamarin.Forms XAML and set/bind the properties of these controls.

In order to include native views, first the namespaces for the native views should be declared:

```
xmlns:ios="clr-namespace:UIKit;assembly=Xamarin.iOS;targetPlatform=iOS"
xmlns:androidWidget="clr-
namespace:Android.Widget;assembly=Mono.Android;targetPlatform=Android"
xmlns:formsandroid="clr-
namespace:Xamarin.Forms;assembly=Xamarin.Forms.Platform.Android;targetPlatf
orm=Android"
```

Once the namespace is declared, we can, for instance, replace `Label` in our `ItemView.xaml` and use its native counterpart directly:

```
<!-- <Label Text="{Binding Description}" /> -->
<ios:UILabel Text="{Binding Description}" View.HorizontalOptions="Start"/>
<androidWidget:TextView Text="{Binding Description}" x:Arguments="{x:Static
formsandroid:Forms.Context}" />
```

Now the view will include a different native control for each platform. Additionally, the `UILabel.Text` and `TextView.Text` properties now carry the binding to the `Description` field.

> It is important to note that, for native view references to work, the view in question should not be included in `XamlCompilation`. In other words, the view should carry the `[XamlCompilation(XamlCompilationOptions.Skip)]` attribute.

It is also possible to further customize the native fields using native types and properties. For instance, in order to add a drop-shadow on the `UILabel` item, we can use the `ShadowColor` and `ShadowOffset` values:

```
<ios:UILabel
    Text="{Binding Description}"
    View.HorizontalOptions="Start"
    ShadowColor="{x:Static ios:UIColor.Gray}">
        <ios:UILabel.ShadowOffset>
            <iosGraphics:CGSize>
```

```
            <x:Arguments>
                <x:Single>1</x:Single>
                <x:Single>2</x:Single>
            </x:Arguments>
          </iosGraphics:CGSize>
       </ios:UILabel.ShadowOffset>
   </ios:UILabel>
```

The outcome of this declaration is as follows (compare this to the Xamarin.Forms `Label` field defined earlier):

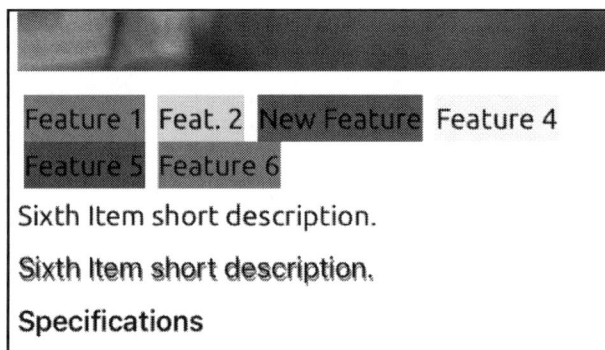

Creating data-driven views

MVVM architecture, as you saw in the `Chapter 4`, *Developing Mobile Applications with Xamarin*, mainly concentrates on data and how to decouple data from views. However, this decoupling does not mean the views and controls created should not respond to data content changes, either as a result of user input or state data being updated. In order to facilitate the propagation of data models, from the view model to view, as well as between views, data bindings and other data-related Xamarin.Forms mechanisms are crucial tools.

Data-binding essentials

The simplest data binding in Xamarin.Forms is comprised of the path of the property we want to link to the current view property. In this type of declaration, we assume that the `BindingContext` of the whole and/or the parent view is set to use the target source view model.

If we take a look at the navigation implementation from our ListItemView to ItemView, you will notice that the selected item from the list is set as the binding context for the ItemView:

```
private void Handle_ItemTapped(object sender,
Xamarin.Forms.ItemTappedEventArgs e)
{
    var itemView = new ItemView();
    itemView.BindingContext = (ItemViewModel) e.Item;
    Navigation.PushAsync(itemView);
}
```

Once BindingContext is set, we can move on to using the property model of ItemViewModel, given that ItemViewModel is set to trigger PropertyChangedEvent (from INotifyPropertyChanged) for the Title property:

```
<Label Text="{Binding Title}" FontSize="Large" />
```

Data binding does not always need to be related to a value property (for example, Text, SelectedItem, and so on) but it can also be used to identify the visual properties of a view.

For instance, the chips that we previously added to ItemView define whether certain features are supported for the currently selected item. Let's assume that we have Boolean properties on the view model side to show or hide these values. The bindings would look similar to the following:

```
<Label x:Name="Feat1" Text="Feature 1" IsVisible="{Binding HasFeature1}"
BackgroundColor="Gray" />
  <Label x:Name="Feat2" Text="Feat. 2" IsVisible="{Binding HasFeature2}"
BackgroundColor="Lime"/>
```

In both of these binding scenarios, we are binding a value from the view model to a specific view element. Another valid scenario is where the change of view affects another view (that is, View-to-View binding). Let's assume that, on ItemView, the visibility of our specs depends on the visibility of the label, with x:Name set to Feat1:

```
<Grid IsVisible="{Binding Path=IsVisible,Source={x:Reference Feat1}}">
```

It is important to note that, in a real-world project, the View-to-View binding would generally be utilized to reflect the user input in one view on another view. In this example, it would be much more appropriate for the binding to use the same view model property (that is, HasFeature1).

The bindings we have outlined so far do not really depend on any change being reflected in the UI once the visual tree is created. In such a setup, it would be an avoidable performance compromise to listen for any change event on the view model properties. In order to remedy this overhead, we could have set the binding mode to `OneTime`:

```
<Label Text="{Binding Title, Mode=OneTime}" FontSize="Large" />
```

This way, the binding is executed only when the `BindingContext` changes. If we wanted the changes in the `ViewModel` (generally referred to as the source) to be reflected in the `View` (referred to as the target), we could have used `OneWay` binding. If the direction of this unidirectional data flow provided by the OneWay binding is other way round, we could also utilize `OneWayToSource`. `TwoWay` bindings provide the infrastructure to support the bi-directional flow of data.

Although the runtime tries to convert the source type to the target type while establishing a binding, the outcome might not always be desirable (for example, the `ToString` method of a different type might not provide the correct display value). In these types of situations, developers can resort to using value converters.

Value converters

Value converters can be described as simple translation tools that implement the `IValueConverter` interface. This interface provides two methods, which allow the translation of the source to the target, as well as from the target to the source to support various binding scenarios.

For instance, if we were to display the release date of an item from our inventory, we would need to bind to the respective property on `ItemViewModel`. However, once the page is rendered, the result is less than satisfactory:

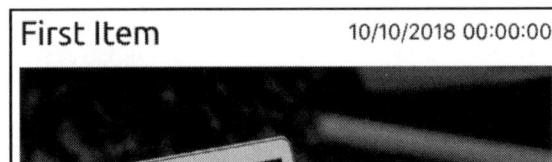

In order to format the date, we can create a value converter, which is responsible for converting the `DateTime` value to a string:

```
public class DateFormatConverter : IValueConverter
{
    public object Convert(object value, Type targetType, object
```

```
    parameter, CultureInfo culture)
    {
        if(value is DateTime date)
        {
            return date.ToShortDateString();
        }

        return null;
    }

    public object ConvertBack(object value, Type targetType, object
    parameter, CultureInfo culture)
    {
        // No Need to implement ConvertBack for OneTime and OneWay
            bindings.
        throw new NotImplementedException();
    }
}
```

And it is also responsible for declaring this converter in our XAML:

```
<ContentPage
    ...

xmlns:converters="using:FirstXamarinFormsApplication.Client.Converters"
    x:Class="FirstXamarinFormsApplication.Client.ItemView">
    <ContentPage.Resources>
        <ResourceDictionary>
            <converters:DateFormatConverter x:Key="DateFormatConverter" />
        </ResourceDictionary>
    </ContentPage.Resources>
    <ContentPage.Content>
        <!-- Removed for brevity -->
        <Label Text="{Binding ReleaseDate, Converter={StaticResource
DateFormatConverter}}" />
        <!-- Removed for brevity -->
    </ContentPage.Content>
</ContentPage>
```

Now, the display would use the short date format, which is culture-dependent (for example, M/d/yyyy for the EN-US region):

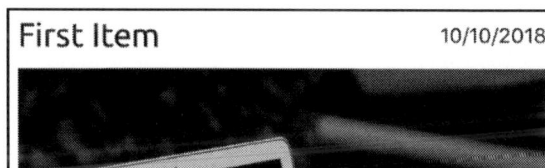

We can take this implementation one step further by using a binding to pass the date format string (for example, M/d/yyyy) to use a fixed date format.

Xamarin.Forms also provides the use of formatted strings to handle simple string conversions, so that simple converters such as DateFormatConverter could be avoided. The same implementation with a fixed date format could have been set up as follows:

```
<Label Text="{Binding ReleaseDate, StringFormat='Release {0:M/d/yyyy}'}}"
/>
```

The outcome would look like this:

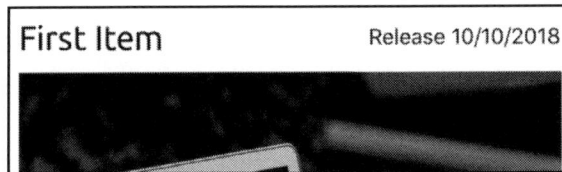

Additionally, we may like to handle scenarios where the release date is set to null (that is, when the ReleaseDate property is set to Nullable<DateTime> or simply DateTime). For this scenario, we can resort to the use of TargetNullValue:

```
<Label Text="{Binding ReleaseDate, StringFormat='Release {0:M/d/yyyy}',
TargetNullValue='Release Unknown'}" />
```

TargetNullValue, as the name suggests, is a replacement value when the binding target is resolved but the value found was null. Similarly, FallbackValue could be used when the runtime cannot resolve the target property on the binding context.

Expanding this implementation, we may want to display the Label with a different color if the release date is unknown. In order to achieve this, we could potentially create a converter to return a certain color depending on the release value, but we could also use a property trigger to set the font color depending on the label's Text property value. In this situation, the use of a trigger is a better choice, since using the converter would mean hardcoding the color value, whereas the trigger can use dynamic or static resources and can be applied with styles for the target view.

Triggers

Triggers can be defined as declarative actions that need to be executed. The different types of triggers are as follows:

- Property trigger: Property changes for a view
- Data trigger: Data value changes for a binding
- Event trigger: The occurrence of certain events on the target view
- Multi-trigger: Using this, it's also possible to implement a combination of triggers

To illustrate the use of triggers, we can use our previous example, where `ReleaseDate` for a certain item does not exist. In this scenario, because of `TargetNullValue` attribute being defined, the text of the label would be set to `Release Unknown`. Here, we can make use of a property trigger, which set the font color:

```
<Label x:Name="ReleaseDate" Text="{Binding ReleaseDate,
StringFormat='Release {0:M/d/yyyy}', TargetNullValue='Release Unknown'}">
    <Label.Triggers>
        <Trigger TargetType="Label" Property="Text" Value="Release
        Unknown">
            <Setter Property="TextColor" Value="Red" />
        </Trigger>
    </Label.Triggers>
</Label>
```

Here, the target type defines the containing element (that is, the target of the trigger action), and the property and value define the cause of the trigger. Multiple setters then can be applied to the target modifying the values of the view.

In a similar fashion, we could have created a data trigger to set the color of the title depending on the release date label's value:

```
<Label Text="{Binding Title, Mode=OneTime}" FontSize="Large">
    <Label.Triggers>
        <DataTrigger TargetType="Label"
            Binding="{Binding Source={x:Reference ReleaseDate},
                            Path=Text}"
            Value="Release Unknown">
            <Setter Property="TextColor" Value="Red" />
        </DataTrigger>
    </Label.Triggers>
</Label>
```

Here, we are setting the binding context of the `DataTrigger` to another view (that is, View-to-View) binding. If we were using the view model as the binding context, we could have used the `ReleaseDate` as well.

Finally, if we have don't have a release date but we have the data to support that an item is, in fact, already released to the public, we can use `MultiTrigger`:

```
<MultiTrigger TargetType="Label">
    <MultiTrigger.Conditions>
        <PropertyCondition Property="Text" Value="Release Unknown" />
        <BindingCondition Binding="{Binding IsReleased}" Value="false"/>
    </MultiTrigger.Conditions>
    <Setter Property="TextColor" Value="Red" />
</MultiTrigger>
```

Event triggers are odd members of the trigger family, since they rely on events being triggered on the target view instead of `Setters`; they use the so-called `Action`.

For instance, in order add a little UX enhancement, we can add a fade animation to the image in the item view. In order to use this animation, we first need to implement it as part of an `Action`:

```
public class AppearingAction : TriggerAction<VisualElement>
{
    public AppearingAction() { }

    public int StartsFrom { set; get; }

    protected override void Invoke(VisualElement visual)
    {
        visual.Animate("FadeIn",
        new Animation((opacity) => visual.Opacity = opacity, 0, 1),
        length: 1000, // milliseconds
        easing: Easing.Linear);
    }
}
```

After `TriggerAction` is created, we can now define an event trigger on the image (that is, using the `BindingContextChanged` event):

```
<Image Source="{Binding Image}" HorizontalOptions="FillAndExpand">
    <Image.Triggers>
        <EventTrigger Event="BindingContextChanged">
            <actions:AppearingAction />
        </EventTrigger>
    </Image.Triggers>
</Image>
```

This will create a subtle fade-in effect, which should coincide with the loading of the image, hence a more pleasant user experience.

Actions can also be used with property and data triggers using `EnterAction` and `ExitAction`, defining the two states according to the trigger condition(s). However, in the context of property and data triggers, in order to create more generalized states, as well as modifying the common states for a control, **Visual State Manager** (**VSM**) can also be utilized. This way, multiple setters can be unified in a single state, decreasing the clutter within the XAML tree, creating a more maintainable structure.

Visual states

Visual states and VSM will be familiar concepts for WPF and UWP developers; however, they were missing from Xamarin.Forms' runtime until recently. Visual states define various conditions that a control can be rendered according to certain conditions. For instance, an `Entry` element can be in `Normal`, `Focused`, or `Disabled` states and each state defines a different visual setter for the element. Additionally, custom states can also be defined for a visual element and, depending on triggers or explicit calls to `VisualStateManager`, can manage the visual state of elements.

In order to demonstrate this, we can create three different states for our label (for example, `Released`, `UnReleased`, and `Unknown`) and deal with states using our triggers.

First, we need to define states for our label control (which can then be moved to a resource dictionary as part of a style):

```
<Label x:Name="ReleaseDate" ...>
    <Label.Triggers>
        <!-- Removed for Brevity -->
    </Label.Triggers>
    <VisualStateManager.VisualStateGroups>
        <VisualStateGroup x:Name="CommonStates">
            <VisualState x:Name="Released">
                <VisualState.Setters>
                    <Setter
                        Property="BackgroundColor"
                        Value="Lime" />
                    <Setter
                        Property="TextColor"
                        Value="Black" />
                </VisualState.Setters>
            </VisualState>
            <VisualState x:Name="UnReleased">
                <VisualState.Setters>
```

```
                    <Setter Property="TextColor" Value="Black" />
                </VisualState.Setters>
            </VisualState>
            <VisualState x:Name="Unknown">
                <VisualState.Setters>
                    <Setter Property="TextColor" Value="Red" />
                </VisualState.Setters>
            </VisualState>
        </VisualStateGroup>
    </VisualStateManager.VisualStateGroups>
</Label>
```

As you can see, one of the defined states is Unknown, and it should set the text color to red. In order to change the state of the label using a trigger, we can implement a trigger action:

```
public class ChangeStateAction : TriggerAction<VisualElement>
{
    public ChangeStateAction() { }

    public string State { set; get; }

    protected override void Invoke(VisualElement visual)
    {
        if(visual.HasVisualStateGroups())
        {
            VisualStateManager.GoToState(visual, State);
        }
    }
}
```

And we can use this action as our EnterAction for the previously defined multi-trigger:

```
<MultiTrigger TargetType="Label">
    <MultiTrigger.Conditions>
        <!-- Removed for brevity -->
    </MultiTrigger.Conditions>
    <MultiTrigger.EnterActions>
        <actions:ChangeStateAction State="Unknown" />
    </MultiTrigger.EnterActions>
</MultiTrigger>
```

We can achieve the same result as using setters. However, it is important to mention that, without defining an ExitAction once the label is set to the given state, it will not revert to the previous state.

Summary

In this chapter, we implemented some simple views using the intrinsic controls of the Xamarin.Forms framework. With the extensive set of layouts, views, and customization options available, developers can create attractive and intuitive user interfaces. Moreover, the data-driven UI options can help developers separate (decouple) and business domain implementation from these views, which in return will improve the maintainability of any mobile development project.

Nevertheless, at times, standard controls might not be enough to meet project requirements. In the next chapter, we will take a closer look at customizing the existing UI views and implementing custom native elements.

6
Customizing Xamarin.Forms

Xamarin.Forms allows developers to modify the UI-rendering infrastructure in various ways. The customizations that are introduced by developers may target a certain platform feature on a certain control element or create a completely new view control. These customizations can be made on the Xamarin.Forms tier or on the target native platform.

This chapter will go through the steps and procedures for customizing Xamarin.Forms without compromising the performance and user experience. Some of the features that will be analyzed include effects, behaviors, extensions, and custom renderers.

The following sections will cover different development domains for Xamarin.Forms customizations:

- Xamarin.Forms development domains
- Xamarin.Forms shared domains
- Customizing the native domains
- Creating custom controls

Xamarin.Forms development domains

As we have seen so far in this book, application development using the Xamarin.Forms framework is executed on multiple domains. While the Xamarin.Forms layer creates a shared development domain that will be used to target native platforms, the target platforms can still be utilized for platform-specific implementation.

If we were to separate a Xamarin.Forms application into four quadrants by development strategy and application domain category, it will look like this:

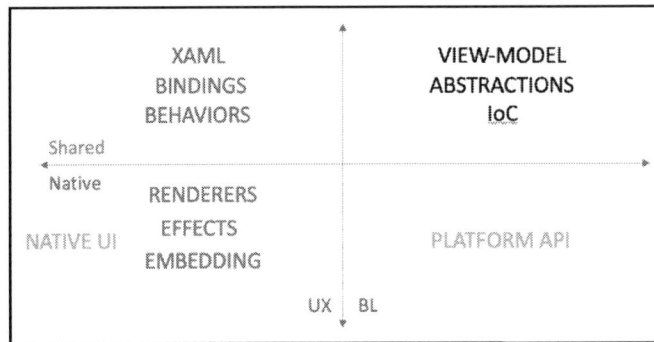

In this setup, quadrant I (that is, the shared business logic) would represent the core logic implementation of the application. This domain will contain the view models, domain data descriptions, and service client implementation. Most importantly, the abstractions for platform-specific APIs (that is, the interfaces that will be implemented on the native platform) should be created in this domain so that each other domain, as well as the view models within this domain, can make use of them.

Quadrant II and III represent the UI customizations that we will need to implement to create the desired UX for the application. Up until now, we have been creating our visual trees using only quadrant II. Simple data-driven applications and **line of business (LOB)** applications can solely utilize this domain. However, if we were to create a consumer-facing application, complying with the branded UX requirements and creating an intuitive UI should be our main goals. In this case, we can resort to creating customizations for Xamarin.Forms views with quadrant III.

In this paradigm, quadrant I only connects with quadrant II using data binding and converter implementations. Quadrant II is responsible for propagating the delivered data to quadrant III.

In quadrant II, the customization options for developers are mostly related to using the extensibility options provided by the out-of-the-box views offered by the Xamarin.Forms framework. Compositions of these views and behavioral modifications can provide highly maintainable cross-platform source code. By using styling options, visual states, and data-driven templates, the UX can meet these requirements.

Moving from the shared to the native platform (that is, crossing from quadrant II to quadrant III), the developers are blessed with platform specifics, as well as Xamarin.Forms effects. Using these extensibility points, we, as developers, can modify the behavior of native controls, as well as modify the rendered native UI, creating a bridge between the Xamarin.Forms view abstractions and the target native controls. A combination of these extensibility features with Xamarin.Forms behaviors can improve the maintainability of the application.

Quadrant-III-specific development is comprised of custom renderers and native controls. Native controls can be created and combined under Xamarin.Forms compositions, thereby decreasing the complexity of the Xamarin.Forms XAML trees (that is, the composite controls).

Finally, quadrant IV represents the platform-specific APIs, such as geolocation, the usage of peripherals, such as Bluetooth or NFC, or SaaS integrations/SDKs that require native implementation.

Xamarin.Forms shared domain

In the previous chapters, we used intrinsic Xamarin.Forms controls and their styling attributes to create our user interface. By using data binding and data triggers, we created data-driven views. The extensibility options are of course not limited to the control attributes that are available on this layer. Both the behavior and the look and feel of rendered controls can be modified using standard customization and extensibility options. Let's take a look at the different customization options in the shared Xamarin.Forms domain.

Using styles

Previously, we created a simple chips container to display the various features of an item that is currently being offered through our application.

In the previous setup, we were only utilizing the `Margin` property
and `VerticalTextAlignment` for the labels:

```
<FlexLayout Direction="Row" Wrap="Wrap">
    <Label Text="Feature 1" Margin="4"
VerticalTextAlignment="Center" BackgroundColor="Gray" />
    <Label Text="Feat. 2" Margin="4" VerticalTextAlignment="Center"
BackgroundColor="Lime"/>
    <!-- Additional Labels -->
</FlexLayout>
```

> The background property is specific for each feature element, and so
> we will not be modifying it for now.

Let's modify the look of these items to make the labels look more like chips in order to
improve the user experience:

1. We will start by wrapping up the label in a frame and styling the frame:

```
<Frame
    IsVisible="{Binding HasFeature1}"
    BackgroundColor="Gray"
    CornerRadius="7"
    Padding="3"
    Margin="4"
    HasShadow="false">
  <Label x:Name="Feat1" Text="Feature 1"
      VerticalTextAlignment="Center"
      HorizontalTextAlignment="Center" />
</Frame>
```

While this creates a more desirable look, adding these properties to each feature
would create a completely redundant XAML structure:

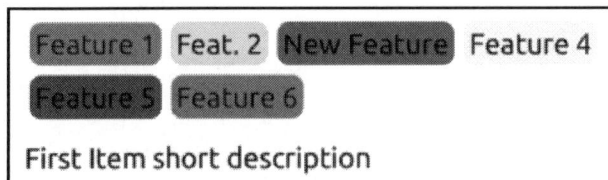

2. We can now create create two styles (that is, one for the feature label and one for the frame itself) that will be applied to each element, thereby decreasing the redundancy:

```
<Style TargetType="Frame">
    <Setter Property="HasShadow" Value="false" />
</Style>
<Style TargetType="Frame" x:Key="ChipContainer">
    <Setter Property="CornerRadius" Value="7" />
    <Setter Property="Padding" Value="3" />
    <Setter Property="Margin" Value="3" />
</Style>
<Style TargetType="Label" x:Key="ChipLabel">
    <Setter Property="VerticalTextAlignment" Value="Center" />
    <Setter Property="HorizontalTextAlignment" Value="Center" />
    <Setter Property="TextColor" Value="White" />
</Style>
```

3. Next, we need to apply these implicit (that is, the HasShadow="false" setter will be applied to all the frames on the application level) and explicit styles (note the x:Key declaration on the ChipContainer and ChipLabel styles):

```
<FlexLayout Direction="Row" Wrap="Wrap" FlowDirection="LeftToRight"
AlignItems="Start">
    <Frame IsVisible="{Binding HasFeature1}"
            BackgroundColor="Gray" Style="{StaticResource
            ChipContainer}">
        <Label x:Name="Feat1" Text="Feature 1" Style="
        {StaticResource ChipLabel}" />
    </Frame>
    <Frame IsVisible="{Binding HasFeature2}"
BackgroundColor="Lime"
        Style="{StaticResource ChipContainer}">
            <Label x:Name="Feat2" Text="Feat. 2"
Style="{StaticResource
            ChipLabel}" />
    </Frame>
    <!-- Additional Labels -->
</FlexLayout>
```

In doing so, we will be decreasing the clutter and redundancy in our XAML tree. Styles can be declared at the application level (as in this case) as global styles using App.xaml. Additionally, they can also be declared at page and view levels using local resource dictionaries.

Another approach to styling controls would be to use CSS style sheets. While the style sheets currently do not support the full extent of the XAML control styles, can prove powerful, especially when utilizing the CSS selectors. Let's get started:

1. If we were to recreate the styles for our chip views, the style declarations would be similar to the following:

```
.ChipContainerClass {
    border-radius: 7;
    padding: 3;
    margin: 3;
}

.ChipLabelClass {
    text-align: center;
    vertical-align: central;
    color: white;
}
```

For those of you who are not familiar with CSS, here, we have created two style classes named `ChipContainerClass` and `ChipLabelClass`.

2. Now, we can use these classes with our controls using the `StyleClass` attribute:

```
<Frame IsVisible="{Binding HasFeature1}"
        BackgroundColor="Gray" StyleClass="ChipContainerClass">
    <Label x:Name="Feat1" Text="Feature 1"
      StyleClass="ChipLabelClass" />
</Frame>
```

3. We can take our style declaration one step further and apply the style directly to the child label within the frame with the `ChipContainerClass` style class (that is, we will remove the style class from the `Label` element):

```
.ChipContainerClass {
    border-radius: 7;
    padding: 3;
    margin: 3;
}

.ChipContainerClass>^label {
    text-align: center;
    vertical-align: central;
    color: white;
}
```

> **TIP**
>
> The difference between `.ChipContainerClass>label` and `.ChipContainerClass>^label` is that by using the `^` (base class) notation, we can make sure that even if we modify the view using a custom control deriving from label, we can make sure that the styles are applied in the same way.

Styles can also be used in conjunction with Xamarin.Forms behaviors to not only modify the visualization, but also the behavior of elements.

Creating behaviors

Behaviors are an eloquent use of the decorator pattern, allowing the developers to modify the Xamarin.Forms controls without having to create derived controls.

A simple example of creating a behavior would be to implement a validation behavior in our `LoginView`. As you may remember, we actually used the `Command.CanExecute` delegate to validate our fields. In this example, we will separate the validators for the email field and the password field. This way, we can allow the UI to give feedback to the user as a result of an incorrect entry. This would be more user-friendly than only disabling the login window. To set this up, follow these steps:

1. First, we need to create a validation rule infrastructure, starting with the validation interface:

```
public interface IValidationRule<T>
{
    string ValidationMessage { get; set; }
    bool Validate (T value);
}
```

2. A simple implementation of this rule would be required so that we can check whether we have a short validation message stating that the field is a required field:

```
public class RequiredValidationRule : IValidationRule<string>
{
    public string ValidationMessage { get; set; } = "This field is
    a required field";
    public bool Validate (string value)
    {
        return !string.IsNullOrEmpty(value);
    }
}
```

3. Now, we can create our validation behavior for the `Entry` field, which will make use of any given validation rule (starting with `RequiredValidationRule`, which we just implemented):

```
public class ValidationBehavior : Behavior<Entry>
{

    protected override void OnAttachedTo(Entry bindable)
    {
        base.OnAttachedTo(bindable);

        bindable.TextChanged += ValidateField;
    }

    protected override void OnDetachingFrom(Entry bindable)
    {
        base.OnDetachingFrom(bindable);

        bindable.TextChanged -= ValidateField;
    }

    private void ValidateField(object sender, TextChangedEventArgs
    args)
    {
        if (sender is Entry entry)
        {
            // TODO:
        }
    }
}
```

In this implementation, the `OnAttachedTo` and `OnDetachingFrom` methods are the crucial access points and the teardown logic implementations. In this case, when the behavior is attached to a target control, we are subscribing to the `TextChanged` event, and when the behavior is removed, we are unsubscribing from the event so that undesired memory leak issues are avoided.

4. The next order of business will be to implement a bindable property for the validation rule so that the validation rules are dictated by the view model (or another business logic module), decoupling it from the view:

```
public static readonly BindableProperty ValidationRuleProperty =
        BindableProperty.CreateAttached("ValidationRule",
typeof(IValidationRule<string>), typeof(ValidationBehavior), null);


public static readonly BindableProperty HasErrorProperty =
```

```
            BindableProperty.CreateAttached("HasError", typeof(bool),
    typeof(ValidationBehavior), false, BindingMode.TwoWay);

    public IValidationRule<string> ValidationRule
    {
        get { return this.GetValue(ValidationRuleProperty) as
    IValidationRule<string>; }
        set { this.SetValue(ValidationRuleProperty, value); }
    }

    public bool HasError
    {
        get { return (bool) GetValue(HasErrorProperty); }
        set { SetValue(HasErrorProperty, value); }
    }
```

5. Now that we have an outlet for the validation rule and an output field (so that we can attach additional UX logic to it), we can implement the `validate` method:

```
    private void ValidateField(object sender, TextChangedEventArgs
    args)
    {
        if (sender is Entry entry && ValidationRule != null)
        {
            if (!ValidationRule.Validate(args.NewTextValue))
            {
                entry.BackgroundColor = Color.Crimson;
                HasError = true;
            }
            else
            {
                entry.BackgroundColor = Color.White;
                HasError = false;
            }
        }
    }
```

6. After adding the appropriate rule to the view model property (in this case, `UserNameValidation`), we can bind the behavior to the validation rule that's exposed from the view model and observe the entry field behavior according to the text input:

```
<Entry x:Name="usernameEntry" Placeholder="username" Text="{Binding
UserName, Mode=OneWayToSource}" >
    <Entry.Behaviors>
        <behaviors:ValidationBehavior x:Name="UserNameValidation"
            ValidationRule="{Binding
            BindingContext.UserNameValidation,
                Source={x:Reference RootView}}" />
    </Entry.Behaviors>
</Entry>
```

Here, the main benefit is that we do not have to modify the `Entry` field, and the implemented behavior can be maintained as a separate module.

> It is important to note that the binding context for a behavior is not the same as the page layout or the view, which is why the source of the binding value for the validation rule has to reference the page itself and use `BindingContext` as part of the binding path.

7. To extend this implementation, we can add a validation error message label that will display in accordance with the `HasError` bindable property (anywhere on the page layout, as long as the `UserNameValidation` element is accessible):

```
<Label Text="UserName is required" FontSize="12" TextColor="Gray"
    IsVisible="{Binding HasError, Source={x:Reference
UserNameValidation}}"/>
```

8. The outcome would look similar to the following:

Attached properties

Another way to implement behaviors in order to modify default control behavior is to use attached properties to declare a bindable extension to existing controls. This approach is generally used for small behavioral adjustments, such as enabling/disabling other behaviors, as well as adding/removing effects. Let's get started:

1. In order to implement such a behavior, we need to create a bindable property that will be used on other controls:

```
public static class Validations
{
    public static readonly BindableProperty
ValidateRequiredProperty =
        BindableProperty.CreateAttached(
            "ValidateRequired",
            typeof(bool),
            typeof(RequiredValidation),
            false,
            propertyChanged: OnValidateRequiredChanged);

    public static bool GetValidateRequired(BindableObject view)
    {
        return (bool)view.GetValue(ValidateRequiredProperty);
    }

    public static void SetValidateRequired(BindableObject view,
    bool value)
    {
        view.SetValue(ValidateRequiredProperty, value);
    }

    private static void OnValidateRequiredChanged(
            BindableObject bindable, object oldValue, object
            newValue)
    {
        // TODO:
    }
}
```

2. In the case of attached behaviors, the static class can be directly accessed so that it sets the attached property on the current control (instead of creating and adding a behavior):

```
<Entry x:Name="usernameEntry" Placeholder="username"
    Text="{Binding UserName, Mode=OneWayToSource}"
    behaviors:Validations.ValidateRequired="true" >
```

3. By implementing the `ValidateRequired` property changed handler, we can have the attached property insert and remove the required validation to various `Entry` views:

```
private static void OnValidateRequiredChanged(
    BindableObject bindable,
    object oldValue,
    object newValue)
{
    if(bindable is Entry entry)
    {
        if ((bool)newValue)
        {
            entry.Behaviors.Add(new ValidationBehavior()
            {
                ValidationRule = new RequiredValidationRule()
            });
        }
        else
        {
            var behaviorToRemove = entry.Behaviors
                .OfType<ValidationBehavior>()
                .FirstOrDefault(
                    item => item.ValidationRule is
                    RequiredValidationRule);

            if (behaviorToRemove != null)
            {
                entry.Behaviors.Remove(behaviorToRemove);
            }
        }
    }
}
```

XAML markup extensions

Up until now, when we created XAML views, we resorted to several markup extensions that are supported either by the Xamarin.Forms framework or the XAML namespace itself. Some of these extensions are as follows:

- `x:Reference`: Used to refer to another view on the same page
- `Binding`: Used throughout the view model implementations
- `StaticResource`: Used to refer to styles

These are all markup extensions that are resolved by the associated service implementation within the Xamarin.Forms framework.

For specific needs in your application, custom markup extensions can be implemented to create a more maintainable XAML structure. In order to create a markup extension, the `IMarkupExtension<T>` class needs to be implemented. This depends on the type that needs to be provided.

For instance, in our previous example, the error label and the field descriptors were hard coded into the XAML view. This would create issues if the application needs to support multiple localizations. This can be resolved by doing the following:

1. First, we need to create a markup extension that will translate the associated text values:

```
[ContentProperty("Text")]
public class TranslateExtension : IMarkupExtension<string>
{
    public string Text { get; set; }

    public string ProvideValue(IServiceProvider serviceProvider)
    {
        // TODO:
    }

    object IMarkupExtension.ProvideValue(IServiceProvider
    serviceProvider)
    {
        return (this as
        IMarkupExtension<string>).ProvideValue(serviceProvider);
    }
}
```

2. Note that the `Text` property is set as `ContentProperty`, which allows developers to provide a value for this extension simply by adding a value for the extension. Let's incorporate it into the XAML structure:

```
<Label Text="{behaviors:Translate LblUsername}" />
<Entry x:Name="usernameEntry" Placeholder="username"
    Text="{Binding UserName, Mode=OneWayToSource}" >
    <Entry.Behaviors>
        <behaviors:ValidationBehavior x:Name="UserNameValidation"
ValidationRule="{Binding BindingContext.UserNameValidation,
Source={x:Reference RootView}}" />
    </Entry.Behaviors>
</Entry>
```

```
<Label Text="{behaviors:Translate LblRequiredError}"
    FontSize="12" TextColor="Gray"
    IsVisible="{Binding HasError, Source={x:Reference
UserNameValidation}}"/>
```

3. `ProvideValue` method would therefore need to translate the `LblUsername` and `LblRequiredError` keys:

```
public string ProvideValue(IServiceProvider serviceProvider)
{
    switch (Text)
    {
        case "LblRequiredError":
            return "This a required field";
        case "LblUsername":
            return "Username";
        default:
            return Text;
    }
}
```

This completes the quadrant II customizations. Now, we will move on to quadrant III and the customize native controls.

Customizing native domains

Native customizations of UI controls can vary from simple platform-specific adjustments to completely creating a custom native control to replace the existing platform renderer.

Platform specifics

While the UI controls offered by Xamarin.Forms are customizable enough for most UX requirements, additional native behaviors may be needed. For certain native control behaviors, platform-specific configuration can be accessed using the `IElementConfiguration` interface implementation of the target control. For instance, in order to change the `UpdateMode` picker (that is, `Immediately` or `WhenFinished`), you can use the `On<iOS>` method to access the platform-specific behavior:

```
var picker = new Xamarin.Forms.Picker();
picker.On<iOS>().SetUpdateMode(UpdateMode.WhenFinished);
```

The same can be implemented in XAML using the
`Xamarin.Forms.PlatformConfiguration.iOSSpecific` namespace:

```
<ContentPage
    ...
    xmlns:ios="clr-
namespace:Xamarin.Forms.PlatformConfiguration.iOSSpecific;assembly=Xamarin.
Forms.Core">
        <!-- ... -->
            <Picker ios:Picker.UpdateMode="WhenFinished">
            <!-- Removed for brevity -->
            </Picker>
        <!-- ... -->
</ContentPage>
```

Similar platform configurations are available for other controls and platforms within the
same namespace (that is, `Xamarin.Forms.PlatformConfiguration`).

Xamarin.Forms effects

Xamarin.Forms effects are an elegant bridge between the cross-platform domain and the
native domain. Effects are generally used to expose a certain platform behavior or
implementation of a given native control through the shared domain so that a completely
new custom native control is not needed for the implementation.

Similar to the Xamarin.Forms views/controls, effects exist on both the shared domain and
the native domain with their abstraction and implementation, respectively. While the
shared domain is used to create a routing effect, the native project is responsible for
consuming it.

For instance, let's assume the details that we receive for our product items actually contain
some HTML data that we would like to present within the application. In this case, we are
aware of the fact that the `Label` element on Xamarin.Forms is rendered with a `UILabel` in
iOS and `TextView` in Android. While `UILabel` provides the `AttributedString` property
(which can be created from HTML), the Android platform offers the intrinsic module for
parsing HTML. We can expose these platform-specific features using an effect and enable
the Xamarin.Forms abstraction to accept HTML input. Let's get started:

1. Create the routing effect that will provide the data for the platform effects:

    ```
    public class HtmlTextEffect: RoutingEffect
    {
        public HtmlTextEffect():
    base("FirstXamarinFormsApplication.HtmlTextEffect")
    ```

```
        {
        }

        public string HtmlText { get; set; }
    }
```

2. Now, we can use this effect in our XAML:

```
<Label Text="{Binding Description}">
 <Label.Effects>
 <effects:HtmlTextEffect
HtmlText="<b>Here</b> is some
<u>HTML</u>" />
 </Label.Effects>
</Label>
```

Without the platform implementation of this routing effect, the label will still display the binding data (in this case, encoded HTML text).

3. Now, we need to implement the iOS effect that will parse the `HtmlText` property of our effect. Platform effects are mainly composed of two main components: registration and implementation:

```
[assembly: ResolutionGroupName("FirstXamarinFormsApplication")]
[assembly: ExportEffect(typeof(HtmlTextEffect), "HtmlTextEffect")]
namespace FirstXamarinFormsApplication.iOS.Effects
{
    public class HtmlTextEffect: PlatformEffect
    {
        protected override void OnAttached()
        {
        }

        protected override void OnDetached()
        {
        }
    }
}
```

The effect that's registered with `ResolutionGroupName` of the `ExportEffect` attributes will be used in the runtime environment to resolve the routing effect that was implemented in the first step. In order to modify the native control, you can use the `Control` property of `PlatformEffect`. The `Element` property refers to the Xamarin.Forms control that requires this effect.

4. Now, we need to implement the `OnAttached` method (which will be executed when `PlatformEffect` is resolved):

```
protected override void OnAttached()
{
    var htmlTextEffect = Element.Effects
            .OfType<Client.Effects.HtmlTextEffect>
            ().FirstOrDefault();

    if(htmlTextEffect != null && Control is UILabel label)
    {
        var documentAttributes = new
NSAttributedStringDocumentAttributes();
        documentAttributes.DocumentType = NSDocumentType.HTML;
        var error = new NSError();

        label.AttributedText = new
NSAttributedString(htmlTextEffect.HtmlText, documentAttributes, ref
error);
    }
}
```

5. A similar implementation for the Android platform will create the HTML rendering of the controls:

```
protected override void OnAttached()
{
    var htmlTextEffect = Element.Effects
            .OfType<Client.Effects.HtmlTextEffect>
            ().FirstOrDefault();

    if (htmlTextEffect != null && Control is TextView label)
    {
        label.SetText(
            Html.FromHtml(htmlTextEffect.HtmlText,
            FromHtmlOptions.ModeLegacy),
            TextView.BufferType.Spannable);
    }
}
```

While we have managed to display HTML content on our Xamarin.Forms view, the value we have used is still not bindable. With a little restructuring, and by using attached properties (that is, attached behavior), we can use the data binding and effects together.

Composite customizations

Behaviors and effects, when used together, can create eloquent solutions to common native element requirements without having to resort to custom controls and renderers. Let's see how we can do this:

1. Picking up from where we left off with the `HtmlText` effect, let's create an attached behavior that will allow us to switch the HTML rendering on/off:

```
public static class HtmlText
{
    public static readonly BindableProperty IsHtmlProperty =
        BindableProperty.CreateAttached("IsHtml",
            typeof(bool), typeof(HtmlText), false,
            propertyChanged: OnHtmlPropertyChanged);

    private static void OnHtmlPropertyChanged(
        BindableObject bindable, object oldValue, object newValue)
    {
        var view = bindable as View;
        if (view == null)
        {
            return;
        }

        if (newValue is bool isHtml && isHtml)
        {
            view.Effects.Add(new HtmlTextEffect());
        }
        else
        {
            var htmlEffect = view.Effects.FirstOrDefault(e => e is
            HtmlTextEffect);

            if (htmlEffect != null)
            {
                view.Effects.Remove(htmlEffect);
            }
        }
    }
}
```

The behavior will be the addition or removal of the HTML effect, depending on the `IsHtml` property declaration.

2. Now, we will modify our HTML effect so that it uses the existing text assignment on the forms view to create `NSAttributedText` and `ISpannable` for iOS and Android platforms, respectively:

```
public class HtmlTextEffect: PlatformEffect
{
    protected override void OnAttached()
    {
        SetHtmlText();
    }

    protected override void OnDetached()
    {
        // TODO: Remove formatted text
    }

    protected override void
OnElementPropertyChanged(PropertyChangedEventArgs args)
    {
        base.OnElementPropertyChanged(args);

        if (args.PropertyName == Label.TextProperty.PropertyName)
        {
            SetHtmlText();
        }
    }

    // Removed for brevity
}
```

Note that we have also used the `OnElementPropertyChanged` method to listen for `Text` property value changes. This would be the main access point for binding data.

3. Now, we will add the behavior to our XAML:

```
<Label Text="{Binding Description}"
    effects:HtmlText.IsHtml="{Binding IsHtml}" />
```

We can now control the displayed text attributes on both platforms using the `IsHtml` attached property.

If none of these customization options provide what really required for the desired UI, a complete custom control implementation can be considered an option.

Creating custom controls

Just like any other development platform, it is also possible to create custom views/controls that look, behave, and render differently compared to the out-of-the-box Xamarin.Forms controls; however, creating a custom control doesn't mean that the complete Xamarin.Forms render infrastructure needs to be implemented for target platforms as well as the shared domain. Depending on the UX and platform requirements, the following can occur:

- Custom controls can be created solely as a composition of other Xamarin.Forms controls
- Existing Xamarin.Forms controls can be modified with custom renderers on different platforms
- Custom Xamarin.Forms controls can be created with custom renderers

Creating a Xamarin.Forms control

A Xamarin.Forms control can be created for various reasons, one of which is to decrease the clutter in your XAML tree and create reusable view blocks. Let's begin:

1. First, we will take a step back and take a look at the validatable entries we created previously for the login screen:

```
<Label x:Name="lblUserName" Text="..." />
<Entry x:Name="txtUserName" Placeholder=".." Text="..." >
    <Entry.Behaviors>
        <behaviors:ValidationBehavior x:Name="UserNameValidation"
                ValidationRule="..." />
    </Entry.Behaviors>
</Entry>
<Label x:Name="errUserName" Text="..." IsVisible="..."/>
```

The control block is composed of a label that is associated with the entry and the error label, which is only visible if there is a validation error within the label. A similar structure is used with the password field.

2. Just by exposing a couple of the binding data points, this block can easily be converted into a custom control. In order to create the base control, we will use `ContentView`:

```
<ContentView
    xmlns="http://xamarin.com/schemas/2014/forms"
    xmlns:x="http://schemas.microsoft.com/winfx/2009/xaml"
```

```
        xmlns:behaviors="clr-
    namespace:FirstXamarinFormsApplication.Client.Behaviors"

    x:Class="FirstXamarinFormsApplication.Client.Controls.ValidatableEn
    try"
        x:Name="RootView">
        <ContentView.Content>
            <StackLayout>
                <!-- TODO: // Insert Controls -->
            </StackLayout>
        </ContentView.Content>
    </ContentView>
```

Here, it is important to note that a name declaration is used to create a reference to the control itself, since we will create bindable properties on the control and bind them to the children values we previously identified.

3. Now, we will create our bindable properties within the `ValidatableEntry.xaml.cs` file:

```
    public static readonly BindableProperty LabelProperty =
        BindableProperty.CreateAttached("Label", typeof(string),
        typeof(ValidatableEntry), string.Empty);

    public static readonly BindableProperty PlaceholderProperty =
        BindableProperty.CreateAttached("Placeholder", typeof(string),
        typeof(ValidatableEntry), string.Empty);

    public static readonly BindableProperty ValueProperty =
        BindableProperty.CreateAttached("Value", typeof(string),
        typeof(ValidatableEntry), string.Empty, BindingMode.TwoWay);

    public static readonly BindableProperty ValidationRuleProperty =
        BindableProperty.CreateAttached("ValidationRule",
    typeof(IValidationRule<string>),
        typeof(ValidationBehavior), null);
```

4. We can also create accessors for these properties, like so:

```
    public string Label
    {
        get
        {
            return (string)GetValue(LabelProperty);
        }
        set
        {
            SetValue(LabelProperty, value);
```

```
            }
        }
```

5. Next, we will wire up these properties to the children attributes:

```
<StackLayout>
    <Label Text="{Binding Label, Source={x:Reference RootView}}" />
    <Entry Placeholder="{Binding Placeholder, Source={x:Reference
RootView}}" Text="{Binding Value, Mode=OneWayToSource,
Source={x:Reference RootView}}" >
        <Entry.Behaviors>
            <behaviors:ValidationBehavior
x:Name="ValidationBehavior"
                      ValidationRule="{Binding ValidationRule,
Source={x:Reference RootView}}" />
        </Entry.Behaviors>
    </Entry>
    <Label Text="{Binding ValidationRule.ValidationMessage,
Source={x:Reference RootView}}" FontSize="12" TextColor="Gray"
IsVisible="{Binding HasError, Source={x:Reference
ValidationBehavior}}"/>
</StackLayout>
```

6. Finally, we will replace the original `LoginView.xaml` file:

```
<controls:ValidatableEntry
    Label="{behaviors:Translate LblUsername}"
    Placeholder="{behaviors:Translate LblUsername}"
    ValidationRule="{Binding UserNameValidation}"
    Value="{Binding UserName, Mode=OneWayToSource}"/>
```

Here, we have created our custom `ContentView`, which will bundle a node of the visual tree in a single control. This control can also be used for other entry fields that require validation.

Next, we will look at how we can create a custom renderer for Android so that we can make use of the built-in validation displays, as well as the floating label design concept.

Creating a custom renderer

At times, a target platform can offer out-of-the-box functionality that exceeds the expected requirements via the use of customized controls from Xamarin.Forms. In these types of situations, it could be a good idea to replace the Xamarin.Forms implementation on a specific platform.

For instance, the form entry fields that we were trying to achieve with our custom implementation in the previous section would look much more platform appropriate if they were implemented with a TextInputLayout that followed material design guidelines:

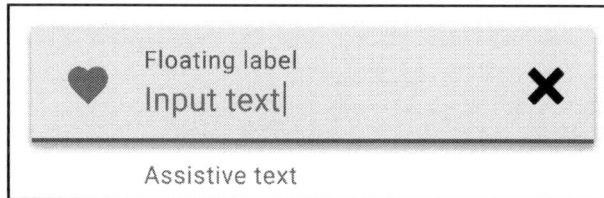

In this layout, we can bind the label to the floating label and the error text to the helper text area of the floating label edit text; however, by default, Xamarin.Forms uses FormsEditText (a derivative of EditText) rather than TextInputLayout for Android. In order to remedy this, we can implement our own custom renderer. Let's see how we can do this:

1. The first step of creating a renderer is to decide whether to create a renderer deriving from ViewRenderer<TView, TNativeView>, or the actual render implementation. For EntryRenderer, the Xamarin.Forms base class is ViewRenderer<Entry, FormsEditText>. Unfortunately, this means that we won't be able to make use of the base class implementation since our renderer will need to return TextInputLayout. Therefore, the renderer declaration will look similar to the following:

```
public class FloatingLabelEntryRenderer : ViewRenderer<Entry,
TextInputLayout>
    {
        public FloatingLabelEntryRenderer(Context context) :
        base(context)
        {
        }

        private EditText EditText => Control.EditText;

        protected override TextInputLayout CreateNativeControl()
        {
            var textInputLayout = new TextInputLayout(Context);
            var editText = new EditText(Context);
            editText.SetTextSize(ComplexUnitType.Sp,
            (float)Element.FontSize);
            textInputLayout.AddView(editText);
            return textInputLayout;
```

```
        }

        protected override void OnElementPropertyChanged(object
sender,
        PropertyChangedEventArgs e)
        {
            // TODO:
        }

        protected override void
OnElementChanged(ElementChangedEventArgs<Entry> e)
        {
            base.OnElementChanged(e);

            if (e.OldElement == null)
            {
                var textView = CreateNativeControl();
                // TODO:
                SetNativeControl(textView);
            }

            // TODO:
        }
    }
```

In this declaration, we should be initially be dealing with several override methods:

- CreateNativeControl: Responsible for creating the native control using the Element properties
- OnElementChanged: Similar to the OnAttached method on behaviors and effects
- OnElementPropertyChanged: Used to synchronize changes from the Xamarin.Forms element to the native element

2. As a first step, we are interested in the `Placeholder` property and the associated `OneWay` binding (that is, from `Element` to `Native`). Therefore, we will be using the `Placeholder` value as the hint text for the `EditText` field:

```
protected override void OnElementPropertyChanged(object sender,
PropertyChangedEventArgs e)
{
    if (e.PropertyName == Entry.PlaceholderProperty.PropertyName)
    {
        Control.Hint = Element.Placeholder;
    }
}
```

3. Secondly, we want to update the placeholder when the `Element` is attached to the renderer (the initial synchronization):

```
protected override void
OnElementChanged(ElementChangedEventArgs<Entry> e)
{
    base.OnElementChanged(e);

    if (e.OldElement == null)
    {
        var textView = CreateNativeControl();
        // textView.EditText.AddTextChangedListener(this);
        SetNativeControl(textView);
    }

    Control.Hint = Element.Placeholder;
    EditText.Text = Element.Text;
}
```

Another value we would like to keep in sync is the actual `Text` value; however, the synchronization in this case should be able to support `TwoWay` binding.

4. In order to listen for input text changes, we will implement the `ITextWatcher` interface:

```
void ITextWatcher.AfterTextChanged(IEditable @string)
{
}

void ITextWatcher.BeforeTextChanged(ICharSequence s, int start, int
count, int after)
{
}
```

```
void ITextWatcher.OnTextChanged(ICharSequence s, int start, int
before, int count)
{
    if (string.IsNullOrEmpty(Element.Text) && s.Length() == 0)
    {
        return;
    }

    ((IElementController)Element)
        .SetValueFromRenderer(Entry.TextProperty, s.ToString());
}
```

5. Once the renderer is complete, we will also need to register the renderer so that the Xamarin.Forms runtime is aware of the association between the Entry control and this new renderer:

```
[assembly: ExportRenderer(typeof(Entry),
typeof(FloatingLabelEntryRenderer))]
namespace FirstXamarinFormsApplication.Droid.Renderers
```

6. Now that the renderer is going to be handling both the label and the placeholder, we won't need the additional label within ValidatableEntry, so we will be using them only for iOS:

```
<ContentView>
    <OnPlatform x:TypeArguments="View">
        <On Platform="iOS">
            <Label Text="{Binding Label, Source={x:Reference
RootView}}" />
        </On>
    </OnPlatform>
</ContentView>
<Entry Placeholder="{Binding Placeholder, Source={x:Reference
RootView}}" Text="{Binding Value, Mode=OneWayToSource,
Source={x:Reference RootView}}" >
    <Entry.Behaviors>
        <behaviors:ValidationBehavior x:Name="ValidationBehavior"
                    ValidationRule="{Binding ValidationRule,
Source={x:Reference RootView}}" />
    </Entry.Behaviors>
</Entry>
```

> **TIP**
>
> The reason why we have wrapped the OnPlatform declaration is that even though it's syntactically correct, adding a view to a parent with multiple children cannot be rendered because of the way reflection is implemented. In order to remedy this issue, the platform-specific declaration needs to be wrapped into a benign view with a single child.

7. This is what the final outcome looks like:

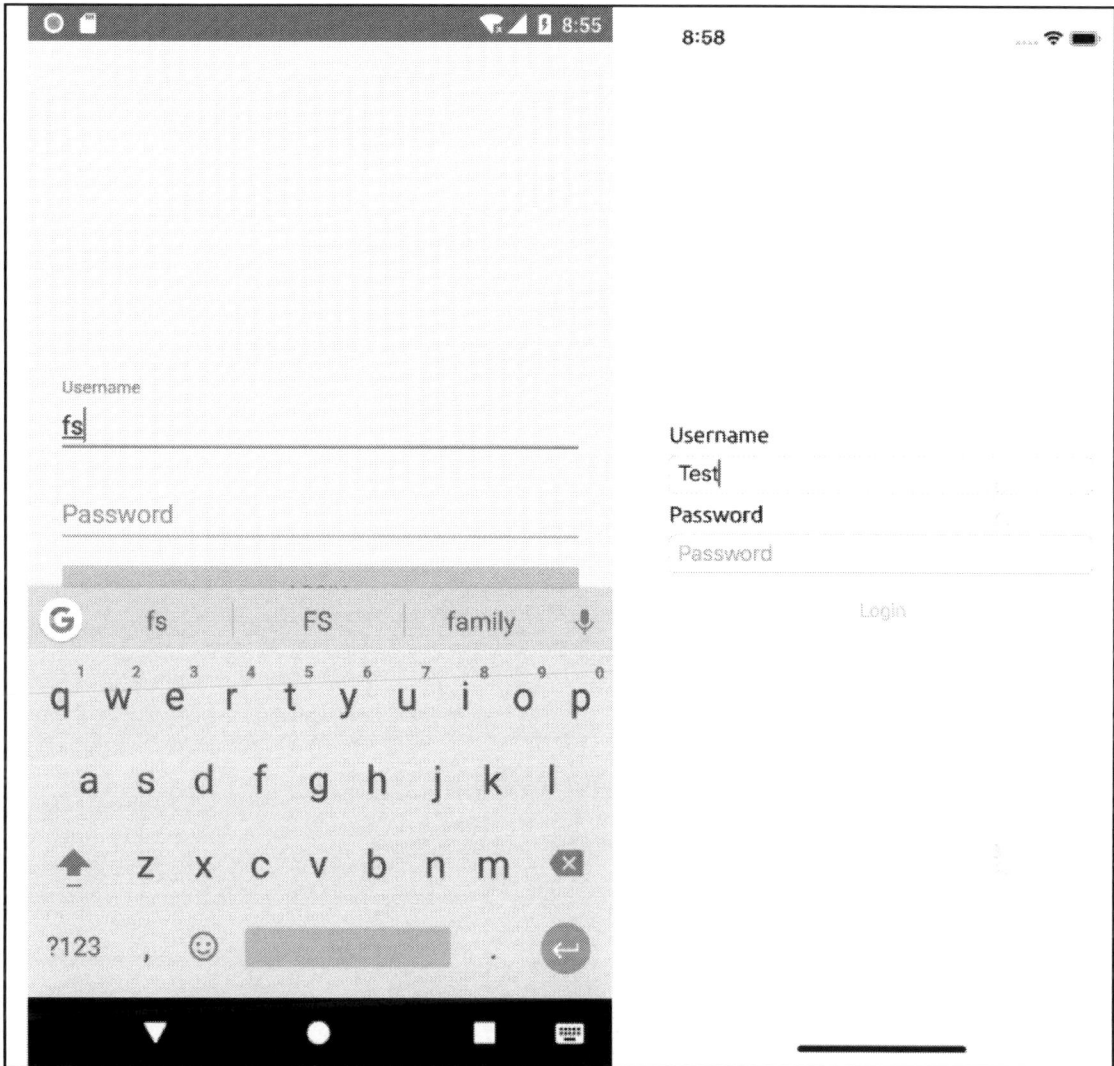

We can further expand this implementation to include the error indicator within the custom control, though this would mean that we would need to create a custom control and attach the custom renderer to it.

Creating a custom Xamarin.Forms control

For a completely custom control, the implementation starts with the Xamarin.Forms view abstraction. This abstraction provides the integration with XAML, as well as the view models (that is, the business logic) that are associated with that specific view.

For the floating label entry, we would, therefore, need to create a control with the required bindable properties exposed. For our use case, in addition to the Entry control attributes, we would need the validation error description and a flag identifying whether there is such an error. Let's begin the implementation of our custom control:

1. We will start by deriving our custom control from the Entry itself and adding the additional properties:

```
public class FloatingLabelEntry : Entry
{
    public static readonly BindableProperty ErrorMessageProperty =
        BindableProperty.CreateAttached("ErrorMessage",
typeof(string), typeof(FloatingLabelEntry), string.Empty);

    public static readonly BindableProperty HasErrorProperty =
    BindableProperty.CreateAttached("HasError", typeof(bool),
typeof(FloatingLabelEntry), false);

    public string ErrorMessage
    {
        get
        {
            return (string)GetValue(ErrorMessageProperty);
        }
        set
        {
            SetValue(ErrorMessageProperty, value);
        }
    }

    public bool HasError
    {
        get
        {
            return (bool)GetValue(HasErrorProperty);
```

```
            }
        set
        {
            SetValue(HasErrorProperty, value);
        }
    }
}
```

2. We can now modify our `FloatingLabelRenderer` to use the new control as the `TElement` type parameter:

```
[assembly: ExportRenderer(typeof(FloatingLabelEntry),
typeof(FloatingLabelEntryRenderer))]
 namespace FirstXamarinFormsApplication.Droid.Renderers
 {
     public class FloatingLabelEntryRenderer :
ViewRenderer<FloatingLabelEntry, TextInputLayout>, ITextWatcher
```

3. In the renderer, we would need to listen for `HasErrorProperty` changes and set the error description and error indicator accordingly:

```
protected override void OnElementPropertyChanged(object sender,
PropertyChangedEventArgs e)
{
    . . . .
    else if (e.PropertyName ==
FloatingLabelEntry.HasErrorProperty.PropertyName)
    {
        if (!Element.HasError ||
string.IsNullOrEmpty(Element.ErrorMessage))
        {
            EditText.Error = null;
            Control.ErrorEnabled = false;
        }
        else
        {
            Control.ErrorEnabled = true;
            EditText.Error = Element.ErrorMessage;
        }
    }
    . . . .
}
```

4. Using this control within `ValidatableEntry` instead of the `Entry` control would create a pleasant material design layout:

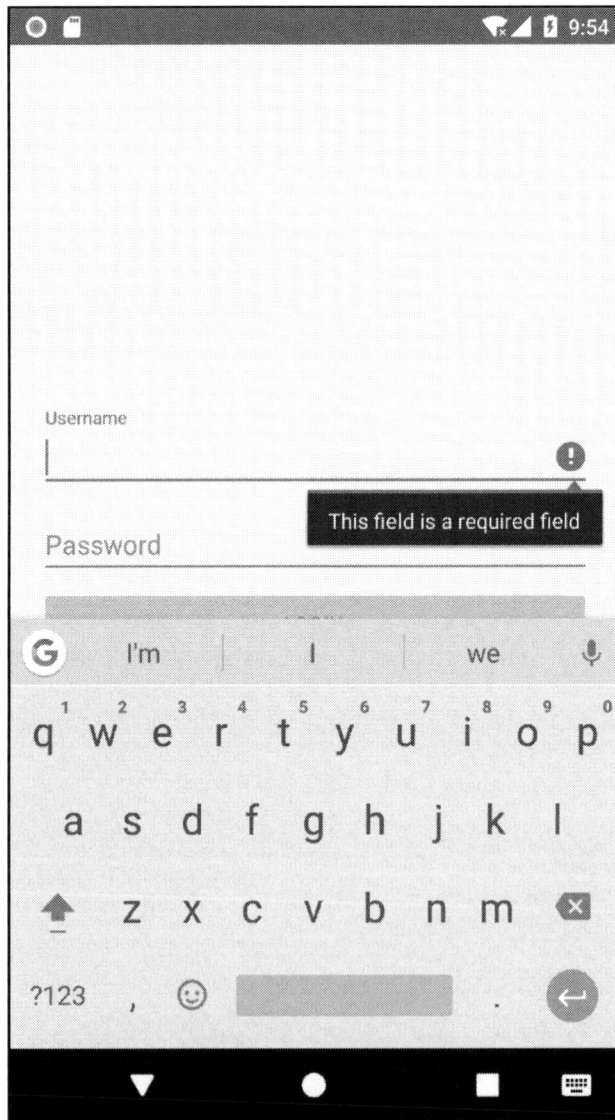

It is important to note that, even though we have created and used this custom control for both Android and iOS, since the iOS renderer is not implemented, iOS would still display the next best thing from the inheritance tree (that is, `EntryRenderer`).

Summary

Overall, Xamarin.Forms has many extensibility points for various scenarios. Nevertheless, we, as developers, should be careful about using these extensibility points sensibly in order to create robust, simple, yet sophisticated user interfaces. In this chapter, in order to understand the customization options, we identified the implementation domains/quadrants of our Xamarin.Forms application and went over different customization options for each quadrant. Finally, we created a custom control to demonstrate the complete implementation of a user control in both shared and native domains.

This chapter finalizes the Xamarin side of the development effort of our project. In the next few chapters, we will continue developing a cloud infrastructure using .NET Core for our mobile application.

3
Section 3: Azure Cloud Services

Creating a modern mobile application generally requires a robust service backend and infrastructure. Azure Cloud infrastructure provides a wide spectrum of services and development platforms for developers to create .NET Core components that can be used with mobile applications. These services vary from simple **Platform as a Service (PaaS)** hosting components to sophisticated multi-model persistence stores.

The following chapters will be covered in this section:

- Chapter 7, *Azure Services for Mobile Applications*
- Chapter 8, *Creating a Datastore with Cosmos DB*
- Chapter 9, *Creating Microservices Azure App Services*
- Chapter 10, *Using .NET Core for Azure Serverless*

7
Azure Services for Mobile Applications

Whether you are dealing with a small startup application or handling lots of data for an enterprise application, Microsoft Azure is always a convenient choice because of its cost-effective subscription model and the scalability that it offers. There are a number of services on offers such as **Software as a Service (SaaS)**, **Platform as a Service (PaaS)**, and **Infrastructure as a Service (IaaS)**. These include Notification Hubs, Cognitive Services, and Azure Functions, which can change the impression of the user regarding your application with few or no additional development hours. This chapter will give you a quick overview on how to use some of these services whilst developing .NET Core applications.

In this chapter, we will be designing our service backend using the available service offerings on the Azure platform:

- An overview of Azure services
- Data store
- Azure serverless
- Development services

An overview of Azure services

We are living in the age of cloud computing. Many of the software paradigms that we learned and applied to our applications 10 years ago are now completely obsolete. The good old n-tier applications and development team simplicity have been replaced with distributed modules for the sake of maintainability and performance.

Without further ado, let's start preparing the scope of our application by setting up the architecture and exploring the concepts of the Azure platform.

An introduction to distributed systems

We have started the development of our client application; it will require some additional views and modifications. In order to continue with the development, we first need to set up our backend. For our application, we will need a service backend that will do the following:

- Provide static metadata about products
- Manage user profiles and maintain user-specific information
- Allow users to upload and publicly share data
- Index and search user uploads and shares
- Notify a set of users with real-time updates

Now, putting these requirements and our goal of creating a cloud infrastructure aside, let's try to imagine how we could implement a distributed system with an on-premise n-tier application setup:

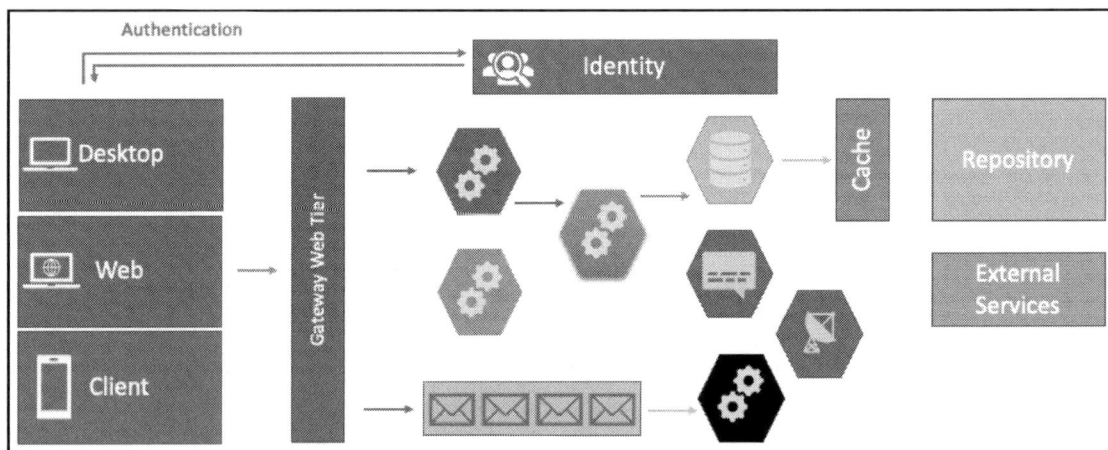

In this setup, we have a web tier that exposes the closed logical n-tier structure to the client. Notice that the system is divided into logical tiers and there is no over-the-wire communication involved. We will maintain this structure with an on-premise server. Multiple servers with a load-balanced implementation could still work if there is a need for scaling. In most cases, synchronization and normalization will occur in the data tier. From a deployment and management perspective, each deployment will result in a complete update (on multiple servers). Additionally, each logical module's requirements will have to be maintained separately even though it is a monolith implementation, and applying updates to the on-premise server should not be taken lightly because of these requirements. Deployments to distinct servers will also have to be handled with care.

Knight Capital Group was an American global financial services firm specializing in the electronic execution of sales and trading. In August 2012, the company went from $400 million in assets to bankruptcy overnight because of a new deployment that was only released on seven of the eight servers that the company operated. The 45-minute nightmare, where the correct deployments were competing with the old code on a single server, resulted in a $460 million loss.

We can easily move the complete web application to a cloud IaaS **virtual machine** (**VM**). However, this migration will only help with maintenance, and scaling would still have to be on the system level rather than the component level; the bottleneck, in this case, would most likely be the data tier.

For this n-tier setup, we would use a SQL database for data storage, a message queue such as Rabbit MQ or MSMQ, and an ASP.NET web API implementation for the web tier. Identity management would probably be an integrated solution such as ASP.NET Identity. Notifications could be a polling implementation from the client side or, alternatively, a SignalR implementation can be considered within the ASP.NET web application. Search functionality would probably be have to be on the SQL Server level for better performance. All of this is with the assumption that we are using the Microsoft .NET stack and the target hosting platform is a Microsoft IIS server on a Windows host.

Next, let's break down our logical modules into smaller services that can communicate with each other within a **Service-Oriented Architecture** (**SOA**) ecosystem:

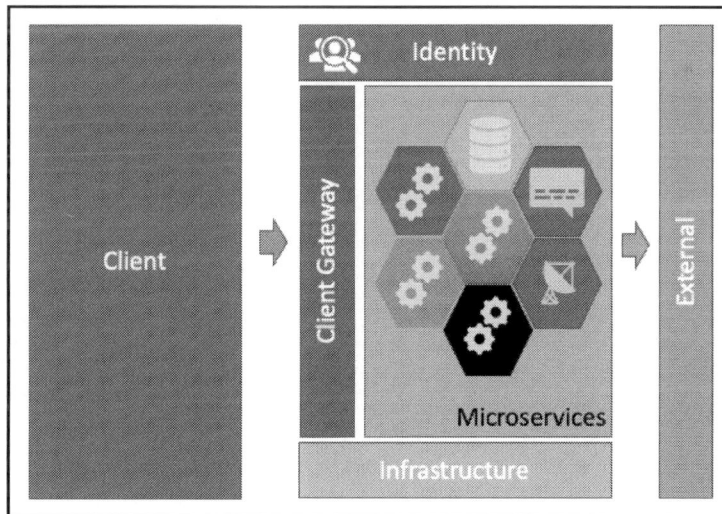

This setup is a little lighter, where each component can be independently developed and deployed (that is, decoupled from the other elements in the system). From a maintenance perspective, each service can be deployed onto separate servers or VMs. In return, they can be scaled independently from each other. Moreover, each of these services can now be containerized so that we can completely decouple our services from the operating system. After all, we only need to have a web server in which our set of services can be hosted and served to the client. At this point, .NET Core will turn our application into a cross-platform web module, allowing us to use both Windows and Unix containers. This whole endeavor could be label as migrating from an IaaS strategy to a PaaS approach. Additionally, the application can now implement an **Infrastructure as Code (IaC)** structure, where we don't need to worry about the current state of the servers that the application is running on.

Well, this sounds great, but how does it relate to cloud architecture and Azure? The main purpose of creating a cloud-ready application is to create an application with functionally independent modules that can be hosted by appropriate, maintainable, and scalable cloud resources. At this point, we are no longer talking about a single application, but a group of resources working hand-in-hand for various application requirements:

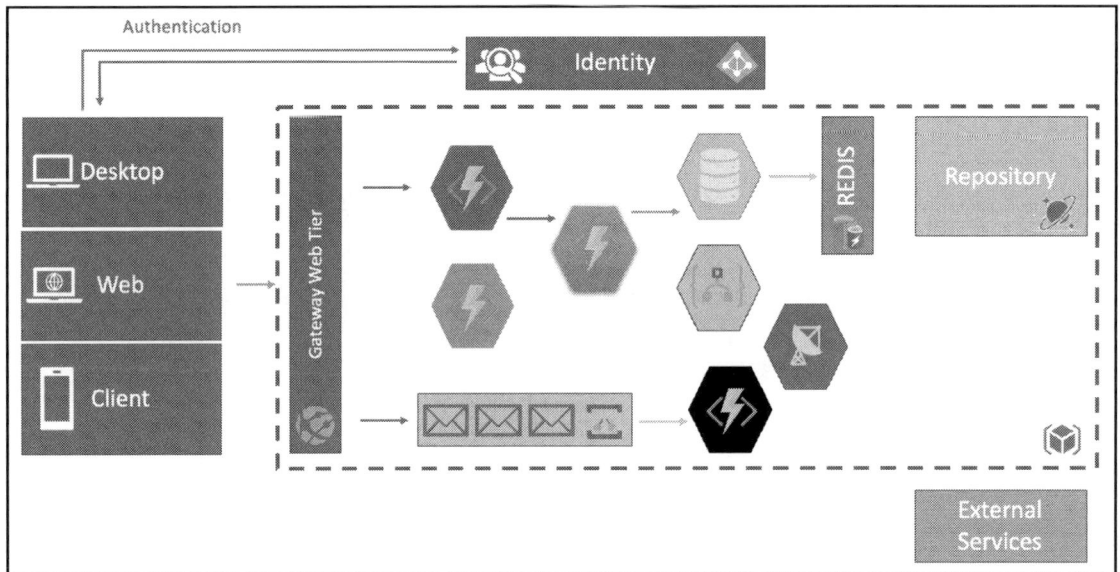

In this distributed model, each component is made up of simple PaaS services where there is no direct dependency among them. The components are completely scalable, and are replaceable as long as the system requirements are satisfied. For instance, if we begin with a small web API application service, it would probably reside within an application service plan. However, if the requirements are satisfied, then we can replace this microservice with an Azure function implementation that would change the deployment model and the execution runtime, but still keep the system intact. Overall, in a cloud model, the replaceable nature of individual components (as long as the overall system is in check) minimizes risks and the effort of maintenance.

Going back to our requirements, we are free to choose between a relational database such as a SQL Server PaaS or a NoSQL setup with Cosmos DB. Additionally, we can improve performance by using a Redis Cache between the data stores and the web gateway. Search functions could be executed using Azure Search indices, and App Services and Azure Functions can be employed for the API layer. Additionally, a simple ESB implementation or Azure Durable Functions can help with the long-running asynchronous operations. Finally, notifications can be done by using Azure SignalR or Notification Hubs.

Of course, the choice of resources will largely depend on the architectural approach that is chosen.

Cloud architecture

In the cloud platform, the design of the system consists of individual components. While each component should be designed and developed separately, the way in which these components are composed should follow certain architectural patterns that will allow the system to provide resilience, maintainability, scalability, and security.

Particularly for mobile applications, some of the following compositional models can help to contribute to the success of the application.

Gateway aggregation

In a microservices setup, the application is made up of multiple domains, and each domain implements its own microservice counterpart. Since the domain is segregated, the data that is required for the client application view can be constructed by executing multiple calls for the backend services. In an evolving application ecosystem, this will, in time, push all the complexities of the business tier into the client application. While this could still be acceptable for a web application, a mobile application's performance will degrade in time as the complexity of the system grows. In order to avoid this problem, a gateway service facade can be placed in between the client application and the microservices:

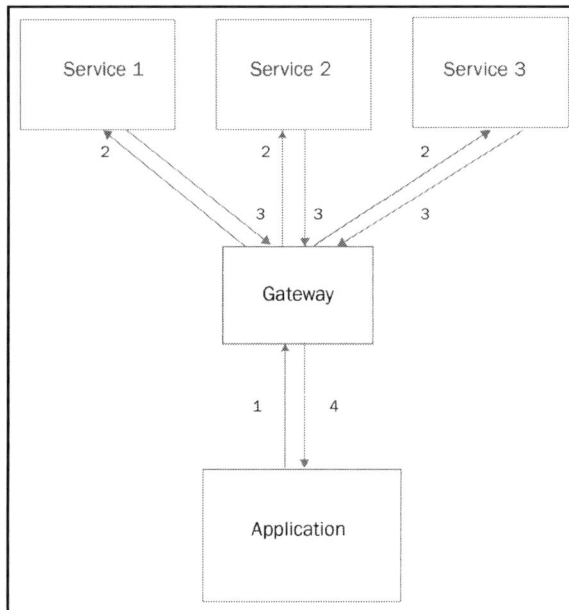

Let's consider applying the same logic within our application. Let's assume that an auctioned vehicle data is handled by one API (business item), the vehicle metadata is handled by another API (static data), the user information is served through yet another API, and, finally, we have a bidding API. While this setup provides the necessary segregation for a microservice setup, it requires the client application to execute multiple service calls to view and/or create a single posting. In such a scenario, the gateway can be used to orchestrate the microservices so that the client application can be relieved of this responsibility.

If, in fact, we are planning to support a web application as a client, then the data models and service orchestration might differ from the mobile application. In this case, we would consider creating separate gateways for each client app, thus decreasing the maintainability costs of a single super gateway.

Backends for frontends

In a multiclient system, each client might require the data to be aggregated in a certain way. This will depend on the target platform resources, technical feasibility, and use cases. In this type of a scenario, the gateway API would be required to expose multiple service endpoints for different microservice and data compositions. Instead, each client app can be served data through a separate gateway, decreasing the complexities of supporting multiple client applications through a single facade:

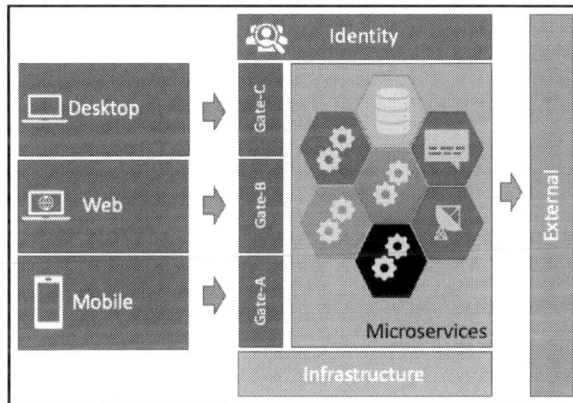

For instance, let's assume that our application development team implements a UWP application on top of an original mobile application that is targeting iOS and Android platforms. In this case, UWP views would be viewed on a larger design real estate and the data requirements would be different than the mobile applications. For a simple solution, gateway API endpoints can now be extended with parameters to limit or extend the object tree that is returned in the response (that is, info, normal, or extended), or additional `Get{Entity}Extended` endpoints can be introduced. Nevertheless, in this way, while minimizing the complexity of the client applications, we are causing the gateway to grow and are decreasing the maintainability of this tier. If we introduce separate gateways, we will be separating the life cycle of these APIs for clients that already have separate application life cycles. This could help in creating a more maintainable system.

However, what if we have certain compositions or aggregations that repeat throughout the execution of client applications? These repeating patterns can be construed as data design problems, where the segregation of data results in a degradation of performance. If the microservice setup, in fact, requires these domain separations, we will need to come up with a data composition on the data store level.

A materialized view

The aggregation of certain data dimensions can be done on the data store level. In fact, as developers with a SQL background, we are familiar with SQL views that can be composed of multiple relational tables and can be indexed on the data store level. Similar strategies can be applied to NoSQL databases such as Cosmos.

For instance, this data denormalization process can be executed on Cosmos DB using the Azure Cosmos DB change feed. Changes on one document collection can be synchronized across multiple collections, which are optimized for executing various searches or aggregating data operations:

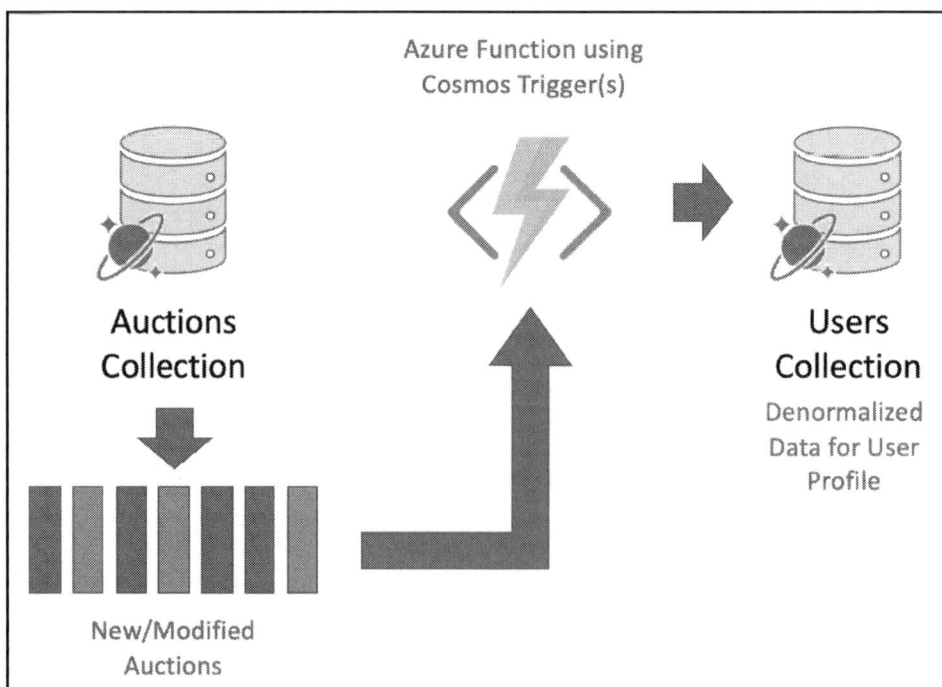

For instance, going back to our auction functionality, when we are dealing with the search function, we will be executing a search on multiple document collections; that is, the user will need to search by vehicle, auction data, bids, and profile data. In other words, the data points on different dimensions should all be available through an inner-join for the search execution. This can be achieved using a summary table for the vehicle posts, allowing searchable fields to be synchronized across collections.

The cache-aside pattern

Caching is yet another factor that can help improve the performance of the application, that is, the type of data we are caching and the application layers that we are caching this information on. The cache-aside pattern is the implementation of a multiplexer that will handle data consistency between the cache store and the data store depending on the incoming requests and the data lifespan:

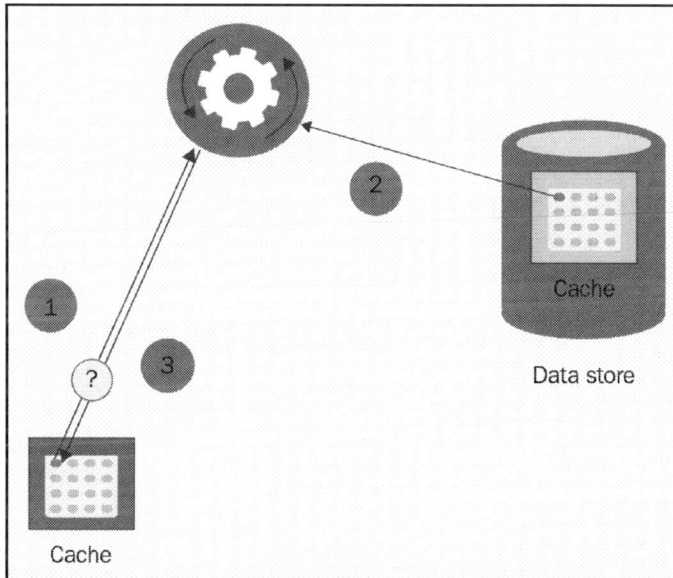

In this setup, an incoming request, branded with a certain unique identifier (for example, {EntityName}_{EntityId}), is first searched for within the cache store and, if not there, it is retrieved from the data store and inserted into the cache. In this way, the next request will be able to retrieve the data from the cache.

In a to-cache or not-to-cache dilemma, data entropy can be a fundamental decision factor. For instance, caching data for static reference items can be beneficial; however, caching the auction information, where the data is impure and the recurrence of requests for the same data points is less likely than static references, will not provide added value to the system.

The cache-aside strategy can also be implemented on the client side using local storage such as SQLite. At times, a certain document collection that it would not make sense to cache on the server side can beneficially be cached on the client side. For instance, the vehicle metadata for a certain make and model for the current user might be a repeating request pattern; however, considering the entropy of this data and the access frequency of other users to the same item, it would not be a server cache dimension.

Queue-based load leveling

Message queues are neither a new concept nor exclusive to the cloud architecture. However, in a distributed system with microservices, they can help with the decoupling of services and allow you to throttle resource utilization. Serverless components that are designed for scalability and performance, such as Azure Functions, can provide excellent consumers for work queues within the cloud infrastructure.

For instance, let's consider an application use case where the registered user is creating an auction item. They have selected the make and model, added additional information, and have even added several photos for the vehicle. At this point, if we allow the posting of this auction item to be a synchronous request, we will be locking certain modules in the pipeline to a single request. Primarily, the request would need to create a document in the data store; however, additional functions would also be triggered within the system to process images, notify subscribed users, and even start an approval process for the content administrator. Now, imagine this request is executed by multiple users of the application (for example, multiple registered users creating multiple posts). This would result in resource utilization peaks which, in turn, would put the resilience and availability of the application at risk.

As a solution to this problem, we can create a message queue that will be consumed by an Azure function, which will orchestrate the creation of the auction data. The message queue can be either an Enterprise Service Bus or an Azure storage queue:

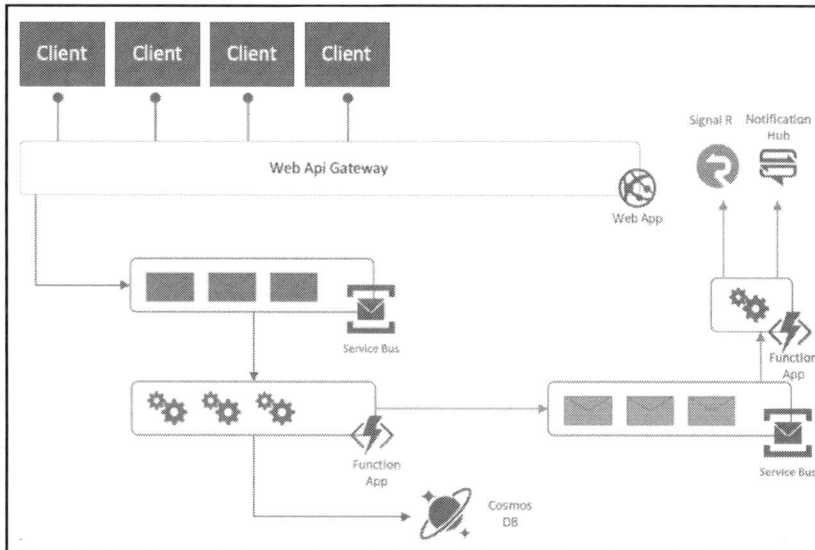

Well, this sounds great, but how does this implementation a effect the client implementation? Well, the client application will need to implement either a polling strategy to retrieve the status of the asynchronous job, or it can be notified using a push-pull mechanism, where the server would first send the auction ID before the process is even queued. Then, when it is finalized, the server can notify the client with the same ID, allowing it to pull the completed server data. At this point, the local version of the data can be stored and served to the user until the actual server data is available. For this type of notification, notification mechanisms such as Azure SignalR or Notifications Hub can be employed.

Competing consumers

In the previous example, we used Azure functions as the consumers of a message queue. This approach could already be accepted as an implementation of competing consumers where the provided message queue is handled by multiple worker modules.

While this would provide scaling requirements and allow for performant execution, as a product owner we would not have any control over the function instances created to consume the events from the message queue. In order to be able to throttle and manage the queue, a message broker mechanism can be introduced, which will control the flow of messages into the queue. Once the messages are pushed into the queue, multiple consumers can retrieve, process, and complete the messages.

The publisher/subscriber pattern

Let's assume that we have completed our brokered queue implementation and dispatched a consumer to finalize a long-running operation. At this point, as previously mentioned, our application is expecting a *done* signal so that it can get rid of any transient data.

In an open system, like the one we are in the process of implementing, where each service can communicate with each other (rather than a closed system where the execution is done sequentially down the stream), we are no longer dealing with a deterministic synchronous model, and yet the consumers of the system still expect results. In order to allow the source system (that is, the publisher) to propagate the output of an operation to the interested parties (that is, the subscribers), an output channel can be established. This implementation pattern can be attributed to the publisher/subscriber pattern (which is also known as the pub/sub pattern)

Going back to our asynchronous web request, the output channel would then deliver the result to the notification module and deliver the results to the client application.

The same pattern implementation can be established with another message queue using Service Bus or the actual implementation of the pub/sub pattern on the Azure infrastructure; EventGrid. Either of these services can allow the output from a long-running process to be fanned out to interested parties, such as an Azure function, which will push the notification message or trigger a message on Azure SignalR.

The circuit breaker and retry patterns

In a cloud system, where there are multiple moving pieces involved, it is hard to avoid failure. In this context, the resilience of a system is determined by how fast and how often it can recover from a failure. The circuit breaker and retry patterns are complementary patterns that are generally introduced in a microservice ecosystem. The circuit break pattern can be used to decrease the time and resources of a system if a failure is imminent. In these types of situations, it is better to allow the system to fail sooner rather than later so that the failure can be handled by a secondary process or a failover mechanism can be initiated.

For instance, if we have a service that is prone to timeouts (for example, under heavy load or due to an external service failure), a circuit breaker can be implemented to continuously monitor the incoming requests in the closed circuit state. Failures can be retried seamlessly with regard to the client application. When consequent failures occur, the circuit can be put into a half-open or open state temporarily, so that the following requests are immediately dropped without trying the execution (knowing that it will probably fail until the issue is fixed). In this state, the client app can disable the feature or, if there is a failover/workaround implemented, then this implementation can be used. Once the circuit open state expires, the system can reintroduce this endpoint, first, in a half-open state and, finally, in a closed state, and the system is said to be healed.

Built-in Azure-monitoring functions accompanied by application telemetry can provide alerts and notifications, which can help with the maintenance of Azure applications.

Azure service providers and resource types

The Azure ecosystem, along with the ever-growing set of services it provides, allows developers to create various distributed cloud applications with ease. As we have seen in the *Cloud architecture* section, many PaaS and SaaS offerings create a catalog of solutions for everyday problems by designing scalable and resilient applications.

By quickly looking at the (incomplete) catalog of services, you will notice that each service is provided as part of a provider category:

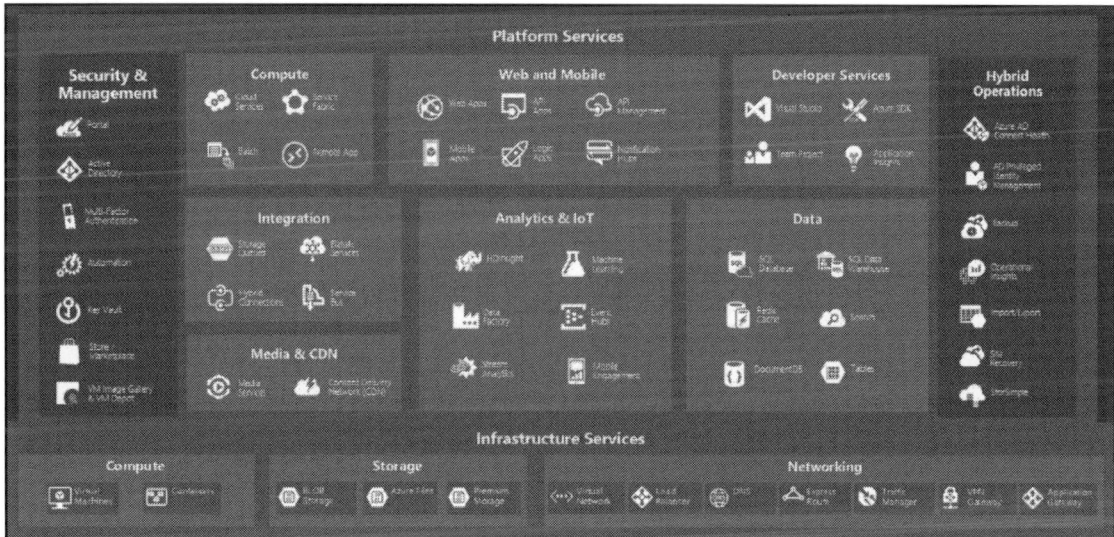

The services in each category are provided by one or multiple service providers. Each service in these catalogs is versioned so that the provisioning of these services can be handled by Azure Resource Manager.

In order to visualize the number of providers available under the same roof, you can use the Az PowerShell module:

```
Get-AzResourceProvider -ListAvailable | Select-Object ProviderNamespace,
RegistrationState
```

This will return a set of providers that are available for your subscription. These providers can be Microsoft-provided modules or third-party offers:

Microsoft Azure Documentation provides helpful Azure PowerShell commands, which can be directly executed in the Cloud Shell without having to use PowerShell (on Windows) or bash (on Linux or macOS). Additionally, a cross-platform version, PowerShell (Core), which utilizes .NET Core, is available on non-Windows operating systems.

If you dive into a specific namespace, for instance, the Microsoft.Compute provider namespace, you can get a better overview of the services offered and the geographical regions these resources are available at:

```
Azure:/
PS Azure:\> Get-AzResourceProvider -ProviderNamespace Microsoft.Compute `
>> | Select-Object ResourceTypes, Locations

ResourceTypes                                                    Locations
------------                                                     ---------
{availabilitySets}                                              {East US, East US 2, ...
{virtualMachines}                                              {East US, East US 2, ...
{virtualMachines/extensions}                                   {East US, East US 2, ...
{virtualMachineScaleSets}                                      {East US, East US 2, ...
{virtualMachineScaleSets/extensions}                           {East US, East US 2, ...
{virtualMachineScaleSets/virtualMachines}                      {East US, East US 2, ...
{virtualMachineScaleSets/networkInterfaces}                    {East US, East US 2, ...
{virtualMachineScaleSets/virtualMachines/networkInterfaces}    {East US, East US 2, ...
{virtualMachineScaleSets/publicIPAddresses}                    {East US, East US 2, ...
{locations}                                                    {}
{locations/operations}                                         {East US, East US 2, ...
{locations/vmSizes}                                            {East US, East US 2, ...
{locations/runCommands}                                        {East US, East US 2, ...
{locations/usages}                                             {East US, East US 2, ...
{locations/virtualMachines}                                    {East US, East US 2, ...
{locations/publishers}                                         {East US, East US 2, ...
{operations}                                                   {East US, East US 2, ...
{restorePointCollections}                                      {Southeast Asia, East...
{restorePointCollections/restorePoints}                        {Southeast Asia, East...
{virtualMachines/diagnosticSettings}                           {East US, East US 2, ...
{virtualMachines/metricDefinitions}                            {East US, East US 2, ...
```

In an Azure resource group, the `Resource` type defines which resource we are really after as well as the version of this resource. These resource definitions, if prepared as part of a resource group **Azure Resource Manager (ARM)** template, make up our declarative IaC.

ARM is the platform service that allows provisioning resources within a subscription. It exposes a web API that can be consumed using PowerShell, Azure CLI, and CloudShell, as well as the Azure portal itself. The declarative syntax used in resource manager templates provides a consistent, idempotent deployment experience, which allows developers and automation engineers to manage the infrastructure life cycle with confidence.

Data stores

Defining domains, and creating the architecture that our distributed system is going to be built upon, inherently starts with deciding on the persistence store. In return, data domains can be defined and access models can be designated. In most cases, this decision does not need to be limited to a single data store, but the system can make use of multiple data types and different data stores. The Azure platform offers various resources with different data management concepts and feature sets. It is important to choose a data store model that is best suited to the application requirements and take account of cost and management. Let's take a look at these different models and when to use them.

Relational database resources

Relational databases are probably the most prominent applications of a data store. Transactional consistency that implements the **Atomic, Consistent, Isolated, Durable (ACID)** principles offers developers a strong consistency guarantee. Nevertheless, from a scalability and performance perspective, common SQL implementations such as MSSQL or MySQL are, in most scenarios, outperformed by NoSQL databases such as Mongo and Cosmos DB. Azure SQL Database, Azure Database for MySQL, and PostgreSQL are available both as IaaS and PaaS offerings on the Azure platform.

In the PaaS resource model, the operational costs and scalability of the databases are handled through a unit called **Database Transaction Unit (DTU)**. This unit is an abstract benchmark that is calculated using the of CPU, memory, and data I/O measures . In other words, DTU is not an exact measure but a normalized value depending on the aforementioned measures. Microsoft offers a DTU Calculator, which can provide estimates on DTU usage based on performance counters collected on a live database.

From a security perspective, several advanced features are available for Azure SQL Databases. These security features are available on various levels of data accessibility:

- Network security is maintained by firewalls and access is granted explicitly by using IP and Virtual Network firewall rules.
- Access management implementation consists of SQL Authentication and Azure Active Directory authentication. The security permissions can be as granulized as data tables and rows.
- Threat protection is available through log analytics and data auditing as well as threat detection services.
- Information protection through data masking and encryption on various levels protects the data itself.

Azure storage

The Azure storage model is one of the oldest services in the cloud ecosystem. It is a NoSQL store and provides developers with a durable and scalable persistence layer. Azure storage is made up of four different data services, and each of these services is accessed over HTTP/HTTPS with a well-established REST API.

Let's take a closer look at these data services available within Azure storage.

Azure blobs

Azure blob storage is the cloud storage offering for unstructured data. Blobs can be used to store any kind of data chunks such as text or binary data. Azure blob storage can be accessed through the URL provided for the storage account created:

```
http://{storageaccountname}.blob.core.windows.net
```

Each storage account contains at least one container, which is used to organize the blobs that are created. Three types of blobs are used for different types of data chunk to be uploaded:

- **Block blobs**: These are designed for large binary data. The size of a block blob can go up to 4.7 TB. Each block blob is made up of smaller blocks of data, which can be individually managed. Each block can hold up to 100 MB of data. Each block should define a block ID, which should conform to a specific length within the blob. Block blobs can be considered as discrete storage objects such as files in a local operating system. They are generally used for storing individual files such as media content.
- **Page blobs:** These are used when there is a need for random read/write operations. These blobs are made up of pages of 512 bytes. A page blob can store up to 8 TB of data. In order to create a page blob, a maximum size should be designated. Then, the content can be added in pages by specifying an offset and a range that aligns with the 512-byte page boundaries. VHDs stored in the cloud are a perfect fit for page blob usage scenarios. In fact, the durable disks provided for Azure VM are page blob-based (that is, they are Azure IaaS disks).
- **Append blobs:** As the name suggests, these are append-only blobs. They cannot be updated or deleted and the management of individual blocks is not supported. They are frequently used for for logging information. An append blob can grow up to 195 GB.

As you can see, blob storage, especially block blobs, are ideal for storing image content for our application. Azure Storage Client library methods provide access to CRUD operations for blobs and can be directly used in the client application. However, it is generally a security-aware approach to use a backend service to execute the actual upload to blob storage so that the Azure security keys can be kept within the server rather than the client.

Azure files

Azure files can be considered a cloud-hosted file sharing system. It is accessible through the **Server Message Block (SMB)**, also known as Samba, and allows storage resources to be used on hybrid (that is, on-premise and cloud) scenarios. Legacy applications that are using network shared folders (or even local files) can be easily pointed to the Azure files network storage. Azure files, just like other Azure storage data service, are accessible through the REST API and Azure storage client libraries.

Azure queues

In order to implement asynchronous processing patterns, if you are not after advanced functionalities and queue consistency, Azure queues can be a cost-effective alternative to Service Bus. Azure queues can be larger and easier to implement and manage for simpler use cases. Similar to Service Bus, Azure queue messages can also be used with Azure Functions, where each message triggers an Azure function that handles the processing. If triggers are not used, then only a polling mechanism can handle the message queue. This is because, unlike Service Bus, they don't provide blocking access or an event trigger mechanism such as `OnMessage` on Service Bus.

Azure tables

Azure tables are a NoSQL-structured cloud data store solution. The implementation of Azure Table storage follows a **key-value pair (KVP)** approach, where structured data without a common schema can be stored in a table store. Azure Table storage data can be easily visualized on the Azure portal and data operations are supported through Azure Storage client libraries such as other Azure storage services. Nevertheless, Azure Table storage is now part of Azure Cosmos DB and can be accessed using the Cosmos DB table API and SDK.

Cosmos DB

Cosmos DB is the multifacade, globally distributed database service offering of Microsoft on the Azure cloud. With its key benefits being scalability and availability, Azure Cosmos DB is a strong candidate for any cloud-based undertaking. Being a write-optimized database engine, it guarantees less than 10 ms latency on read/write queries at the 99th percentile globally.

Cosmos DB provides developers with five different consistency models to allow for an optimal compromise between performance and availability, depending on requirements. The so-called consistency spectrum defines various levels between strong consistency and eventual consistency or, in other words, higher availability and higher throughput.

In spite of the fact that it was designed as NoSQL storage, it does support various storage model protocols including SQL. These storage protocols support the use of existing client drivers and SDKs, and can replace existing NoSQL data stores seamlessly. Each API model can also be accessed through the use of the available REST API:

API	Model	Containers	Items
SQL API	Document	Collections	Documents
MongoDB	Document	Collections	Documents
Gremlin	Graph	Graphs	Nodes and Edges
Cassandra	Column Family	Table	Rows
Azure Table Storage	KVP Store	Table	Items

Azure Cache for Redis

The Azure Cache for Redis resource is a provider that implements an in-memory data structure store that is similar to Redis. It helps improve the performance and scalability of distributed systems by decreasing the load on the actual persistence store. With Redis, data is stored as KVPs, and because of its replicated nature, it can also be used as a distributed queue. Redis supports the execution of transactions in an atomic manner.

We will be using the cache-aside pattern with the help of Azure Cache for Redis to implement in our application backend.

Azure serverless

As you might have noticed, in modern cloud applications, PaaS components are more abundant than IaaS resources. Here, application VMs are replaced with smaller application containers, and the database as a platform replaces the clustered database servers. Azure Serverless takes infrastructure and platform management one step further. In a serverless resource model, such as Azure Functions, event-driven application logic is executed on-demand on a platform that is provisioned, scaled, and managed by the platform itself. In the Azure serverless platform, event triggers can vary from message queues to Webhooks, with intrinsic integration into various resources within the ecosystem.

Azure functions

Azure functions are managed event-driven logic implementations that can provide lightweight ad hoc solutions for the cloud architecture. In Azure functions, the engineering team is oblivious not only of the execution infrastructure but also of the platform, since Azure functions are implemented cross-platform with .NET Core, Java, and Python.

The execution of an Azure function starts with the trigger. Various execution models are supported for Azure functions, including the following:

- **Data triggers**: CosmosDBTrigger and BlobTrigger
- **Periodic triggers**: TimerTrigger
- **Queue triggers**: QueueTrigger, ServicesBusQueueTrigger, and ServiceBusTopicTrigger
- **Event triggers**: EventGridTrigger and EventHubTrigger

The trigger for an Azure function is defined within the function manifest/configuration: `function.json`.

Once the trigger is realized, the function runtime executes the run block of an Azure function. The request parameters (that is, the input bindings) that are passed to the run block are determined by the trigger that was used.

For instance, the following function implementation is triggered by a message entry in an Azure storage queue:

```
public static class MyQueueSample
{
    [FunctionName("LogQueueMessage")]
    public static void Run(
        [QueueTrigger("%queueappsetting%")] string queueItem,
        ILogger log)
    {
        log.LogInformation($"Function was called with: {queueItem}");
    }
}
```

The output parameters (that is, the output binding) can also be defined as an `out` parameter within the function declaration:

```
public static class MyQueueSample
{
    [FunctionName("LogQueueMessage")]
    public static void Run(
        [QueueTrigger("%queueappsetting%")] string queueItem,
        [Queue("%queueappsetting%-out")] string outputItem
        ILogger log)
    {
        log.LogInformation($"Function was called with: {queueItem}");
    }
}
```

Note that the `[Queue]` attribute is used from the queue storage bindings for Azure functions as an output binding, which will create a new message entry in another queue. Further similar binding types are available out-of-the-box for Azure functions.

> We have used C# and .NET Standard in these examples for creating compiled Azure functions. Script-based C#, Node.js, and Python are also options for creating functions using a similar methodology.

Conceptually, Azure functions can be treated as per-call web services. Nevertheless, Durable Functions, an extension of Azure Functions, allows developers to create durable (that is, stateful) functions. These functions allow you to write stateful functions with checkpoints, where the orchestrator function can dispatch stateless functions and execute a workflow.

Azure functions can be used either as individual modules executing business logic on certain triggers or as a bundle of imperative workflows (using durable functions); alternatively, they can be used as processing units of **Azure Logic Apps**.

Azure Logic Apps

Azure Logic Apps are declarative workflow definitions, which are used to orchestrate tasks, processes, and workflows. Similar to functions, they can be integrated with many other Azure resources as well as external resources. Logic Apps are created, versioned, and provisioned using a JSON app definition schema. Designers are available both on the Azure portal as well as Visual Studio:

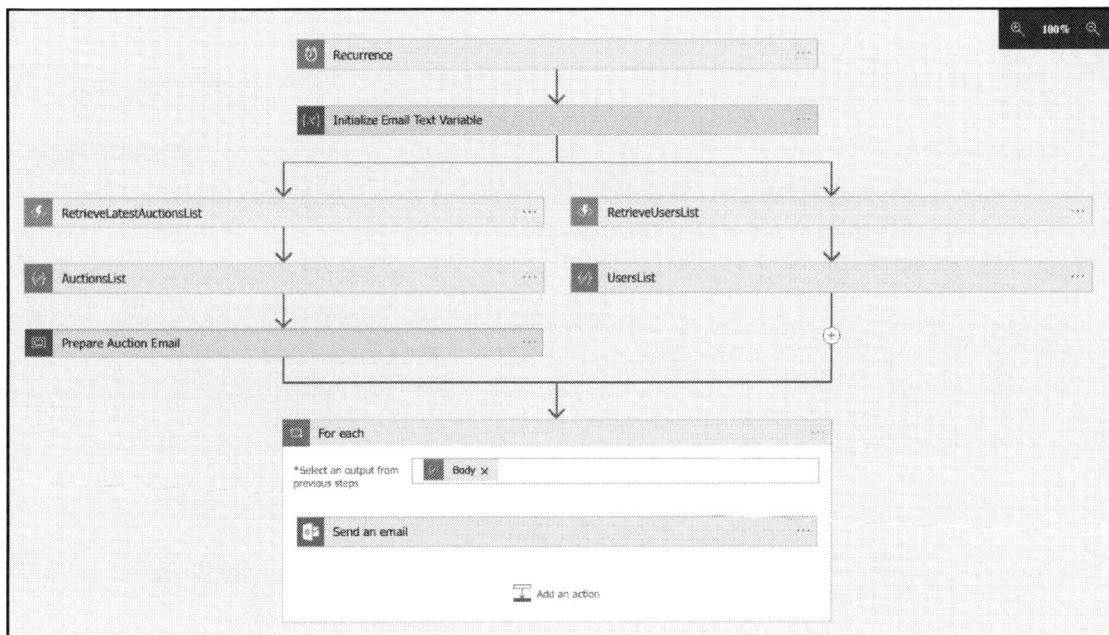

Logic Apps tasks are not only limited to Azure functions but also include so-called commercial and/or third-party connectors (for example, sending an SMS with Twilio, using SendGrid to send an email, posting a Tweet). In addition to this, the **Enterprise Integration Pack (EIP)** provides industry-standard messaging protocols.

Just like with Azure functions, the execution of a logic app starts with the trigger, and each step output is stored within the execution context. Processing blocks such as conditionals, switches, and foreach loops are available within the app flow. Additionally, logic app workflows can be dispatched by **Azure Event Grid** events.

Azure Event Grid

Azure Event Grid is a cloud-based event aggregate implementation that supports the pub/sub event routing strategy. An event grid consists of Event Sources (that is, the publishers) and Event Subscriptions (that is, the consumers):

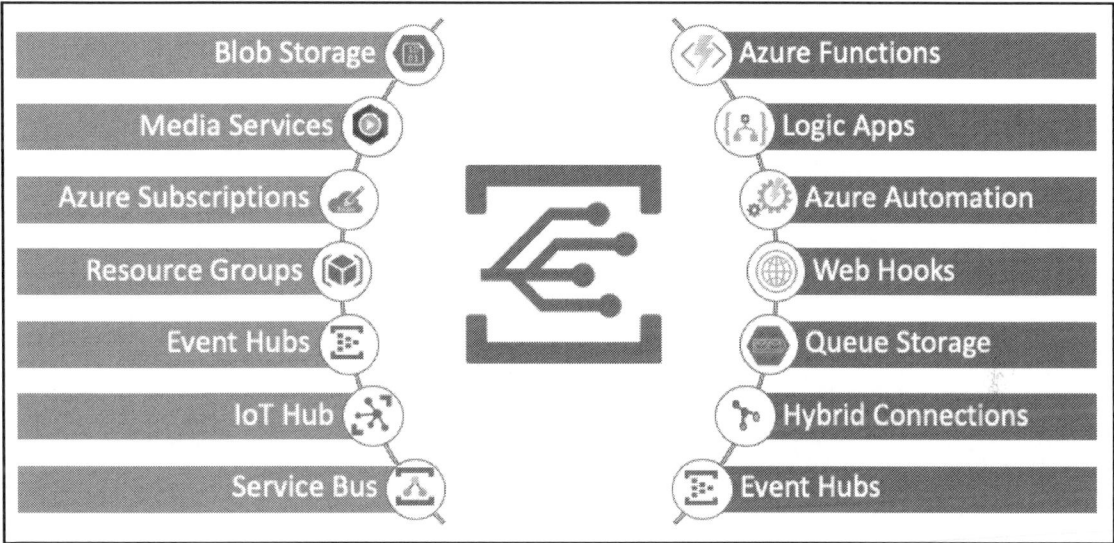

Similar to Azure function triggers, various event sources are available for developers and Azure Event Grid can be used to route or multicast certain events from one Azure resource to another. Events do not need to be triggered by a system resource but can also be created using an HTTP request, allowing custom modules to send events to consumers.

Development services

Azure resources are not just limited to application requirements that are provisioned and maintained with the application life cycle. They also include certain platform services that are used to implement the application life cycle and development pipeline, such as Azure DevOps and Visual Studio App Center. We will be using these resources to manage our application development and deployments throughout the remainder of the book.

Azure DevOps

Azure DevOps (previously known as TFS Online or Visual Studio Team Services), which started as the Microsoft **Application Lifecycle Management** (**ALM**) suite for on-premise product TFS, is now the most utilized freemium management portal. Azure DevOps instances can be created from Azure Portal as well as through the Azure DevOps portal. Let's take a look at how to do this:

1. This process of procurement starts with creating a DevOps organization:

2. Once the organization is created, we can now create a new project to create source control repositories and the backlog. Both the ALM process and the version control options are available under the **Advanced** section of the project settings:

It is important to mention that the TFVC and Git repositories are available at the same time. A single Azure DevOps project may contain multiple repositories. Because of cross-platform support and integration with IDEs (such as Visual Studio Code and Visual Studio for Mac), as well as native IDEs (such as Android Studio), Git is generally the repository type of choice for Xamarin and native mobile developers.

An extensive feature set is available for DevOps implementations on Azure DevOps. The Azure portal (apart from the overview) is divided into five main sections, as follows:

- Boards (project management)
- Repos (source control)
- Pipelines (CI/CD)
- Test plans (test management)
- Artifacts (package management)

According to the freemium subscription model, up to five contributors can be included in a project for free. Additional team members will need to have either a valid Visual Studio or **Microsoft Developer Network (MSDN)** license, or they will be assigned to a read-only stakeholder role.

Visual Studio App Center

Visual Studio App Center is a suite of tools that bundles various development services used by mobile developers (such as Xamarin, Native, and Hybrid) into a single management portal. App Center has tight integration with Azure DevOps and they can be used in conjunction with one another. Multiple application platforms are supported by App Center and various features are available for these platforms:

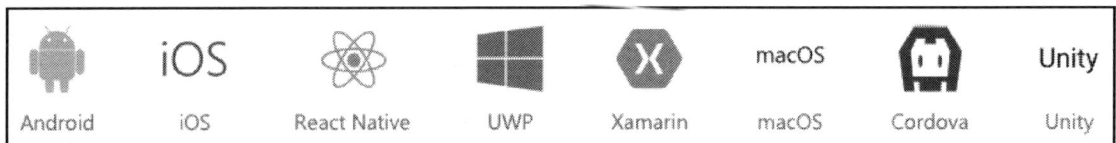

From a CI/CD perspective, App Center allows mobile application builds to be executed with source artifacts from various repository systems such as Azure DevOps and GitHub. Applications can be compiled using out-of-the-box build templates and application packages can be distributed to designated groups, without having to use any other store.

The prepared application packages can also be put through automated acceptance tests by means of UI tests. Multiple testing runtimes are supported for automated UI tests, such as Xamarin.UITest and Appium.

Finally, application telemetry and diagnostic data can be collected both from beta and production versions of mobile applications, and valuable application feedback can be reintroduced into the backlog. Push notifications are another valuable feature that can be used to engage application users.

App Center also uses a freemium subscription model, where the build and test hours are limited by the subscription; however, limited usage of the CI/CD features and unlimited distribution features are available for free.

Summary

Overall, developing a distributed application development is now much easier with tightly integrated Azure modules. Both cloud and hybrid applications can be created using the available resources and modules implemented with the .NET Core stack. It is also important to remember that resources should not define application requirements; rather, an optimal solution should be devised, bearing in mind the available modules, requirements, and costs.

In this chapter, we browsed through available Azure resources that will used in the following chapters to create our application backend. In the next chapter, we will start by creating our data store using Cosmos DB.

8
Creating a Datastore with Cosmos DB

Creating a data store is an essential part of both mobile and web application projects. Scalability, cost-effectiveness, and performance are three key factors that determine which database is appropriate for your application. Cosmos DB, with its wide range of scalability options and subscription models, can provide an ideal solution for mobile applications. Cosmos DB offers a multi-model and multi-API paradigm that allows applications to use multiple data models while storing application data with the most suited API and model for the application, such as SQL, Cassandra, Gremlin, or MongoDB. In this chapter, we will create the data store for our application and implement the data access modules.

In this chapter, we will cover the following topics:

- The basics of Cosmos DB
- Data access models
- Modeling data
- Learning Cosmos DB in depth

The basics of Cosmos DB

From a cloud-based application perspective, Cosmos DB is yet another persistence store that's available that you can include in your resource group. As we discussed previously, the biggest advantages of Cosmos DB also make up the unique feature set of Cosmos DB, namely global distribution, the multi-model, and high availability.

In our example app, we will be using Cosmos DB and creating our data model around the available persistence models. Let's start by adding a Cosmos DB instance to our resource group:

In this screen, we are going to set up the resource group, the account name (which also defines the access URL), and additional parameters related to the data access model, as well as global distribution.

Global distribution

Global distribution is an available option that deals with the global reach of your application. If you are planning to make your application globally available and you expect to have the same latency in each market, it is possible to select Geo-Reduncy to allow distribution on multiple regions. Once the Cosmos DB resource is created, this can be done using the **Replicate data globally** blade:

Great! Now, we have the application present on five different data centers across four continents. In other words, we have enabled multi-homing for our persistence store.

In addition to Geo-Redundancy, you can enable multiple write regions. Multiple write regions allow you to set multiple masters for your distributed data store. This data will not only be replicated across different regions, but will also provide the same write throughput.

When the global distribution regions are configured, you can set one of the read regions, that is, its failover region.

Consistency spectrum

Once data persistence becomes a globally distributed, multi-homed operation, the consistency concept becomes a fundamental subject. In Cosmos DB, there are five well-defined consistency levels that allow the developers to optimize, in simple terms, the trade-off between consistency and performance:

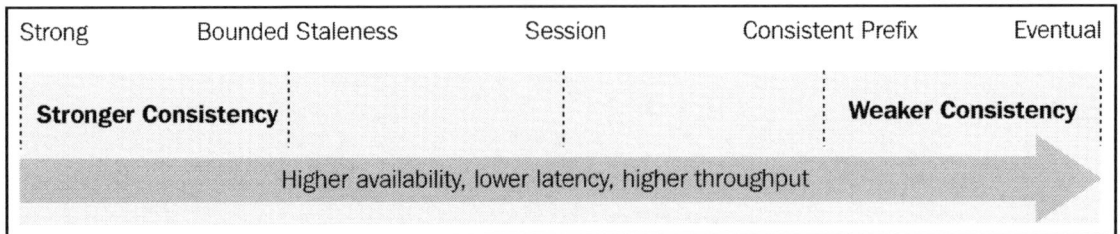

The default consistency level can be set on the database level, as well as within the client session that's consuming the data. However, the client set's consistency cannot be set to a stronger consistency than the default consistency level.

In order to understand this consistency, we will be using the baseball analogy that's used by the Microsoft research paper. In this example, we are considering the various stakeholders in a baseball game and how they read and write the score.

Without a doubt, the person who needs the strongest consistency is the *official scorekeeper*. They would read the current score and increment the score when any of the teams score. Then, they would like to have the assurance that the data they are reading is the latest version. However, since the score keeper is the only person who will be executing a write operation, we might be able to get away with a less consistent read level, such as **session consistency**, which provides the monotonic reads, monotonic writes, read-your-writes and write-follows-reads guarantees given that there is a single writer session.

Following the official scorekeeper, another stakeholder with a **strong consistency** requirement is the umpire, the person who officiates the baseball game behind the home plate. The umpire's decision to end the game in the second half of the ninth inning depends on whether the home team is ahead (that is, if the visiting team has no way of equalizing the score, there is no need for the home team to bat). In this decision, they would require a linearizability guarantee (that is, they will be reading the latest version of any given data point). From their perspective, each operation should be happening atomically, and there is a single global state (that is, the source of truth). In this setup, the performance (that is, low latency) is compromised in exchange for a quorum state in a distributed cluster.

Unlike the umpire, a periodic reporter (for example, for 30 minutes) would just like assurance of a *consistent prefix*; in other words, they just rely on the consistent state of the data up until the returned write operation. For them, the data state consistency is more important than the latency of the result since the operation is executed periodically to give an overall update.

Another stakeholder that doesn't care much about the latency but cares about the consistent state would be the sports writer. The writer can receive the final result of the game and provide their commentary the next day, as long as the result they received is the correct one. In a scenario similar to this, **eventual consistency** would probably return the correct final result, but when you want to limit the eventual consistency promise with a delay period, *bounded staleness* can be a solution. In fact, the umpire could have used a similar strategy for their read operations with a shorter delay.

By applying these concepts to our application model, we can decide for ourselves which modules would require which type of a consistence.

Let's assume that our users receive a review or rating from the participants once the transaction is completed. In this scenario, the rating system does not really require an ordered set of writes, nor is the consistency of much importance. We would be able to get away with each review for the same user if we had written and read with the promise of *eventual consistence*.

Next up is the notification system, which sends out the highest bids for an auction item in certain intervals to only interested parties. Here, the read operation only needs to be performed with the promise so that the order in which the bids have been written to our data store is preserved; in other words, with a *consistent prefix*. This becomes especially crucial if we are sending statistics similar to *the value of the item has raised by 30% in the last hour*. Similarly, the period for this consistence can be defined by the read system, making it a *bounded staleness* consistency.

Now, let's assume, that the user would like to keep a set of auctions in a watch list. This watch list would only be written by the user themselves, and the important read assurance would be to read-your-writes. This can be handled by *session consistence*. Additionally, the creation of a new auction item or updates would again be only session consistent.

Finally, probably the most consistent process in the setup would be the actual bidding (that is, *strong consistence*). Here, in order to bid for an auction item, as well as announce the results of an auction, we rely on strong consistency since bidding on the items is a multi-actor operation and we would like to make sure that the incoming bids are executed in a consistent manner.

This, of course, is just a presumptuous setup where the costs and implementation is completely left out of the equation. In a real-world implementation, session consistence would provide the best trade-off between consistency and performance while decreasing costs.

Pricing

The pricing model for Cosmos DB is rather complicated. This calculation involves many factors, such as the global availability, consistence, and another abstract unit called the **Request Unit (RU)**. Similar to the **Data Transaction Unit (DTU)**, it is a measure of the system resources that are used (for example, CPU, memory, and IO) to read a 1 KB item. The number of factors can effect the RU's usage, such as item size, complexity, indexing, consistency level, and executed queries. It is possible to keep track of the RU consumption by using the request charge headers that are returned by the DB.

Let's take a look at the following document DB client execution:

```
var query = client.CreateDocumentQuery<Item>(
    UriFactory.CreateDocumentCollectionUri(DatabaseId, CollectionId),
        new FeedOptions { MaxItemCount = -1 })
                .Where(item => !item.IsCompleted)
                .AsDocumentQuery();
```

This would translate into a SQL query as follows:

```
select * from Items where Items.isCompleted = false
```

We can retrieve the request charge by using the **Query Stats** tab on the **Data Explorer** blade:

```
Query 1      ×      Scale & Sett...      Documents

▷ Execute Query   ⬆ Load Query

1     select * from Items where Items.isCompleted = false
```

Results   Query Stats

METRIC	VALUE
Request Charge	2.860 RUs
Showing Results	0 - 0
Round Trips	1
Activity id	fc1a30e8-2bc0-4f87-95de-2af077017dcc

According to the datasets and application execution, request charges become more and more important. The Cosmos DB costs can be optimized for your application's needs by analyzing the provided telemetry.

Data access models

Probably the most important option to select before creating the Cosmos DB instance is the access model (that is, the API). In our application, we will be using the SQL API since it is inherently the only native access model and allows the usage of additional features such as triggers. Nevertheless, let's take a quick look at the other options that are available.

SQL API

Previously a standalone offer known as Azure Document DB, the SQL API allows developers to query a JSON-based NoSQL data structure with a SQL dialect. Similar to actual SQL implementations, the SQL API supports the use of stored procedures, triggers (that is, change feeds), and user-defined functions. Support for SQL queries allows for the (partial) use of LINQ and existing client SDKs, such as the entity framework.

MongoDB API

The MongoDB API that's provided by Cosmos DB provides a wide range of support for the MongoDB query language (at the time of writing, the MongoDB 3.4 wire protocol is in preview). Cosmos DB instances that are created with the MongoDB API type can be accessed using existing data managers, such as Studio 3T, RoboMongo, and Mongoose. This level of comprehensive support for MongoDB provides developers with the option of seamless migration from existing MongoDB stores. Azure portal data provides both shell and query access to MongoDB resources in order to visualize and analyze the data. In order to demonstrate this, let's execute several MongoDB queries from the MongoDB documentation library:

1. Given that we have a collection called survey, we will start by inserting the collection of survey results:

```
db.survey.insert([
  { "_id": 1, "results": [{ "product": "abc", "score": 10 }, {
"product": "xyz", "score": 5 }]},
  { "_id": 2, "results": [{ "product": "abc", "score": 8 }, {
"product": "xyz", "score": 7 }]},
  { "_id": 3, "results": [{ "product": "abc", "score": 7 }, {
"product": "xyz", "score": 8 }]}
])
```

This will result in an error message similar to the following:

```
ERROR: Cannot deserialize a 'BsonDocument' from BsonType 'Array'.
```

2. This is because the insert command is not fully supported on the web shell. In order to have proper command execution, we need to move on to a local terminal (given that the Mongo toolset is installed):

```
Cans-MacBook-Pro:~ can.bilgin$ mongo
handsoncrossplatformmongo.documents.azure.com:10255 -u
handsoncrossplatformmongo -p {PrimaryKey} --ssl --
sslAllowInvalidCertificates
MongoDB shell version v4.0.3
connecting to:
mongodb://handsoncrossplatformmongo.documents.azure.com:10255/test
WARNING: No implicit session: Logical Sessions are only supported
on server versions 3.6 and greater.
Implicit session: dummy session
MongoDB server version: 3.2.0
WARNING: shell and server versions do not match
globaldb:PRIMARY> show databases
sample  0.000GB
```

```
globaldb:PRIMARY> use sample
switched to db sample
globaldb:PRIMARY> db.survey.find()
globaldb:PRIMARY> db.survey.insert([{"_id":1,
"results":[{"product":"abc", "score":10}, { "product":"xyz",
"score":5}]}, { "_id":2, "results":[{"product":"abc", "score":8}, {
"product":"xyz", "score":7}]}, { "_id":3,
"results":[{"product":"abc", "score":7}, { "product":"xyz",
"score":8}]} ])
BulkWriteResult({
  "writeErrors" : [ ],
  "writeConcernErrors" : [ ],
  "nInserted" : 3,
  "nUpserted" : 0,
  "nMatched" : 0,
  "nModified" : 0,
  "nRemoved" : 0,
  "upserted" : [ ]
})
globaldb:PRIMARY> db.survey.find()
{"_id":1, "results":[{"product":"abc", "score":10},
{"product":"xyz", "score":5}]}
{"_id":2, "results":[{"product":"abc", "score" : 8},
{"product":"xyz", "score":7}]}
{"_id":3, "results":[{ "product":"abc", "score" : 7},
{"product":"xyz", "score":8}]}
```

The Mongo server and the client, `Mongo.exe`, can be downloaded from the MongoDB website. On macOS, the `brew install mongo` command will install Mongo. The personalized connection string or the complete shell connect command can be copied from the Quick Start section on the Cosmos DB resource.

3. Next, we can continue our execution back on the cloud shell or local mongo shell. We will now execute a find query where the product should be `"xyz"` and the score should be greater or equal to 8:

```
db.survey.find(
    { results: { $elemMatch: { product: "xyz", score: { $gte: 8 } } }
} }
)
```

4. Next, we will find all the survey results that contain the product `"xyz"`:

```
db.survey.find(
    { "results.product": "xyz" }
)
```

5. Finally, we will increment the first score where the product is `"abc"`:

```
db.survey.update({
    "results.product" : "abc"
},
{
    $inc : {'results.0.score' : 1}
});
```

6. You can visualize the results on the shell window of the data explorer:

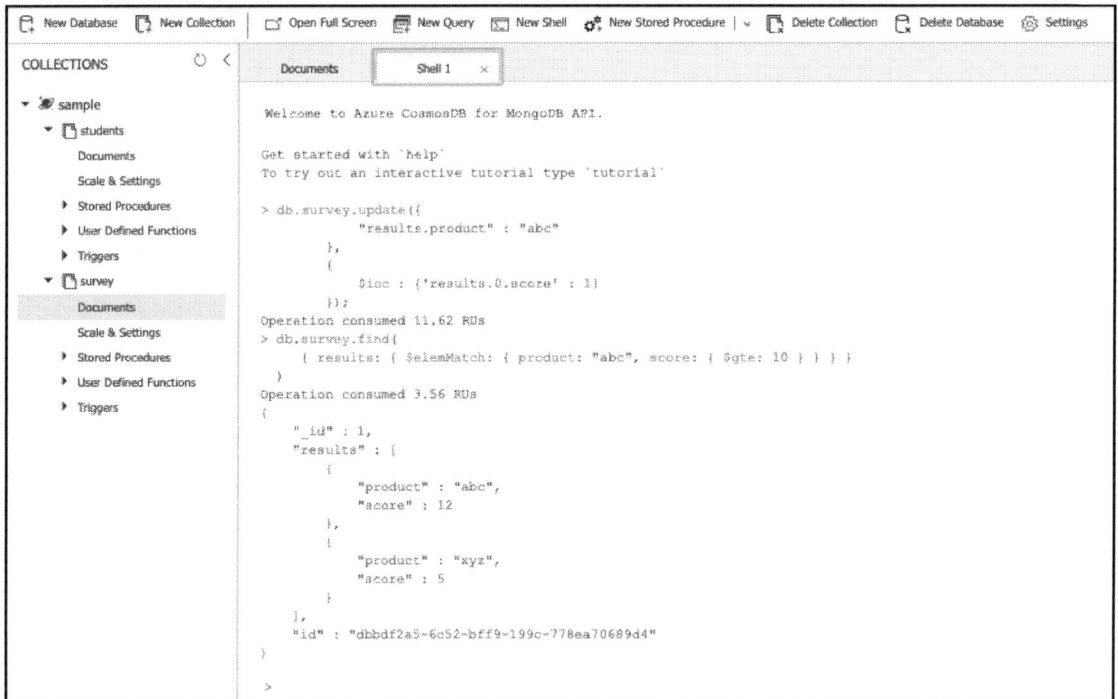

Others

Gremlin, which is a graph data model and Cassandra, which is a column family model, both have wire protocol support. These APIs allow integration with cluster computing and big data analysis platforms such as Spark (GraphX). Apache Spark clusters can be created within Azure HDInsight to analyze streaming and historical data.

The final member of the Cosmos DB, as we mentioned previously, is Azure Table storage, which provides access to a key/value pair store that supports the automatic sharding of data, as well as indexing.

Modeling data

The best way to get accustomed with various data models offered by Cosmos DB would be to implement inherently relational domain models using the provided NoSQL data access APIs. This way, it is easier to grasp the benefits of different data models.

For this exercise, let's create a relational data model for our auction applications.

In this setup, we have three big clusters of data:

- **Vehicles**, which includes the manufacturer, model, year, engine specifications, and some additional attributes describing the car
- **Users**, which consists of the sellers and buyers of the cars sold through auctions
- **Auctions**, which consists of some metadata about the sale that's provided by the selling user, as well as the vehicles and bids provided by the users

We will describe this data using the SQL API.

Creating and accessing documents

The most trivial way of considering the data model design, when dealing with a NoSQL database, would be to imagine the **Data Transformation Object (DTO)** models required for the application. In the case of a non-RBMS data platform, it is important to remember that we are not bound by references, unique keys, or many-2-many relationships.

For instance, let's take a look at the simplest model, namely *User*. User will have basic profile information, which can be used in the remainder of the application. Now, let's imagine what the DTO for the user object would look like:

```
{
    "id": "efd68a2f-7309-41c0-af52-696ebe820199",
    "firstName": "John",
    "lastName": "Smith",
    "address": {
        "addressTypeId": 4000,
        "city": "Seattle",
        "countryCode": "USA",
        "stateOrProvince": "Washington",
```

```
            "street1": "159 S. Jackson St.",
            "street2": "Suite 400",
            "zipCode": "98101"
        },
        "email": {
            "emailTypeId": 1000,
            "emailAddress": "john.smith@test.com"
        },
        "isActive": true,
        "phone": {
            "phoneTypeId": 1000,
            "number": "+1 121-212-3333"
        },
        "otherPhones": [{
            "phoneTypeId": 3000,
            "number": "+1 111-222-3333"
        }],
        "signUpDate": "2001-11-02T00:00:00Z"
}
```

Let's create our collection with the name `UsersCollection` and with the partition key set to `/address/countryCode`.

Next, import this data into our database:

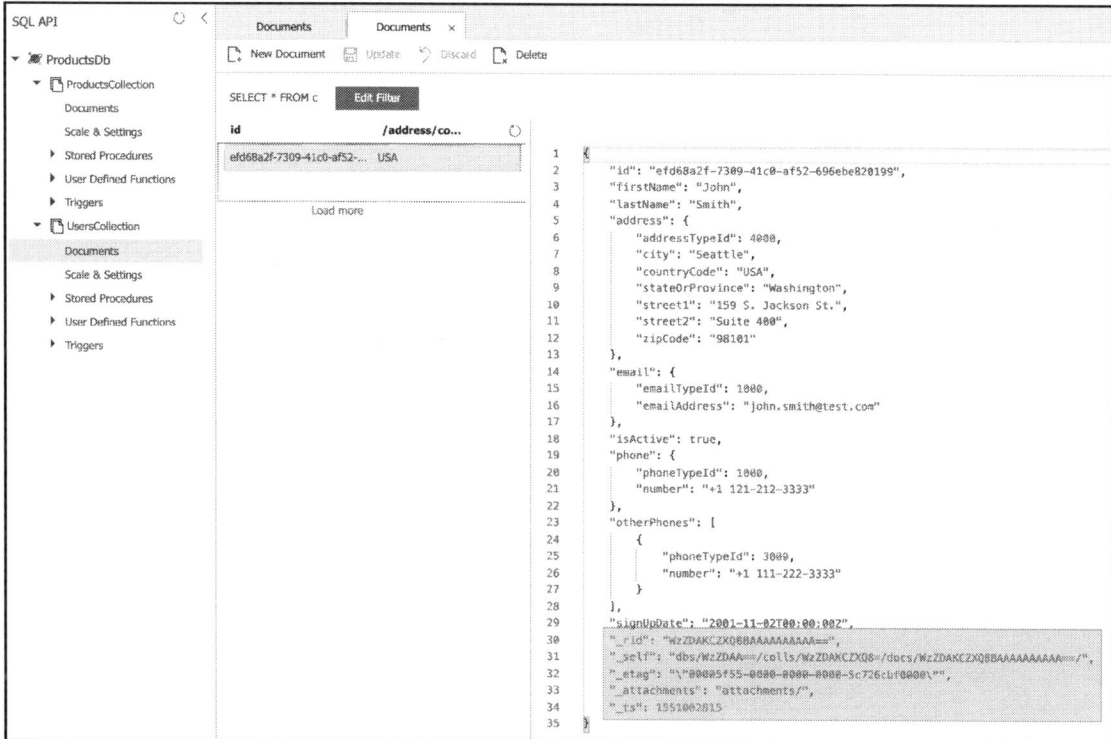

OK; now, we have our first document created. But what are these additional fields that were added by the system? These are references to the container and item that holds the document data for our collection:

`_rid`	System generated	Unique identifier of container
`_etag`	System generated	Entity tag used for optimistic concurrency control
`_ts`	System generated	Last updated timestamp of the container
`_self`	System generated	Addressable URI of the container

Out of these fields, the most important ones are `_etag` and `_ts`, both of which define the state of an entity at a given point in time. Notice that the descriptions do not refer to document, but rather item and entity. The main reason for this is that on Cosmos DB, storage buckets are referred to as containers, and the entities stored within these containers are referred to as items. Collection, table, or graph are the realization of these containers, depending on the API type that is being used.

Now, we can start creating our data access layer, which will be part of the User API, to provide the required data to our mobile application. Let's begin:

1. Create a generic interface that will allow us retrieve the user feed, as well as a single user:

```
public interface IRepository<T> where T : class
{
    Task<T> GetItemAsync(string id);
    Task<IEnumerable<T>> GetItemsAsync();
    //Task<Document> AddItemAsync(T item);
    //Task DeleteItemAsync(string id);
    //Task<Document> UpdateItemAsync(string id, T item);
}
```

2. Create our implementation for Cosmos DB:

```
public class CosmosCollection<T> : IRepository<T> where T : class
{
    // ...Removed for brevity

    private DocumentClient _client;

    public CosmosCollection(string collectionName)
    {
        CollectionId = collectionName;

        _client = new DocumentClient(new Uri(Endpoint), Key);
    }

    // ... Removed for brevity
}
```

3. Implement our repository methods:

```
public class CosmosCollection<T> : IRepository<T> where T : class
{
    // ...Removed for brevity

    public async Task<T> GetItemAsync(string id)
    {
        try
        {
            Document document = await _client.ReadDocumentAsync(
                UriFactory.CreateDocumentUri(DatabaseId,
                CollectionId, id));
            return (T)(dynamic)document;
        }
```

```
            catch (DocumentClientException e)
            {
                // ...Removed for brevity
            }
        }

        public async Task<IEnumerable<T>> GetItemsAsync()
        {
            IDocumentQuery<T> query = _client.CreateDocumentQuery<T>(
                UriFactory.CreateDocumentCollectionUri(DatabaseId,
                CollectionId),
                new FeedOptions { MaxItemCount = -1})
                .AsDocumentQuery();

            List<T> results = new List<T>();

            while (query.HasMoreResults)
            {
                var response = await query.ExecuteNextAsync<T>();
                results.AddRange(response);
            }

            return results;
        }
    }
```

4. Now, we are ready to load the document(s) we have imported into our document collection:

```
var cosmosCollection = new
CosmosCollection<User>("UsersCollection");

var collection = await cosmosCollection.GetItemsAsync()
```

5. This execution would, in fact, result in the following exception:

```
Cross partition query is required but disabled. Please set x-ms-
documentdb-query-enablecrosspartition to true, specify x-ms-
documentdb-partitionkey, or revise your query to avoid this
exception.
ActivityId: c0adcf18-3ea6-4a4a-b033-d946f2000c17,
Microsoft.Azure.Documents.Common/2.2.0.0, Darwin/10.14 documentdb-
netcore-sdk/2.2.2
```

The reason for this is that the collection is set up to use the countryCode as the partition key, but the query isn't executed as a cross-partition query.

6. We can either pass the partition key (that is, `countryCode`) or we can execute it as a cross-partition request (a less performant option):

```
IDocumentQuery<T> query = _client.CreateDocumentQuery<T>(
    UriFactory.CreateDocumentCollectionUri(DatabaseId,
CollectionId),
    new FeedOptions { MaxItemCount = -1, EnableCrossPartitionQuery
    = true })
                                //, PartitionKey = new
PartitionKey("USA") })
    .AsDocumentQuery();
```

7. Now, we will load the complete set of entries for the given collection. However, in most scenarios, we would use a predicate to load the set that is needed. So, let's add a `Where` clause to our query:

```
public async Task<IEnumerable<T>> GetItemsAsync(Expression<Func<T,
bool>> predicate)
{
    IDocumentQuery<T> query = _client.CreateDocumentQuery<T>(
        UriFactory.CreateDocumentCollectionUri(DatabaseId,
        CollectionId),
        new FeedOptions { MaxItemCount = -1,
        EnableCrossPartitionQuery = false })
    .Where(predicate)
    .AsDocumentQuery();

    // ...
}
```

8. Now, we will create the add, update, and remove methods accordingly, which will provide the complete set of CRUD operations for the collections:

```
public async Task<T> AddItemAsync(T item)
{
    return (T)(dynamic)await _client.CreateDocumentAsync(
        UriFactory.CreateDocumentCollectionUri(DatabaseId,
        CollectionId), item);
}

public async Task<T> UpdateItemAsync(string id, T item)
{
    return (T)(dynamic)await _client.ReplaceDocumentAsync(
        UriFactory.CreateDocumentUri(DatabaseId, CollectionId, id),
        item);
}
```

```
public async Task DeleteItemAsync(string id)
{
    await _client.DeleteDocumentAsync(
        UriFactory.CreateDocumentUri(DatabaseId, CollectionId,
id));
}
```

9. Finally, following the code first approach, in order to avoid having the document collection created manually every time, let's initialize the document collection if it does not exist when the client is first created:

```
private async Task CreateCollectionIfNotExistsAsync()
{
    try
    {
        await _client.ReadDocumentCollectionAsync(
            UriFactory.CreateDocumentCollectionUri(DatabaseId,
            CollectionId));
    }
    catch (DocumentClientException e)
    {
        if (e.StatusCode == System.Net.HttpStatusCode.NotFound)
        {
            await _client.CreateDocumentCollectionAsync(
                UriFactory.CreateDatabaseUri(DatabaseId),
                new DocumentCollection { Id = CollectionId,
                PartitionKey = _partitionKey },
                new RequestOptions { OfferThroughput = 400 });
        }
        else
        {
            throw;
        }
    }
}
```

We have now created a complete document collection and basic CRUD functions. Now, we will continue and further extend our domain model via data denormalization.

Denormalized data

Data normalization is the process of structuring a database model by decomposing existing structures and creating replacement references to decrease the redundancy and improve data integrity. However, data normalization inherently applies to relational databases. In the case of a document collection, embedded data is preferred over referential integrity. In addition, data views that cross the boundaries of a single collection should also be replicated on different pivots according to the design requirements.

Let's continue with our data model design with vehicles and auctions. These two data domains are going to be handled with separate APIs and will have separate collections. However, in a general feed (for example, latest auctions), we would need to retrieve data about the auctions, as well as the cars on the auction and bids provided by users for that specific auction. Let's see how we can do this:

1. For the vehicle's declaration, we will need the main product information:

```
{
    "id" : "f5574e12-01dc-4639-abeb-722e8e53e64f",
    "make" : "Volvo",
    "model": "S60",
    "year": 2018,
    "engine": {
        "displacement" : "2.0",
        "power" : 150,
        "torque" : 320,
        "fuelType": { "id": "11", "name": "Diesel" }
    },
    "doors": 4,
    "driveType": { "id" : "20", "name" : "FWD" },
    "primaryPicture" : "",
    "pictures" : [],
    "color": "black",
    "features": [ "Heated Seats",  "Automatic Mirrors",
"Windscreen Defrost",  "Dimmed Mirrors",  "Blind Spot Detection" ]
```

   Notice that the features array contains a list of features that have been selected from a list of reference values, but instead of creating a many-2-many relational table, we chose to embed the data here, which is compromise of the normal form. A similar approach could have been used for the `fuelType` and `driveType` references, but conceptually speaking, we have a many-2-one relationship on these data points, so they are embedded as reference data objects themselves.

2. Moving forward, let's create our auction data:

```
{
    "id" : "7ad0d2d4-e19c-4715-921b-950387abbe50",
    "title" : "Volvo S60 for Sale",
    "description" : "..."
    "vehicle": {
        "id" : "f5574e12-01dc-4639-abeb-722e8e53e64f",
        "make" : "Volvo",
        "model": "S60",
        "year": 2018,
        "engine": {
            "displacement" : "2.0",
            "power" : 150,
            "torque" : 320,
            "fuel": { "id": "11", "name": "Diesel" }
        },
        "primaryPicture" : "",
        "color": "black"
    },
    "startingPrice": {
        "value" : 25000,
        "currency" : {
            "id" : "32",
            "name" : "USD",
            "symbol" : "$"
        }
    },
    "created": "2019-03-01T10:00Z",
    "countryCode": "USA"
    "user": {
        "id" : "efd68a2f-7309-41c0-af52-696ebe820199",
        "firstName": "John",
        "lastName": "Smith"
    }
}
```

3. If this was a relational model, this data would have been enough for identifying an auction. Nevertheless, it would decrease the amount of round trips to load additional data if we embedded the highest bid (or even the most recent or highest bids) within the same structure:

```
"highestBids":[
    {
        "id" : "5d669390-2ba4-467a-b7f9-26fea2d6a129",
        "offer" : {
            "value" : 26000,
        },
```

```
            "user": {
                "id" : "f50e4bd2-6beb-4345-9c30-18b3436e7766",
                "firstName": "Jack",
                "lastName": "Lemon",
            },
            "created" : "2019-03-12T11:00Z"
        }
    ],
```

> **TIP**
>
> In this scenario, we could have also embedded the complete bid's structure within the auction model. While this would decrease the redundancy and dispersement of data across collections, *bids* is not a finite collection like the feature set we saw in the vehicle object, and each new bid would have required a complete document replacement on the auction collection.

We can say that the Auction table is acting as a Materialized View for the listing, while vehicle and bids provide easy access to the required data points for the application views. Here, the responsibility of data integrity falls on the client application rather than the database itself.

Referenced data

In the previous examples, we have used embedding extensively to create an optimized data structure. This, of course, does not mean we haven't used any references. Most of the embedded objects that are used are actually reference descriptions.

In order to visualize the referential data points, let's normalize our Auction data:

```
{
    "id" : "7ad0d2d4-e19c-4715-921b-950387abbe50",
    "description" : "Volvo S60 for Sale",
    "vehicleId": "f5574e12-01dc-4639-abeb-722e8e53e64f",
    "startingPrice": {
        "value" : 25000,
        "currencyId" : "32"
    },
    "highestBids":[
        "5d669390-2ba4-467a-b7f9-26fea2d6a129"
    ],
    "created": "2019-03-01T10:00Z",
    "countryCode": "USA",
    "userId": "efd68a2f-7309-41c0-af52-696ebe820199"
}
```

Here, we have a 1-* relation between the Vehicle and the Auction, a 1-* relation between the currency and the starting price value, a 1-* relation between the Auction and Bids, and a 1-* relation between User and Auctions. All of these references are embedded into the auction object, but what about the reciprocal references? For instance, if we were implementing a user profile view, we might want to show how many bids he/she was involved in and possibly a feedback value from the winning buyer or the seller:

```
{
    "id": "efd68a2f-7309-41c0-af52-696ebe820199",
    "firstName": "John",
    "lastName": "Smith",
    "numberOfAuctions" : 1,
    "auctions" : [
        {
            "auctionId": "7ad0d2d4-e19c-4715-921b-950387abbe50",
            "role" : { "roledId" : "20", "roleName": "seller" },
            "auctionReview" : 1,
            "auctionState" : { "stateId" : "10", "stateName" : "Closed" }
        }
    ]
    ...
}
```

These types of situations completely depend on the application use cases. As we have mentioned previously, we are not bound to foreign keys and constraints in the NoSQL setup, and the design should not necessarily dictate embedding or referencing. Cosmos DB provides features such as stored procedures and triggers to assign the responsibility of data integrity back into the database. Additionally, indexing and partitioning strategies can improve the overall performance of the application.

Cosmos DB in depth

Cosmos DB as a platform is much more than a simple database. The design of your data model, as well as the implementation of the data access layer, depends greatly on the feature being utilized. Partition and indexing setup can help improve performance, while also providing the roadmap for query strategies. Data triggers, stored procedures, and the change feed are extensibility points that allow developers to implement language-integrated transactional JavaScript blocks, which can greatly decrease the system's overall complexity and also compensate for the write transaction compromise in favor of denormalized data.

Partitioning

Cosmos makes use of two types of partitions – namely physical and logical partitions – in order to scale individual containers (that is, collections) in a database. The partition key that's defined at the time of the creation of a container defines the logical partitions. These logical partitions are then distributed into groups to physical partitions with a set of replicas to be able to horizontally scale the database.

In this scheme, the selection of the partition key becomes an important decision that will determine the performance of your queries. With a properly selected partition key, the data would be sharded (that is, data sharding) uniformly so that the distributed system will not show so-called hot partitions (that is, request peaks on certain partitions, while the rest of the partitions are idle). Hot partitions would ultimately result in performance degradation.

In the `UsersCollection`, we have used `/address/countryCode` as the partition key. This means that we are expecting a set of users with a normal distribution across the countries. However, in a real life implementation, the number of users from a certain market really depends on the size of that market. To put it in layman's terms, the number of users from Turkey or Germany could not be the same as Bosnia and Herzegovina, if we were to think about the population count and demand.

> Once a container is created in Cosmos DB, changing other properties of a collection such as the ID or the partition key are not supported. Only the indexing policy can be updated.

The partition key does not necessarily need to be a semantic dissection of data. For instance, in the `UsersCollection` scenario, the partition key could easily be defined according to the first letter of the name or the month that they signed up, as well as a synthetic partition key, such as a generated value from a range (for example, 1-100) that is assigned at the time of creation. Nevertheless, since the ID of an item within a container is only unique in that container, the container and ID combination defines the index of that item. In order to achieve higher throughputs, the queries should be executed within a specific container. In other words, if the partition key can be calculated before a query on the client side, the application would perform better than executing cross-partition queries:

```
IDocumentQuery<T> query = _client.CreateDocumentQuery<T>(
    UriFactory.CreateDocumentCollectionUri(DatabaseId, CollectionId),
    new FeedOptions { MaxItemCount = -1, EnableCrossPartitionQuery =
    true })
                        //, PartitionKey = new PartitionKey("USA")
})
    .AsDocumentQuery();
```

For instance, let's take a look at the following execution on this collection:

```
var cosmosCollection = new CosmosCollection<User>("UsersCollection");

await cosmosCollection.GetItemsAsync((item) =>
item.FirstName.StartsWith("J"));

// Calling with the partition key
await cosmosCollection.GetItemsAsync((item) =>
item.FirstName.StartsWith("J"), "USA");
```

The results from this query (on a collection with only two entries per partition), when using the previous expression, are as follows:

```
Executing Query without PartitionKey
Query: {"query":"SELECT VALUE root FROM root WHERE
STARTSWITH(root[\"firstName\"], \"J\") "}
Request Charge : 2.96 RUs
Partition Execution Duration: 218.08ms
Scheduling Response Time: 26.67ms
Scheduling Run Time: 217.45ms
Scheduling Turnaround Time: 244.65ms

Executing Query with PartitionKey
Query: {"query":"SELECT VALUE root FROM root WHERE
STARTSWITH(root[\"firstName\"], \"J\") "}
Request Charge : 3.13 RUs
Partition Execution Duration: 136.37ms
Scheduling Response Time: 0.03ms
Scheduling Run Time: 136.37ms
Scheduling Turnaround Time: 136.41ms
```

Even with the smallest dataset, the execution results show quite an improvement on the total time required to execute.

> In order to retrieve additional metrics for a query execution, you can enable the PopulateQueryMetrics flag on the FeedOptions. In the FeedResponse<T> object, the QueryMetrics collection can be used to retrieve advanced execution metrics information.

In a similar fashion, we can expand our data models for vehicle and auction, and we can create the collections with the car make or color so that we have evenly distributed partitions.

Indexing

Azure Cosmos DB, by default, assumes that each property in an item should be indexed. When a complex object is pushed into the collection, the object is treated as a tree with properties that make up the nodes and values, as well as the leaves. This way, each property on each branch of the tree is queryable. Each consequent object either uses the same index tree or expands it with additional properties.

This indexing behavior can be changed at any time for any collection. This can help with the costs and performance of the dataset. Index definition uses wildcard values to define which paths should be included and/or excluded.

For instance, let's take a look at the indexing policy of our `AuctionsCollection`:

```
Partition key
/vehicle.color

Indexing Policy
 1    {
 2        "indexingMode": "consistent",
 3        "automatic": true,
 4        "includedPaths": [
 5            {
 6                "path": "/*",
 7                "indexes": [
 8                    {
 9                        "kind": "Range",
10                        "dataType": "Number",
11                        "precision": -1
12                    },
13                    {
14                        "kind": "Range",
15                        "dataType": "String",
16                        "precision": -1
17                    },
18                    {
19                        "kind": "Spatial",
20                        "dataType": "Point"
21                    }
22                ]
23            }
24        ],
25        "excludedPaths": [
26            {
27                "path": "/\"_etag\"/?"
28            }
29        ]
30    }
```

`/*` declaration includes the complete object tree, except for the excluded `_etag` field. These indexes can be optimized using more specialized index types and paths.

For instance, let's exclude all paths and introduce an index of our own:

```
"includedPaths": [
    {
        "path": "/description/?",
        "indexes": [
            {
                "kind": "Hash",
                "dataType": "String",
                "precision": -1
            }
        ]
    },
    {
        "path": "/vehicle/*",
        "indexes": [
            {
                "kind": "Hash",
                "dataType": "String",
                "precision": -1
            },
            {
                "kind": "Range",
                "dataType": "Number",
                "precision": -1
            }
        ]
    }
],
"excludedPaths": [
    {
        "path": "/*"
    }
]
```

Here, we have added two indexes: one hash index for the scalar value of the description field (that is, /?), and one range and/or hash index to the vehicle path and all the nodes under it (that is, /*). The hash index type is the index that's used for equality queries, while the range index type is used for comparison or sorting.

By using the correct index paths and types, query costs can be decreased and scan queries can be avoided. If the indexing mode is set to None instead of Consistent, the database returns an error on the given collection. The queries can still be executed using the EnableScanInQuery flag.

Programmability

One of the most helpful features of Cosmos is its server-side programmability, which allows developers to create stored procedures, functions, and database triggers. These concepts are not too foreign for developers that have created applications on SQL databases, and yet the ability to create stored procedures on NOSQL databases, and what's more, on a client-side scripting language such as JavaScript, is quite unprecedented.

As a quick example, let's implement a trigger to calculate the aggregate value(s) on a user's profile:

1. As you may remember, we added the following reference values to UserProfile for the cross-collection partition:

```
public class User
{
    [JsonProperty("id")]
    public string Id { get; set; }

    [JsonProperty("firstName")]
    public string FirstName { get; set; }

    //...

    [JsonProperty("numberOfAuctions")]
    public int NumberOfAuctions { get; set; }

    [JsonProperty("auctions")]
    public List<BasicAuction> Auctions { get; set; }

    //...
}
```

Now, let's create an aggregate update function that will update the number of auctions whenever there is an update on the user profile. We will use this function to intercept the update requests to the collection (that is, a pre-execution trigger) and modify the object's content.

2. The function should first retrieve the current collection and the document from the execution context:

```
function updateAggregates(){
    // HTTP error codes sent to our callback function by server.
    var ErrorCode = {
        RETRY_WITH: 449,
    }
```

```
var collection = getContext().getCollection();
var collectionLink = collection.getSelfLink();
// Get the document from request (the script runs as trigger,
// thus the input comes in request).
var document = getContext().getRequest().getBody();
```

3. Now, let our function count the auctions that the update is pushing:

```
if(document.auctions != null) {
    document.numberOfAuctions = document.auctions.length;
}

getContext().getRequest().setBody(document);
```

4. We can now add this trigger to the UsersCollection as a **Pre** trigger on **Replace** calls:

5. However, the trigger function will still not execute until we explicitly add the trigger to the client request:

```
var requestOption = new RequestOptions();
requestOption.PreTriggerInclude = new []{ "updateAggregates"};
await _client.ReplaceDocumentAsync(
    UriFactory.CreateDocumentUri(DatabaseId, CollectionId, id),
    item,
    requestOption);
```

Great! The number of auctions the user has participated in is being calculated every time the user's profile is updated. However, in order to insert a new auction (for example, when the user is actually creating an auction, or bidding on one), we would need to update the whole user profile (that is, partial updates are currently not supported on the SQL API).

6. Let's create a stored procedure that will insert an auction item on a specific user profile to push *partial* updates:

```
function insertAuction(id, auction) {
    var collection = getContext().getCollection();
    var collectionLink = collection.getSelfLink();
    var response = getContext().getResponse();
```

7. Next, retrieve the user profile object that we have to insert the auction into:

```
var documentFilter = 'Select * FROM r where r.id = \'' + id + '\'';
var isAccepted = collection.queryDocuments(
    collectionLink,
    documentFilter,
    function (err, docs, options) {
        if (err) throw err;
        var userProfile = docs[0];

        // TODO: Insert Auction
    });
```

8. Now, we can update the document with the new auction:

```
userProfile.auctions[userProfile.auctions.length] = auction;
collection.replaceDocument(userProfile._self, userProfile, function
(err) {
    if (err) throw err;
});
```

9. Finally, we will create an additional function for the `UserProfileRepository`:

```
public async Task InsertAuction(string userId, Auction auction)
{
    try
    {
        RequestOptions options = new RequestOptions { PartitionKey
=
        new PartitionKey("USA") };
        var spLink = UriFactory.CreateStoredProcedureUri(
            DatabaseId,
            CollectionId,
            "insertAuction");

        var result = await
_client.ExecuteStoredProcedureAsync<User>
            (spLink, options, userId, (dynamic)auction);
```

```
        }
        catch (DocumentClientException e)
        {
            throw;
        }
    }
```

Now, the auctions are inserted into the user profile and the aggregate column is updated when the stored procedure is called.

Triggers, functions, and stored procedures are all limited to the collection they are created in. In other words, one collection trigger cannot execute any changes on another collection. In order to execute such an update, we would need to use an external process such as the caller application itself or an Azure Function that's triggered with the change feed on Cosmos DB.

Change feed

Azure Cosmos DB continuously monitors changes in collections, and these changes can be pushed to various services through the change feed. The events that are pushed through the change feed can be consumed by Azure Functions and App Services, as well as stream processing and data management processes.

Insert, update, and soft-delete operations can be monitored through the change feed, with each change appearing exactly once in the change feed. If there are multiple updates being made to a certain item, only the latest change is included in the change feed, thus making it a robust and easy-to-manage event processing pipeline.

Summary

Cosmos DB provides a new perspective to the NOSQL database concept, with a wide range of services for various scenarios. This globally distributed and extremely throughput-oriented data store might be costly for simple applications, but the multi-API access model, as well as server-side programmability, are strong suits of Cosmos DB over other modern persistence stores. Additionally, with Cosmos DB access models, in comparison to relational data models, consumer applications have more responsibility over the referential data integrity. The weak links between the data containers can be used as an advantage by a microservice architecture.

In the next chapter, we will be creating the service layer for our application suite.

Creating Microservices Azure App Services

9

Azure App Services is a platform as a service offering for both mobile and app developers that can host a number of different application models and services. While the developers can create a simple mobile app service to act as a data store access layer within minutes without writing a single line of code, intricate and robust .NET Core applications can also be implemented with intrinsic integration to other Azure Services. In this chapter, we will go through the basics of Azure App Services and create a simple data-oriented backend for our application using ASP.NET Core, with authentication provided by **Azure Active Directory (Azure AD)**.

The following sections will guide you through the process of creating our service backend:

- Choosing the right app model
- Creating our first Microservice
- Integrating with Redis cache
- Hosting the services
- Securing the application

Choosing the right app model

The Azure stack offers multiple ways to host web applications, varying from simple **Infrastructure as a Service (IaaS)** offerings such as VMs to completely managed PaaS hosting services such as App Services. Because of the platform-agnostic nature of .NET Core and ASP.NET Core, even Linux containers and container orchestration services such as Kubernetes are available options.

Azure compute offers can be categorized according the separation of responsibilities in three main categories, namely IaaS, PaaS, and CaaS, as shown in the following diagram:

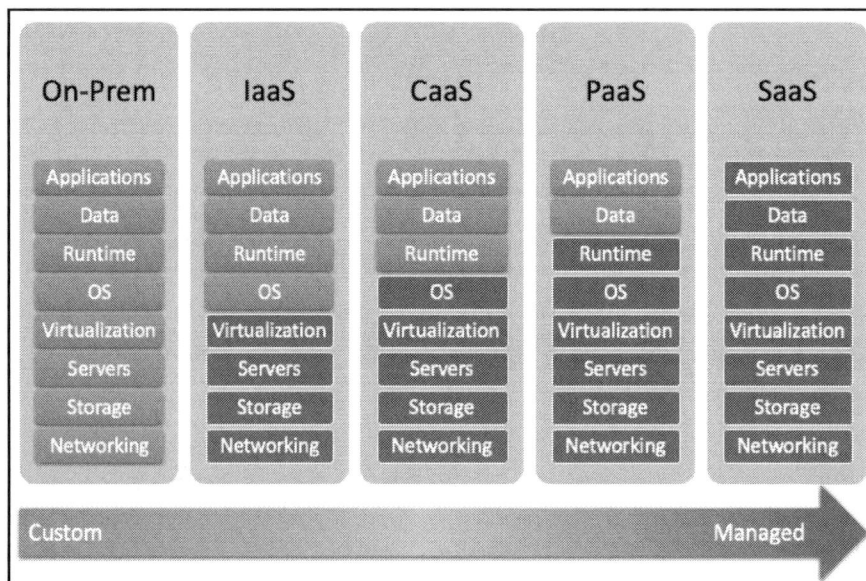

Besides IaaS and PaaS, there are various hosting options, each with their own advantages and use cases. We will now take a closer at these offerings.

Azure virtual machines

Virtual machines (**VMs**), are one of the oldest IaaS offerings on Microsoft Azure Cloud. In simple terms, this offering provides a hosting server on the cloud with complete control. Azure Automation services provides you with the much required tools to manage these servers with **Infrastructure as Code** (**IaC**) principles using **PowerShell Desired State Configuration** (**DSC**) and scheduled runbooks. You can scale and monitor them with ease according to your application requirements.

With both Windows and Linux variants available, Azure VMs can be an easy shallow cloud migration path for already existing applications.

Scaling of VMs is done by adjusting the system configuration (for example, CPU, virtual disk, RAM, and so on) at the VM level or by introducing additional VMs to the infrastructure.

Containers in Azure

If VMs are the virtualization of hardware, containers virtualize the OS. Containers create isolated sandboxes for applications that share the same operating system kernel. This way, application containers that are composed of the application, as well as its dependencies, can be easily hosted on any environment that meets the container demands. Container images can be executed as multiple application containers, and these container instances can be orchestrated with orchestration tools such as Docker Swarm, Kubernetes, and Service Fabric.

Azure currently has two managed container orchestration offerings: Azure Kubernetes Services and Service Fabric Mesh.

Azure Container Services with Kubernetes

Azure Container Services (**ACS**) is one of the managed container orchestration environments where developers can deploy containers without much hassle and enjoy the automatic recovery and scaling experience.

Kubernetes is an open source container orchestration system that was originally designed and developed by Google. Azure Kubernetes is a managed implementation of this service where most the configuration and management responsibilities are delegated to the platform itself. In this setup, as a consumer of this PaaS offering, you are only responsible for managing and maintaining the agent nodes.

AKS provides support for Linux containers as well as Windows operating system virtual containers. Windows container support is currently in private preview.

Service Fabric Mesh

Azure Service Fabric Mesh is a fully managed container orchestration platform where the consumers do not have any direct interaction with the configuration or maintenance of the underlying cluster. So-called polyglot services (that is, any language, any OS) are run in containers. In this setup, developers are just responsible for specifying the resources that the application requires, such as the number of containers and their sizes, networking requirements, and autoscale rules. Once the application container is deployed, Service Fabric Mesh hosts the container in a mesh consisting of clusters of thousands of machines, and cluster operations are completely hidden from the developers. Intelligent message routing through **Software Defined Networking** (**SDN**) enables service discovery and routing between microservices.

Service Fabric Mesh, even though it shares the same underlying platform, differs significantly from Azure Service Fabric. While Service Fabric Mesh is a managed hosting solution, Azure Service Fabric is a complete microservice platform that allows developers to create containerized or otherwise cloud-native applications.

Microservices with Azure Service Fabric

Azure Service Fabric is a hosting and development platform that allows developers to create enterprise grade applications composed of microservices. Service Fabric provides a comprehensive runtime and life cycle management capabilities, as well as a durable programming model that makes use of mutable state containers, such as reliable dictionaries and queues.

Service Fabric applications can be made up of three different hosted service models: containers, services/actors, and guest executables.

Similar to its managed counterpart (that is, Service Fabric Mesh) and Azure Kubernetes, Service Fabric is capable of running containerized applications that are targeting both Linux and Windows containers. Containers can easily be included in Service Fabric Application that are bundled together with other components and scaled across high density shared compute pools, called clusters, with predefined nodes.

Reliable services and reliable actors are truly cloud-native services that make use of the Service Fabric programming model. Ease of development on local development clusters and available SDKs for .NET Core, as well as Java, allow developers to create both stateful and stateless microservices in a platform-agnostic manner.

> Development environments are available for Windows, Linux, and OSX. Linux and OSX development setup relies on running the Service Fabric itself in a container. .NET Core is the preferred language for creating applications that target each of these platforms. The available Visual Studio Code extensions makes it easy to develop .NET Core Service Fabric applications on each operating system.

Finally, guest executables can be an application that's been developed on a variety of languages/frameworks, such as Node.js, Java, or C++. Guest executables are managed and treated as stateless services and can be placed on cluster nodes, side by side with other Service Fabric services.

Azure App Service

Azure App Service is a fully managed web application hosting service. App Service can be used to host applications, regardless of the development platform or operating system. App Services provide first-class support for ASP.NET (Core), Java, Ruby, Node.js, PHP, and Python. It has out of the box continuous integration and deployment with DevOps platforms such as Azure DevOps, GitHub, and BitBucket. Most of the management functionality of a usual hosting environment is integrated into the Azure Portal on the App Service blade, such as scaling, CORS, application settings, SSL, and so on. Additionally, WebJobs can be used to create background processes that can be executed periodically as part of your application bundle.

Web App for Containers is another Azure App Service that allows containerized applications to be deployed as App Services, and orchestrated with Kubernetes.

App Mobile Services is/was an easy way to integrate all the necessary functionality for a simple mobile application, such as authentication, push notifications, and offline synchronization. Nevertheless, App Mobile Services are currently being transitioned to App Center and do not support ASP.NET Core.

Finally, Azure Functions provide a platform to create code snippets or stateless on-demand functions and host them without having to explicitly provision or manage infrastructure. Now that we have covered our basics in regards to choosing a model, we will now look at how to create our first microservice.

Creating our first microservice

For our mobile application, we previously created a simple data access proxy that retrieves data from Cosmos DB. In this exercise, we will be creating small web API components that will expose various methods for CRUD operations on our collections.

Initial setup

Let's begin our implementation:

1. First, create an ASP.NET Core project:

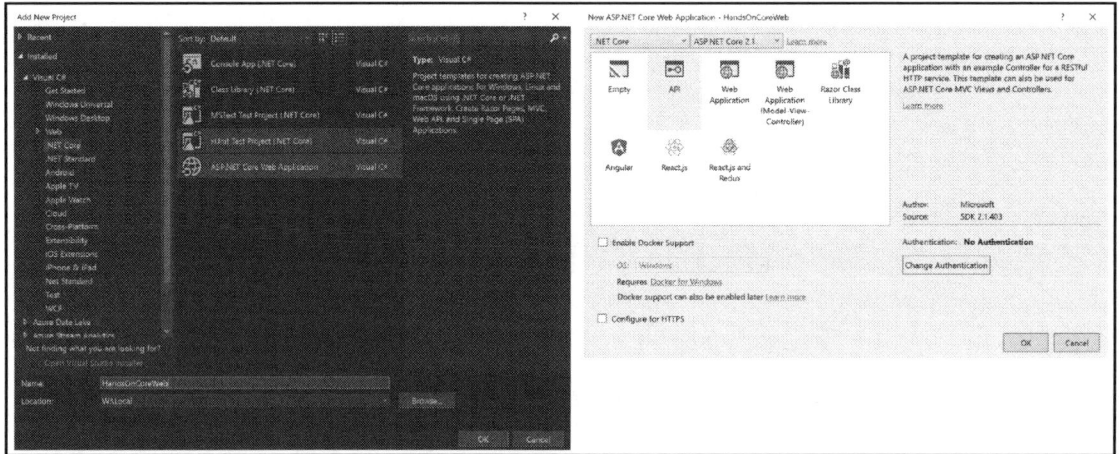

2. Once the project is created, do a quick test to check whether the `dotnet` core components are properly set up. Open a console window and navigate to the project folder. The following commands will restore the referenced packages and compile the application:

```
dotnet restore
dotnet build
```

3. Once the application is built, we can use the run command and execute a GET call to the `api/values` endpoint:

This should result in the output of the values from the `ValuesController.Get` method.

> In the previous example, we used `curl` to execute a quick HTTP request. **Client URL (curl)** is a utility program that is available on Unix-based systems, macOS, as well as Windows 10.

4. Next, we will set up the *Swagger* endpoint, so that we have a metadata endpoint, as well as a UI to execute test requests. For this purpose, we will be using the Swashbuckle NuGet packages to generate the API endpoint metadata. A basic setup of *Swashbuckle* requires three packages, and we can reference them together by adding the `Swashbuckle.AspNetCore` meta package:

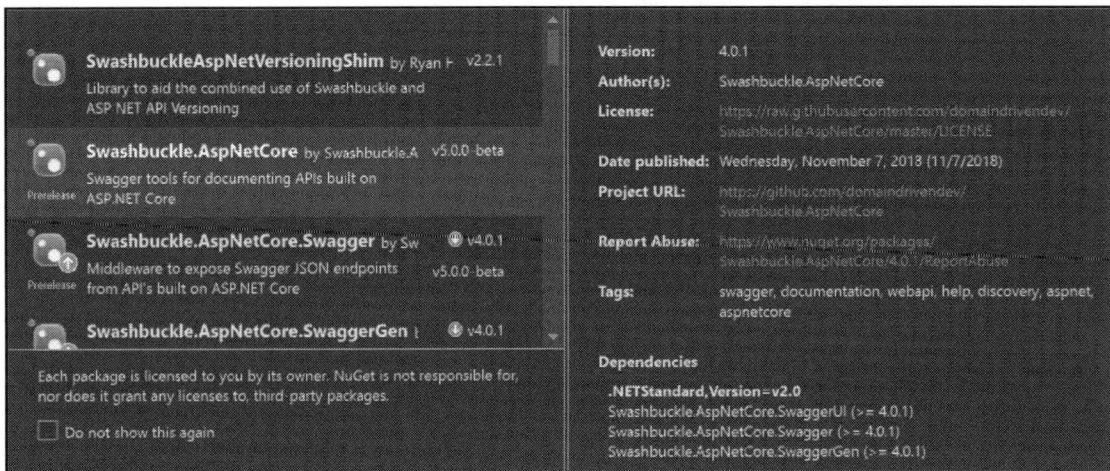

5. After adding the meta package and the dependencies, we will modify our startup class to declare the services:

```
public class Startup
{
    //... <Removed>

    public void ConfigureServices(IServiceCollection services)
    {
        //... <Removed>
        services.AddSwaggerGen(c =>
                {
                    c.SwaggerDoc("v1", new Info{Title = "Auctions
Api", Version = "v1"});
                });
    }

    public void Configure(IApplicationBuilder app,
IHostingEnvironment env)
    {
        //... <Removed>
        app.UseSwagger();
        app.UseSwaggerUI(c =>
```

```
                        {
                c.SwaggerEndpoint("/swagger/v1/swagger.json",
    "Auctions Api");
                });
        }
    }
```

6. Now, when we run the application and navigate to the **{dev host}/swagger** endpoint, we will see the generated Swagger UI and method declarations:

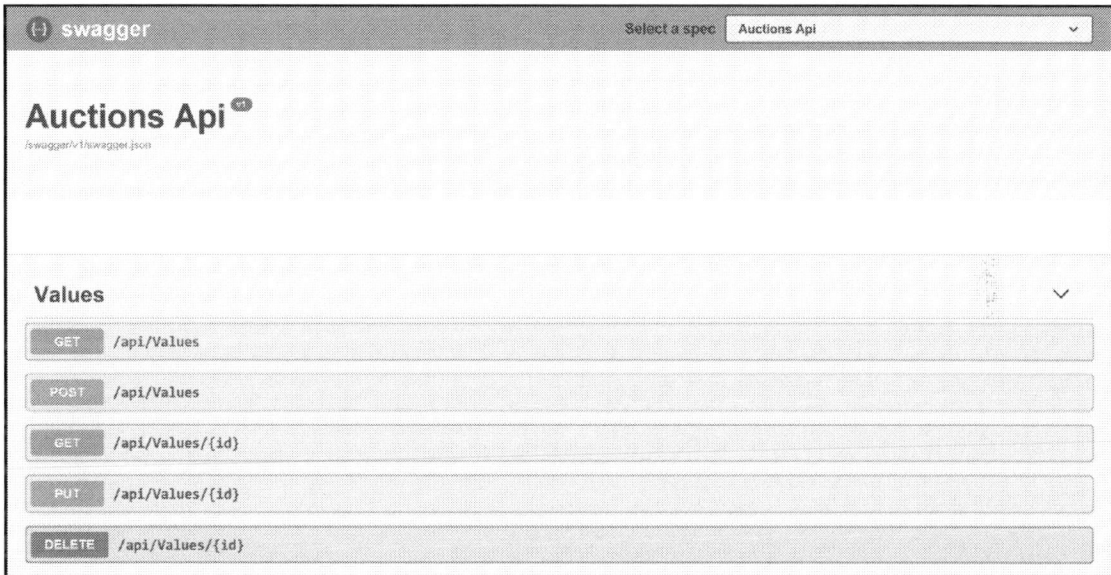

We now have our boilerplate API project ready, which means that we can move on to implementing our services.

Implement retrieval actions

Considering the auctions dataset, since we have created our MVC application, we should include two GET methods in our controller, one for retrieving the complete collection of auctions, and one that will retrieve a specific auction.

Let's see how we can do this:

1. We can simply initialize our repository and return the results for the first GET method:

```
[HttpGet]
public async Task<IEnumerable<Auction>> Get()
{
    var result = Enumerable.Empty<Auction>();

    try
    {
        result = await _cosmosCollection.GetItemsAsync(item =>
        true);
    }
    catch (Exception ex)
    {
        // Log the error or throw depending on the requirements
    }

    return result;
}
```

Notice that the predicate we are using here is targeting the complete set of auctions. We can expand this implementation with query parameters that are filtering the set with additional predicates.

2. The GET method for retrieving a specific item would not be much different:

```
[HttpGet("{id}")]
public async Task<User> Get(string id)
{
    User result = null;

    try
    {
        result = await _cosmosCollection.GetItemAsync(id);
    }
    catch (Exception ex)
    {
        // Log or throw error depending on the requirements
    }

    return result;
}
```

This would suffice the requirements, but to improve the interaction between our mobile application and the cosmos collection, we can also enable **oData** queries and create a transparent query pipeline to the datastore. To do this, we can implement an **oData** controller using the available .NET Core packages, or we can enable the queries on our current MVC controller.

> It is important to note that the Swashbuckle package that we used to generate the Swagger UI, currently does not support **oData** controllers, so these APIs would not be available on the Swagger interface.

3. We begin by setting up our infrastructure in our startup class:

```
public void ConfigureServices(IServiceCollection services)
{
    services.AddOData();
    services.AddODataQueryFilter();
    // ... Removed
}
```

4. We also need to set up the routes for the MVC controller:

```
public void Configure(IApplicationBuilder app, IHostingEnvironment env)
{
    if (env.IsDevelopment())
    {
        app.UseDeveloperExceptionPage();
    }
    app.UseMvc(builder =>
        {
    builder.Count().Filter().OrderBy().Expand().Select().MaxTop(null);
            builder.EnableDependencyInjection();
        });

    // ... Removed
}
```

5. We should also create a new method on the Cosmos repository client, which will be returning a queryable set rather than a result set:

```
public IQueryable<T> GetItemsAsync()
{
    var feedOptions = new FeedOptions {
        MaxItemCount = -1, PopulateQueryMetrics = true,
EnableCrossPartitionQuery = true };
```

```
IOrderedQueryable<T> query = _client.CreateDocumentQuery<T>(
    UriFactory.CreateDocumentCollectionUri(DatabaseId,
CollectionId), feedOptions);

    return query;
}
```

6. Finally, we need to implement our query action:

```
[HttpGet]
[EnableQuery(AllowedQueryOptions = AllowedQueryOptions.All)]
public ActionResult<IQueryable<Auction>>
Get(ODataQueryOptions<Auction> queryOptions)
{
    var items = _cosmosCollection.GetItemsAsync();
    return Ok(queryOptions.ApplyTo(items.AsQueryable()));
}
```

Now, simple **oData** queries can be executed on our collection (for example, first tier **oData** filter queries). For instance, if we were executing a query on the Users endpoint, a simple filter query to retrieve users with one or more auctions would look like this:

```
http://localhost:20337/api/users?$filter=NumberOfAuctions ge 1
```

In order to be able to expand the query options and execute queries on related entities, we would need to create an **entity data model** (**EDM**) and register respective **oData** controllers.

> Another, more appropriate, option to enable an advanced search would be to create an Azure search index and expose the Azure search functionality on top of this index.

For an in-memory EDM and data context, we will be using Microsoft.EntityFrameworkCore's features and functionality. Let's begin with the implementation:

1. Create a DbContext that will define our main data model and the relationships between the entities:

```
public class AuctionsStoreContext : DbContext
{
    public
AuctionsStoreContext(DbContextOptions<AuctionsStoreContext>
options)
        : base(options)
    {
```

```
        }

        public DbSet<Auction> Auctions { get; set; }

        public DbSet<User> Users { get; set; }

        protected override void OnModelCreating(ModelBuilder
        modelBuilder)
        {
            modelBuilder.Entity<User>().OwnsOne(c => c.Address);
            modelBuilder.Entity<User>().HasMany<Auction>(c =>
            c.Auctions);
        }
    }
```

2. Now, let's register this context within the `ConfigureServices` method:

```
    public void ConfigureServices(IServiceCollection services)
    {
        services.AddDbContext<AuctionsStoreContext>(option =>
            option.UseInMemoryData("AuctionsContext"));
        services.AddOData();
        services.AddODataQueryFilter();
        // ... Removed
    }
```

3. Now, we will create our EDM:

```
    private static IEdmModel GetEdmModel()
    {
        ODataConventionModelBuilder builder = new
        ODataConventionModelBuilder();
        var auctionsSet = builder.EntitySet<Auction>("Auctions");
        var usersSet = builder.EntitySet<User>("Users");
        builder.ComplexType<Vehicle>();
        builder.ComplexType<Engine>();
        return builder.GetEdmModel();
    }
```

4. Finally, register an **oData** route so that entity controllers can be served through this route:

```
app.UseMvc(builder =>
    {
    builder.Count().Filter().OrderBy().Expand().Select().MaxTop(null);
        builder.EnableDependencyInjection();
        builder.MapODataServiceRoute("odata", "odata",
GetEdmModel());
        });
```

5. Now that the infrastructure is ready, you can navigate to the `$metadata` endpoint to take a look at the EDM that was generated:

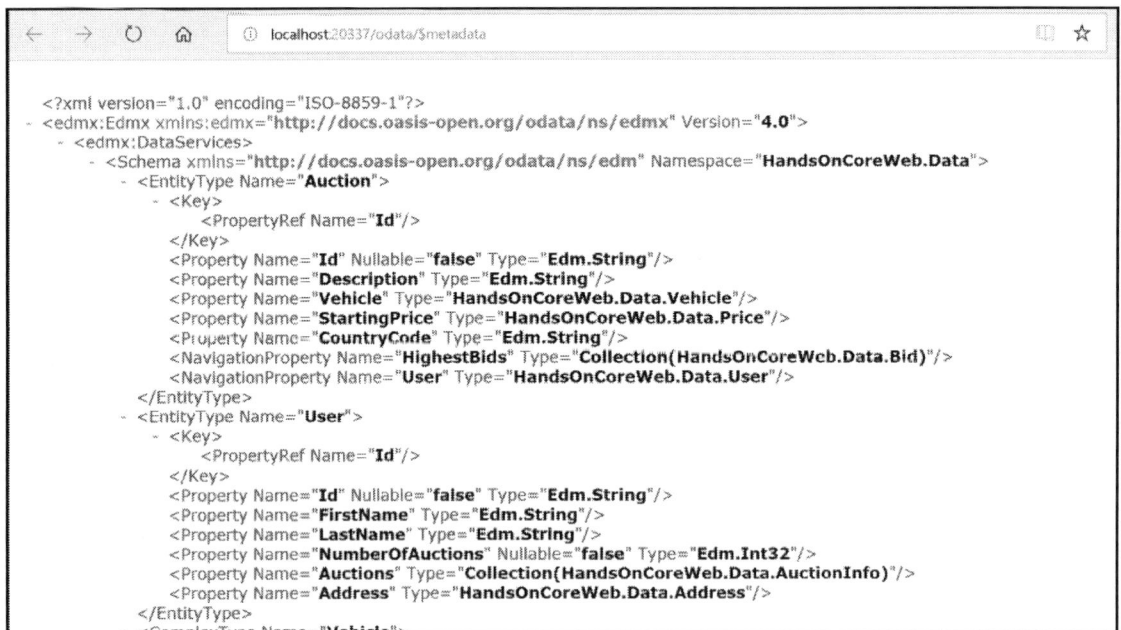

6. Now, by implementing a quick `ODataController`, we can expose the complete set to oData queries:

```
public class AuctionsController : ODataController
{
    private readonly CosmosCollection<Auction> _cosmosCollection;

    public AuctionsController()
    {
        _cosmosCollection = new
```

```
CosmosCollection<Auction>("AuctionsCollection");
    }

    // GET: api/Users
    [EnableQuery]
    public IActionResult Get()
    {
        var items = _cosmosCollection.GetItemsAsync();
        return Ok(items.AsQueryable());
    }

    [EnableQuery]
    public async Task<IActionResult> Get(string key)
    {
        var auction = await _cosmosCollection.GetItemsAsync(item =>
item.Id == key);
        return Ok(auction.FirstOrDefault());
    }
}
```

7. With the given `ODataController` in place and the additional implementation for the users collection, various entity filter expressions can be executed, such as the following:

- `http://localhost:20337/odata/auctions?$filter=Vehicle/Engine/Power%20gt%20120`
- `http://localhost:20337/odata/users?$filter=Address/City%20eq%20'London'`
- `http://localhost:20337/odata/users?$filter=startswith(FirstName,%20'J')`
- `http://localhost:20337/odata/auctions('7ad0d2d4-e19c-4715-921b-950387abbe50')`

With this, we have successfully exposed an **oData** endpoint for a Cosmos Document DB, using ASP.NET Core and EntityFrameworkCore.

Implementing patch updates

Implementing the update endpoint would in fact be different, to say the least, when we are dealing with a NoSQL data store that doesn't support partial updates. According to the application requirements, we can choose two distinct patterns.

In the classic concurrency model, we would be receiving a PUT request with the complete object body and checking whether the update is being executed on the latest version of the object. This concurrency check can be done on the _ts property collection items. The timestamp property is also used by the Cosmos DB containers themselves for handling concurrency issues. Let's begin:

1. In this model, the incoming object body would be verified to check whether it carries the latest timestamp, and, if not, a 409 response signifying a conflict will be sent back as a response. If the timestamp matches the one in the repository, then we are free to upsert the entity:

```
[EnableQuery]
[HttpPut]
public async Task<IActionResult> Put([FromODataUri]string key,
[FromBody] Auction auctionUpdate)
{
    var cosmosCollection = new
CosmosCollection<Auction>("AuctionsCollection");
    var auction = (await cosmosCollection.GetItemsAsync(item =>
item.Id == key)).FirstOrDefault();
    if (auction == null)
    {
        return NotFound();
    }

    if (auction.TimeStamp != auctionUpdate.TimeStamp)
    {
        return Conflict();
    }

    await cosmosCollection.UpdateItemAsync(key, auctionUpdate);
    return Accepted(auction);
}
```

However, with this approach, as the object tree grows in size, and the complexity increases on our container items, the update's requests would get bigger and harder to execute (for example, higher probability of conflicts, unintentional removal of properties, and so on). The following screenshot shows a request that's been sent to update only the description field of an Auction document:

In order to decrease the impact of an update call, at least for the client, we can utilize PATCH methods. In a patch method, only a part of the object tree is delivered as a Delta, or only the partial update operations are delivered as patch operations.

2. Let's implement a PATCH action for the same Auction service and check out the request:

```
[EnableQuery]
[HttpPatch]
public async Task<IActionResult> Patch(
    string key,
    [FromBody] JsonPatchDocument<Auction> auctionPatch)
{
    var cosmosCollection = new
CosmosCollection<Auction>("AuctionsCollection");
    var auction = (await cosmosCollection.GetItemsAsync(item =>
item.Id == key)).FirstOrDefault();

    if (auction == null)
    {
```

```
            return NotFound();
        }

        auctionPatch.ApplyTo(auction);
        await cosmosCollection.UpdateItemAsync(key, auction);
        return Accepted(auction);
    }
```

The request for the given endpoint would look similar to the following:

```
PATCH /odata/auctions('3634031a-1f45-4aa0-9385-5e2c86795c49')

[
    {"op" : "replace", "path" : "description", "value" : "Updated
Description"}
]
```

3. The same optimistic concurrency control mechanism we have implemented with the timestamp value can be implemented with the Test operation:

```
PATCH /odata/auctions('3634031a-1f45-4aa0-9385-5e2c86795c49')

[
    { "op": "test", "path": "_ts", "value": 1552741629 },
    { "op" : "replace", "path" : "vehicle/year", "value" : 2017},
    { "op" : "replace", "path" : "vehicle/engine/displacement",
"value" : "2.4"}
]
```

In this example, if the timestamp value does not match the value in the request, the consequent update operations would not be executed.

Implementing a soft delete

If you're planning on integrating the storage-level operations with triggers and/or change feed, a soft-delete implementation can be used instead of implementing the complete removal of objects. In the soft-delete approach, we can extend our entity model with a specific property (for example, isDeleted) that will define that the document is deleted by the consuming application.

In this setup, the consuming application can make use of the PATCH method that was implemented or an explicit DELETE method, which can be implemented for our entity services.

Let's take a look at the following `PATCH` request:

```
PATCH /odata/auctions('3634031a-1f45-4aa0-9385-5e2c86795c49')

[
    { "op": "test", "path": "_ts", "value": 1552741629 },
    { "op" : "replace", "path" : "isDeleted", "value" : true},
    { "op" : "replace", "path" : "ttl", "value" : "30"}
]
```

With this request, we are indicating that the auction entity with the given ID should be marked for deletion. Additionally, by setting the **Time To Live** (**TTL**) property, we are triggering an expiry on that given entity. This way, both the triggers and the change feed will be notified about this update and, within the given TTL, the entity will be removed from the data store.

> TTL is an intrinsic feature of Cosmos DB. The time to live can be set on the container level, as well as at the item level. If no value has been set on the container level, the value set of the items will be ignored by the platform. However, the container can have -1 default expiry and the items that we want to expire after a certain period can declare a value greater than 0. Time to live does not consume resources and is not calculated as part of the consumed RUs.

With the delete implementation, we have the complete set of functions to create the basic CRUD structure for a microservice in accordance with the Cosmos collections.

Integrating with Redis cache

In a distributed cloud application with a fine-grained microservice architecture, distributed caching can provide much desired data coherence, as well as performance improvements. Generally speaking, the distribution of the infrastructure, the data model, as well as costs, are deciding factors regarding whether to use a distributed cache implementation.

ASP.NET Core offers various caching options, one of which is distributed caching. The available distributed cache options are as follows:

- Distributed memory cache
- Distributed SQL server cache
- Distributed Redis caches

While the memory cache is not a production-ready strategy, SQL and Redis can be viable options for a cloud application that's been developed with .NET Core. However, in the case of a NoSQL database and semi-structured data, Redis would be an ideal choice. Let's see how we can introduce a distributed cache and make it ready for use:

1. In order to introduce a distributed cache that can be used across controllers, we would need to use the available extensions so that we can inject an appropriate implementation of the `IDistributedCache` interface. `IDistributedCache` would be our main tool for implementing the cache, aside from the pattern we mentioned previously:

```
public interface IDistributedCache
{
    byte[] Get(string key);

    Task<byte[]> GetAsync(string key, CancellationToken token =
default(CancellationToken));

    void Set(string key, byte[] value, DistributedCacheEntryOptions
options);

    Task SetAsync(string key, byte[] value,
DistributedCacheEntryOptions options, CancellationToken token =
default(CancellationToken));

    void Refresh(string key);

    Task RefreshAsync(string key, CancellationToken token =
default(CancellationToken));

    void Remove(string key);

    Task RemoveAsync(string key, CancellationToken token =
default(CancellationToken));
}
```

As you can see, using the injected instance, we would be able to set and get data structures as byte arrays. The application would first reach out to the distributed cache to retrieve the data. If the data we require does not exist, we would retrieve it from the actual data store (or a service) and store the result in our cache.

2. Before we can implement the Redis cache in our application, we can head over to the Azure Portal and create a Microsoft Azure Redis Cache resource within the same resource group we have been using up until now:

3. Once the Redis cache instance is created, we would need to note one of the available connection strings. Connection strings can be found under the **Manage Keys** blade, which is accessed by using the **Show access keys** option on the **Overview** screen. We can now continue with our implementation.

4. We will start our implementation by installing the required Redis extension:

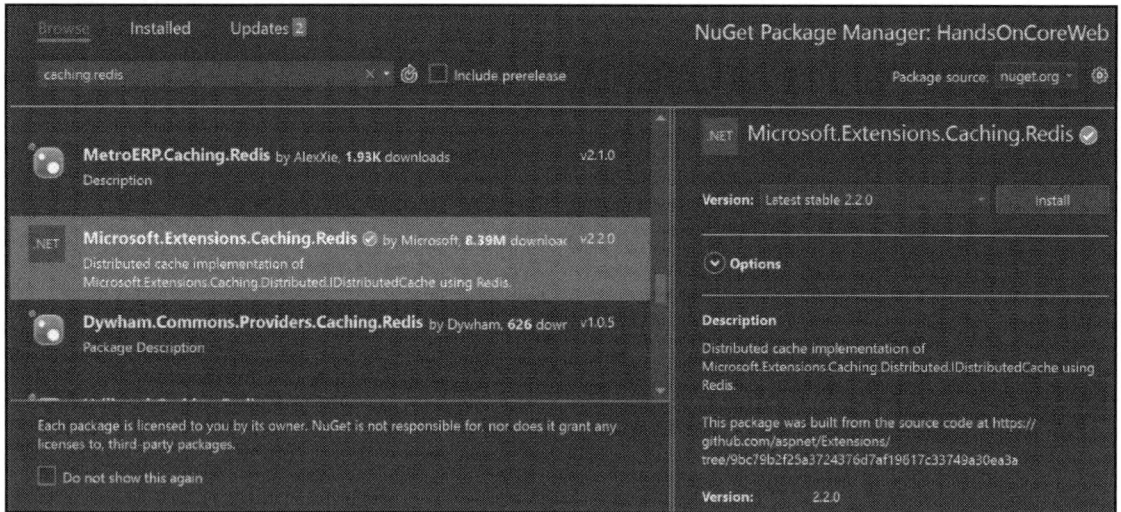

5. After the extension and its dependencies have been installed, we will need to configure our distributed cache service using the extension method:

```
services.AddDistributedRedisCache(
    option =>
        {
            option.Configuration =
Configuration.GetConnectionString("AzureRedisConnection");
            option.InstanceName = "master";
        });
```

6. We can now insert the connection string we retrieved from the Azure Portal in the `appsettings.json` file, and head over to our controller to set up the constructor injection for the `IDistributedCache` instance:

```
private readonly IDistributedCache _distributedCache;

public UsersController(IDistributedCache distributedCacheInstance)
{
    _distributedCache = distributedCacheInstance;
}
```

And that's about it – the distributed cache is ready to use. We can now insert serialized data items as a key/value (where value is a serialized byte array) into our cache without having to communicate with the actual data source.

Combined with the backend for the frontend pattern, the Redis cache can deliver the cached data to the application gateway service without having to contact multiple microservices, providing a simple cost-cutting solution, as well as the promised performance enhancements.

Hosting the services

Since our web implementation makes use of ASP.NET Core, we are blessed with multiple deployment and hosting options that span over Windows and Linux platforms.

Azure Web App for App Service

At this point, without any additional implementation or configuration, our microservices are ready for deployment and they can already be hosted on Azure Cloud as fully managed App Services. In order to deploy the service as an App Service, you can use Visual Studio Azure extensions, which allows you to create a publish profile, as well as the target hosting environment.

Let's see how we can do this::

1. Right-click on the project to be deployed. You will see the **Pick a publish target** selection window:

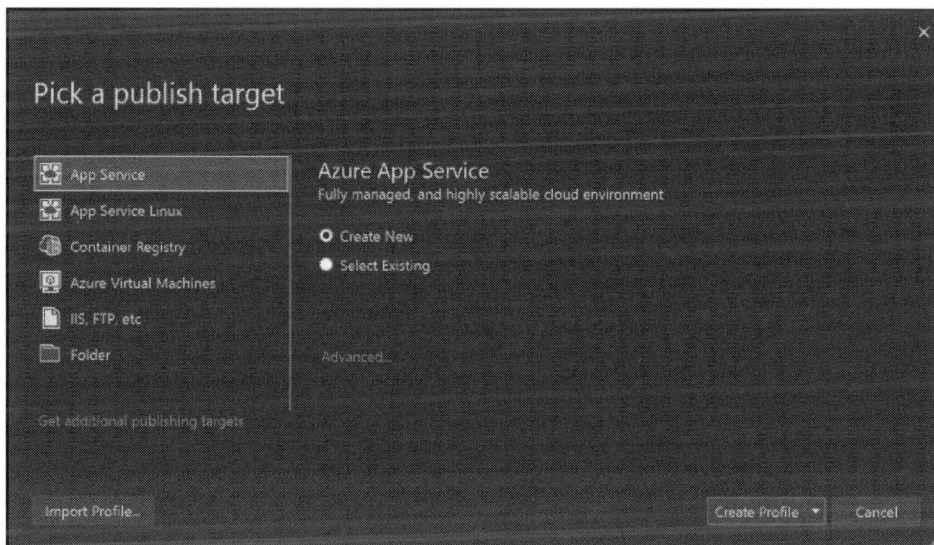

For a complete managed hosting option, we can choose App Service or App Service Linux options and continue to create a new App Service.

2. If we were to choose the App Service option, the application would be hosted on the Windows platform with a full .NET Framework profile, whereas the Linux option would use Linux operating systems with .NET Core runtime. Selecting the **Create New** option allows us to select/create the resource group we want the App Service instance to be added to:

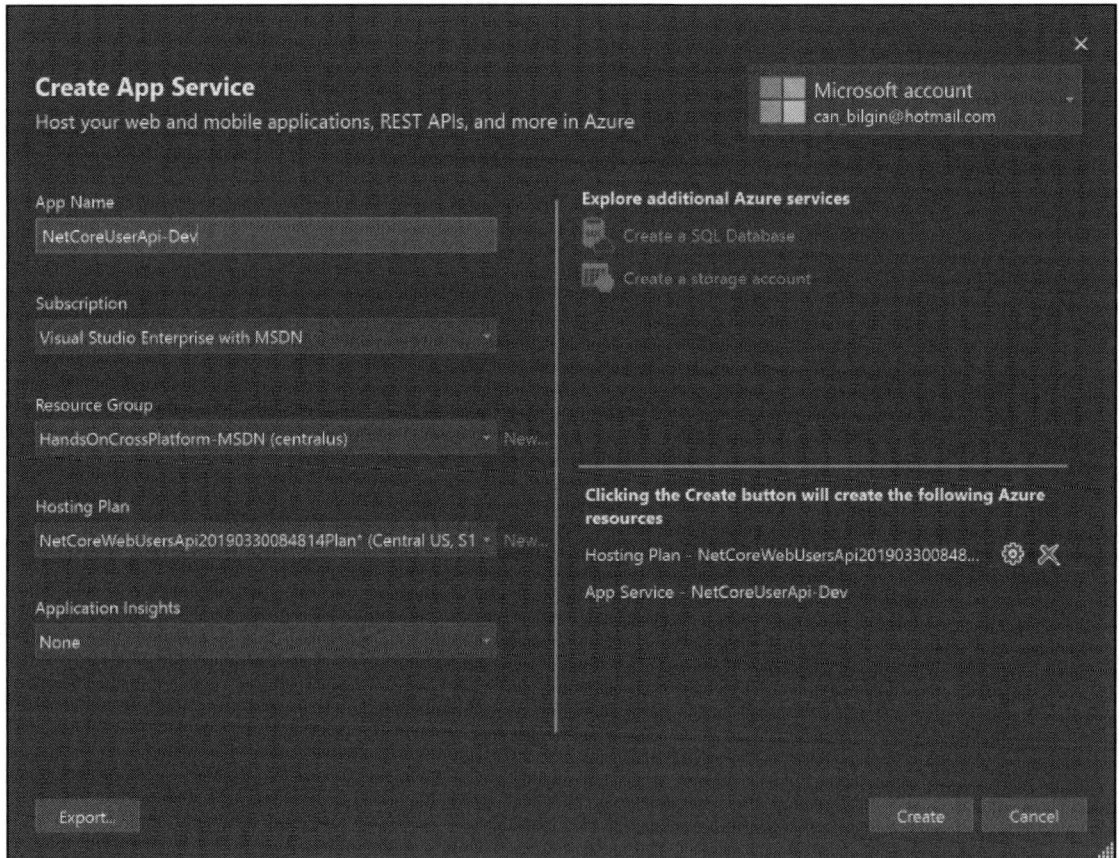

3. Once published, you can copy the URL from the site URL field and execute a query using `curl`:

```
Microsoft Windows [Version 10.0.17134.590] (c) 2018 Microsoft
Corporation. All rights reserved.
```

```
C:\Users\can.bilgin>curl
https://netcoreuserapi-dev.azurewebsites.net/odata/users?$filter=Fi
rstName%20eq%20%27Jane%27

{"@odata.context":"https://netcoreuserapi-dev.azurewebsites.net/oda
ta/$metadata#Users","value":[{"Id":"7aa0c870-cb90-4f02-
bf7e-867914383190","FirstName":"Jane","LastName":"Doe","NumberOfAuc
tions":1,"Auctions":[],"Address":{"AddressTypeId":4000,"City":"Seat
tle","Street1":"23 Pike St.","CountryCode":"USA"}}]}

C:\Users\can.bilgin>
```

Containerizing services

Another option for hosting would be to containerize our application(s), which would come with the added benefit of configuration as code principles. In a container setup, each service would be isolated within its own sandbox and easily migrated from one environment to the next with a high level of flexibility and performance. Containers can also provide cost savings compared to web apps if they are deployed to the aforementioned container registries and platforms such as ACS, AKS, Service Fabric, and Service Fabric Mesh.

> Containers are isolated, managed, and portable operating environments. They provide a location where an application can run without affecting the rest of the system, and without the system affecting the application. Compared to virtual machines, they provide much higher server density since, instead of sharing the hardware, they share the operating system kernel.

Docker is a container technology that has become almost synonymous with containers themselves in recent years. The Docker container hosting environment can be created on Windows and macOS using the provided free software. Let's get started:

1. In order to prepare a Windows development machine for Docker containerization, we would need to make sure that we have installed the following:
   - Docker for Windows (for hosting Windows and Linux containers)
   - Docker Tools for Visual Studio (opt-in installation for Visual Studio 2017 v15.8+)

2. Now that we have the prerequisites, we can add a Docker container image definition into each microservice project using the **Add | Docker Support** menu item:

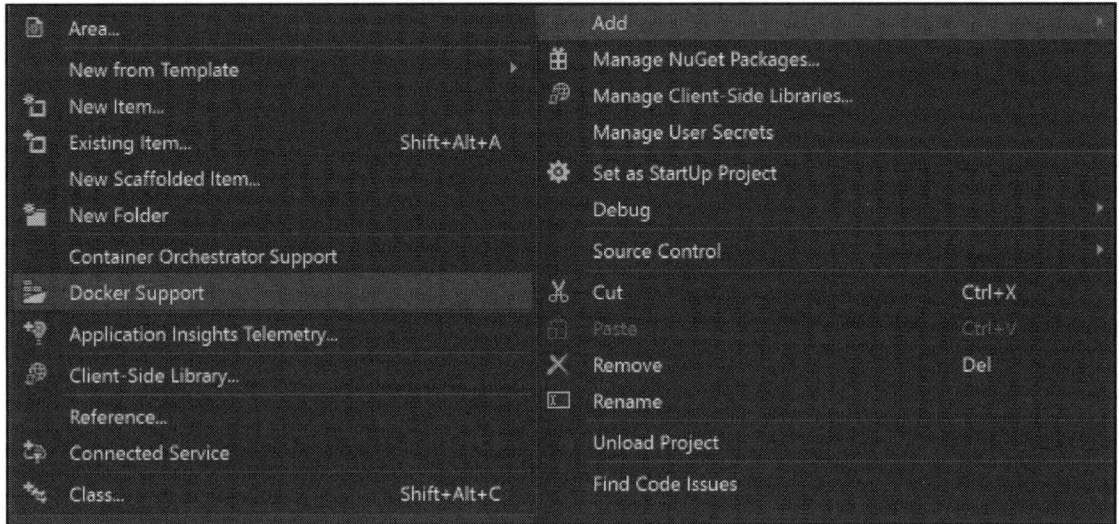

3. This would create a multistage Docker file that will, in simple terms, do the following:
   - Copy the source code for the application into the container image
   - Restore the required .NET Core runtime components, depending on the type of the container (Windows or Linux)
   - Compile the application
   - Create a final container image with the compiled application components

4. Here, the Docker file we have created for `UsersApi` looks like this:

```
FROM microsoft/dotnet:2.2-aspnetcore-runtime AS base
WORKDIR /app
EXPOSE 80
EXPOSE 443

FROM microsoft/dotnet:2.2-sdk AS build
WORKDIR /src
COPY ["NetCore.Web.UsersApi/NetCore.Web.UsersApi.csproj",
"NetCore.Web.UsersApi/"]
COPY ["NetCore.Data.Cosmos/NetCore.Data.Cosmos.csproj",
"NetCore.Data.Cosmos/"]
```

```
COPY ["NetCore.Data/NetCore.Data.csproj", "NetCore.Data/"]
RUN dotnet restore
"NetCore.Web.UsersApi/NetCore.Web.UsersApi.csproj"
COPY . .
WORKDIR "/src/NetCore.Web.UsersApi"
RUN dotnet build "NetCore.Web.UsersApi.csproj" -c Release -o /app

FROM build AS publish
RUN dotnet publish "NetCore.Web.UsersApi.csproj" -c Release -o /app

FROM base AS final
WORKDIR /app
COPY --from=publish /app .
ENTRYPOINT ["dotnet", "NetCore.Web.UsersApi.dll"]
```

As you can see, the first stage that defines the base is a reference to Microsoft managed container images in the public Docker registry. The build image is where we have the source code for our ASP.NET Core application. Finally, the build and final images are the last stages where the application is compiled (that is, dotnet publish) and set up as an entry point on our container.

In other words, the created container image has the final application code, as well as the required components, regardless of the host operating system and other containers running on that host.

5. Now, if you navigate to the parent directory of the UsersApi project on a console or Terminal (depending on the operating system that Docker is installed on) and execute the following command, Docker will build the container image:

```
docker build -f ./NetCore.Web.UsersApi/Dockerfile -t netcore-
usersapi .
```

6. Once the container image is built by the docker daemon, you can check whether the image is available to start as a container using the following command:

```
docker image ls
```

7. If the image is on the list of available container images, you can now run the container, either using the exposed port of 80 or 443 (the following command maps the container port 80 to host port 8000):

```
docker run -p 8000:80 netcore-usersapi
```

8. The container development is, naturally, more integrated with Visual Studio on the Windows platform. In fact, ASP.NET Core applications, once containerized, contain a run/debug profile for Docker that can be directly started from the Visual Studio user interface:

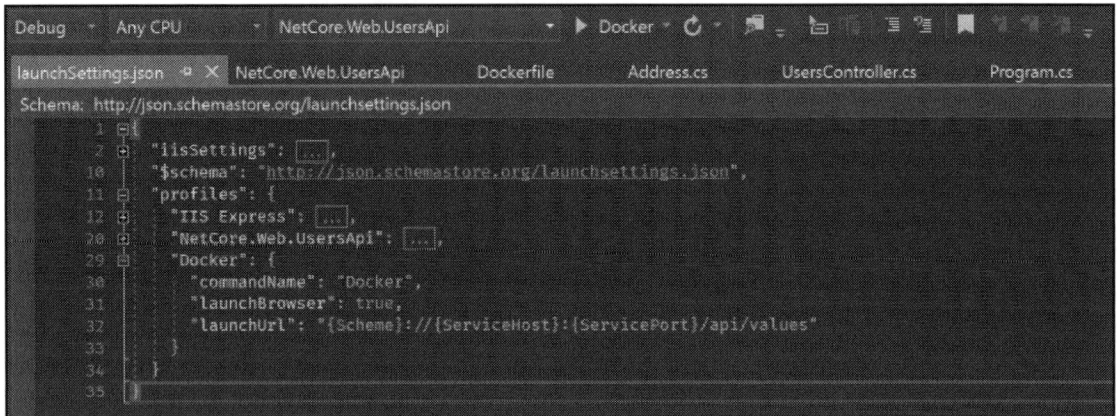

At this stage, our container configuration is ready to be used, and it can already be deployed to Azure Web Applications for containers. Container orchestration support can be added using the available tools on Visual Studio.

Securing the application

In a microservice setup with a client-specific backend, multiple authentication strategies can be used to secure web applications. ASP.NET Core provides the required OWIN middleware components to support most of these scenarios.

Depending on the gateway and downstream services architecture, authentication/authorization can be implemented on the gateway and the user identity can be carried over to the backend services:

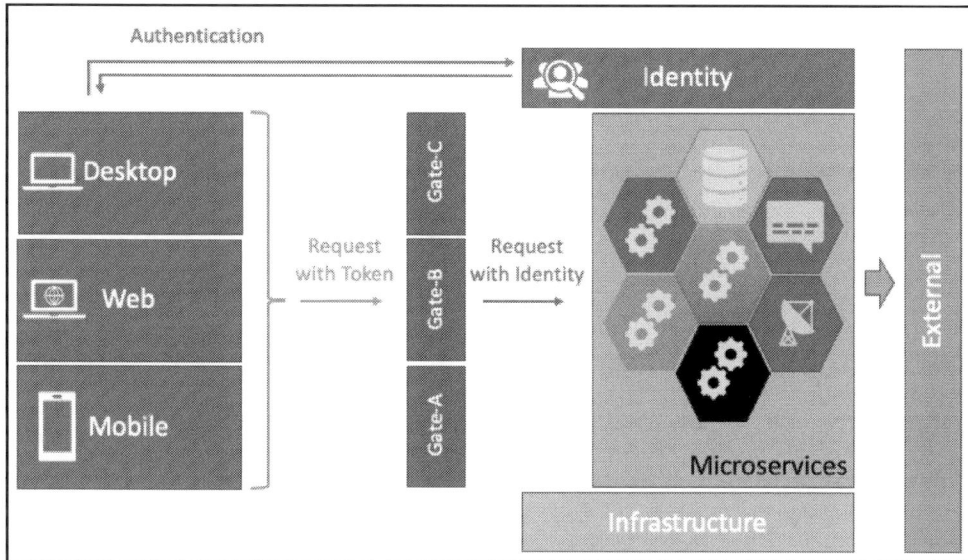

Another approach would be where each service can utilize the same identity provider in a federated setup. In this setup, a dedicated **security token service (STS)** would be used by client applications, and a trust relationship would need to be established between the STS and the app services:

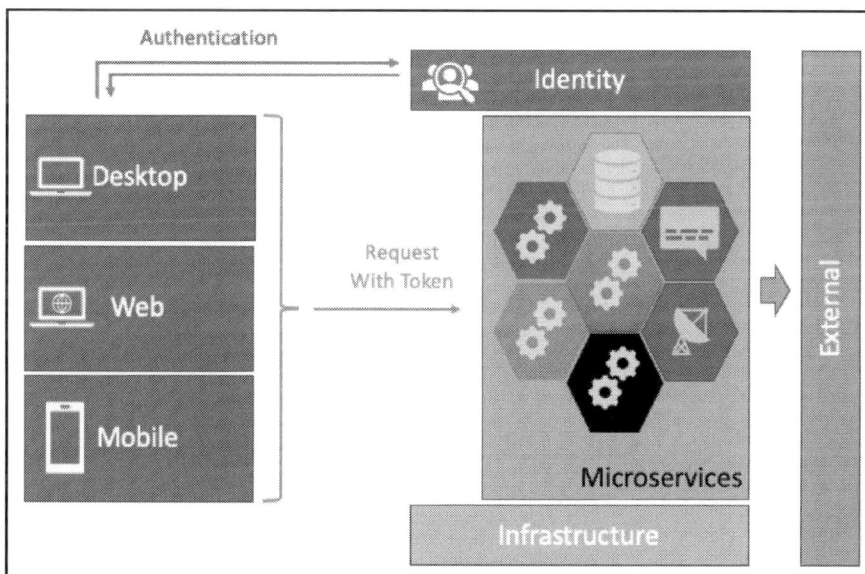

While choosing the authentication and authorization strategy, it is important keep in mind that the identity consumer in this setup will be a native mobile client. When mobile applications are involved, the authentication flow of choice is generally the oAuth2 authorization code flow:

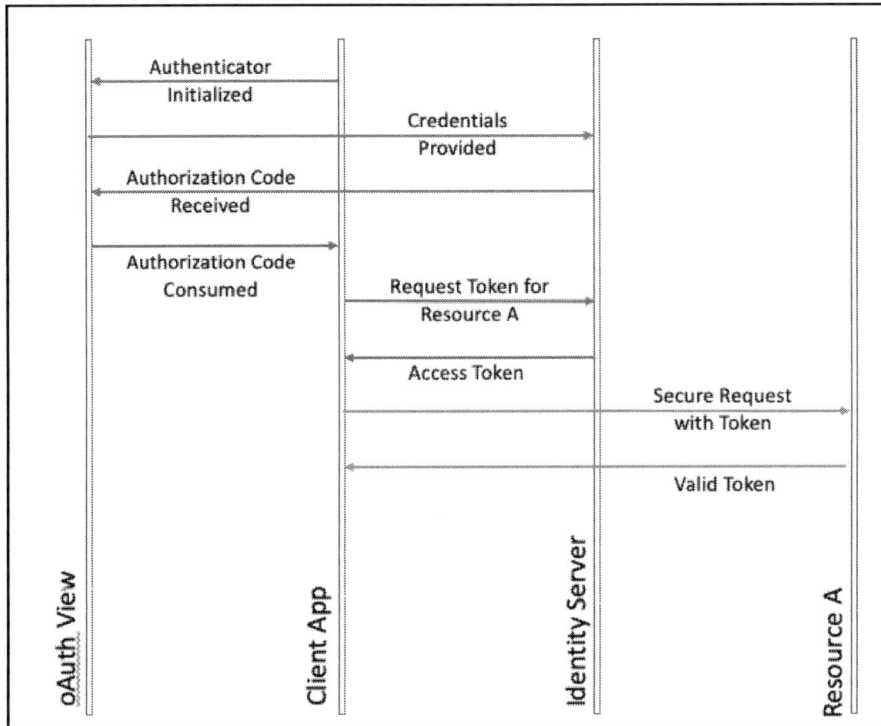

Again, depending on the application you are building, multiple **OpenID Identity Connect (OIDC)** providers, such as Microsoft Live, Facebook, and Google, can be introduced to allow users to choose their preferred identity.

ASP.NET Core Identity

ASP.NET Core Identity is the default membership system that can provide a relatively trivial and yet extensive implementation of an STS as well as login, registration, and management user interfaces. Compared to its predecessor, ASP.NET Core Identity provides developers with a wider range of authentication scenarios, such as oAuth, Two-Factor Authentication, QR Code for **Time-based One-Time Password (TOTP)**, and so on.

ASP.NET Core Identity uses the SQL database as a persistence store by default and can be replaced with other repository implementations. Entity Framework Core is used to implement the standard repository functionality.

External OIDC providers that are supported by ASP.NET Core Identity are Facebook, Twitter, Google, and Microsoft. Additional provider implementations can be found on third-party or community provided packages.

Using ASP.NET Core Identity, the created STS can then be consumed by the Xamarin application through a simple set of HTTP request to register, authenticate, and authorize users. Additionally, Xamarin applications can utilize available identity provider SDKs, as well as cross-provider packages.

While this identity management should suffice, since the requirements are for a solely ASP.NET Core-based solution, once additional Azure resources are included in a distributed application such as Azure Serverless components, a cloud-based identity management could prove to be a better choice.

Azure AD

Azure AD is a cloud-based **identity as a service (IDaaS)** offering and, hitherto, the only authentication and identity management process that's integrated with the resource manager for distributed applications developed on the Azure infrastructure. Azure AD is used to manage access to any SaaS/PaaS resources in resource groups. It supports protocols such as OpenID Connect, oAuth, and SAML to provide SSO and access control to resources within a directory.

The authorization for access to and between resources can be set up using the identity principles defined within the directory. The principle can represent a user with a single organization (possibly with an associated on-premise Active Directory), an application (a resource that's set up within the directory or external application, such as a native mobile app), or an external identity from a different Azure directory or from an external identity provider.

Generally speaking, this setup where user identities are defined within an organization unit and users from other directories are introduced as guests, is referred to as Azure AD **Business to Business (B2B)**.

In order to set up an application wide-authentication scheme using Azure AD B2B, follow these steps:

1. Create a directory that will define the organization that will be using the application suite (that is, mobile application and associated services).

> Any Azure subscription is accompanied with, at a minimum, free tier Azure AD (depending on the subscription type) and, in most cases, the creation of a new directory is not necessary.

   You can create a new directory by using the **Create a Resource** interface on Azure Portal. Once you have selected Azure AD, you will need to declare an organization name and initial domain name (for example: netcorecrossplatform.onmicrosoft.com).

2. Additional custom domains can be added to this declaration at a later time:

3. Once the directory is created, the organization should be available as a domain option, so that you can set up the authentication for an ASP.NET Core web application using Visual Studio:

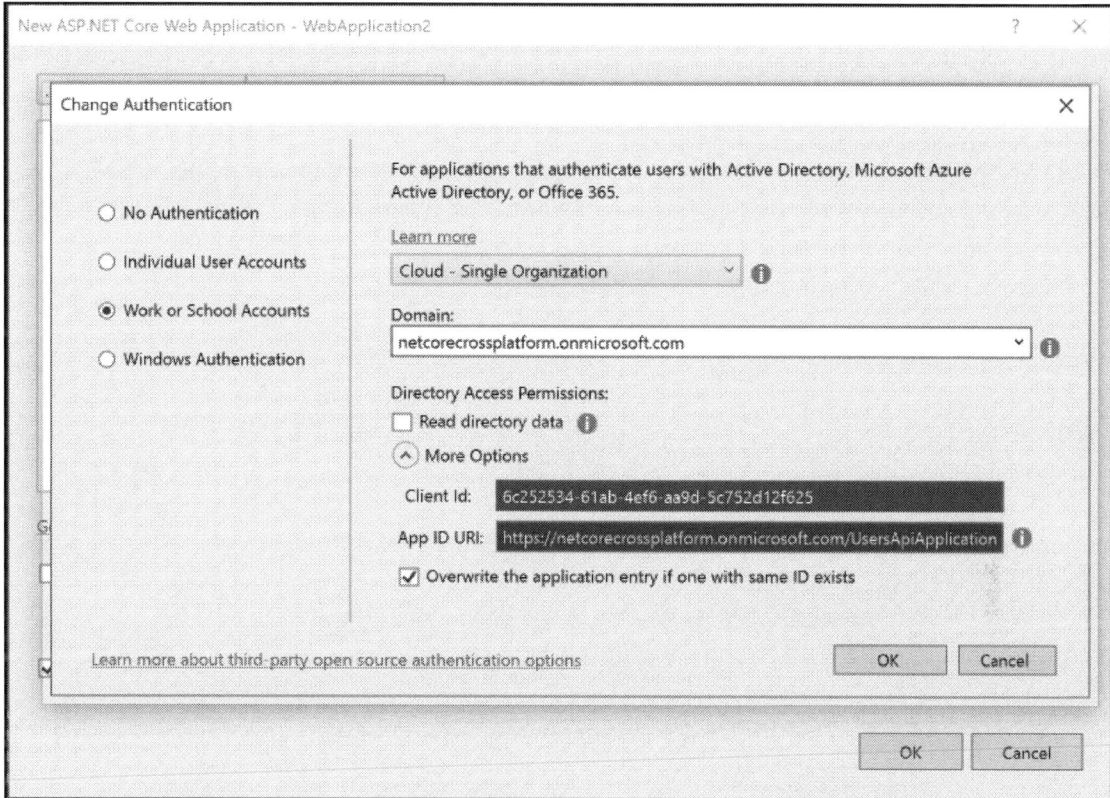

Once the application project is created, it will automatically add an application registry to the Azure AD and add/configure the required middleware for application authentication. The configuration for Azure AD in application startup would look similar to the following:

```
services.AddAuthentication(AzureADDefaults.BearerAuthenticationSche
me)
    .AddAzureADBearer(options => Configuration.Bind("AzureAd",
options));
```

The configuration that was created (matching the Azure AD application registration) is as follows:

```
"AzureAd": {
    "Instance": "https://login.microsoftonline.com/",
    "Domain": "netcorecrossplatform.onmicrosoft.com",
    "TenantId": "f381eb86-1781-4732-9543-9729eef9f843",
    "ClientId": "ababb076-abb9-4426-b7df-6b9d3922f797"
},
```

For Azure AD, authentication on the client application (considering we are using Xamarin and Xamarin.Forms as the development platforms) can be implemented using the **Microsoft Authentication Library (MSAL)**.

4. In order for the client application to be able to use the identity federation within this organization, we would need to register (yet) another application on Azure AD. However, this registration should be declared for a native application:

Home > netcorecrossplatform - App registrations (Preview) > Register an application

Register an application
PREVIEW

* **Name**

The user-facing display name for this application (this can be changed later).

Xamarin Application

Supported account types

Who can use this application or access this API?

◉ Accounts in this organizational directory only (netcorecrossplatform)

○ Accounts in any organizational directory

○ Accounts in any organizational directory and personal Microsoft accounts (e.g. Skype, Xbox, Outlook.com)

Help me choose...

Redirect URI (optional)

We'll return the authentication response to this URI after successfully authenticating the user. Providing this now is optional and it can be changed later, but a value is required for most authentication scenarios.

Public client (mobile & desktop)   ∨   e.g. myapp://auth

By proceeding, you agree to the Microsoft Platform Policies ⊡

Register

5. After the application registration is created, you can use the current directory (that is, tenant) and the client application (that is, application registration) to set up the authentication library. In this setup, the identity flow can be defined simply as follows:

- The native application retrieving an access token using the authorization code flow
- The native application executing an HTTP request to the gateway service (that is, our ASP.NET Core service exposing mobile app-specific endpoints)
- The gateway service verifying the token and retrieving an on-behalf-of token to call the downstream stream services

This way, the user identity can be propagated to each layer and the required authorization procedures can be implemented using the claim principle.

6. In order to allow identity propagation, the gateway service application registration (that is, service principle) should be given the required identity delegation permissions to the downstream service registrations:

7. Now, the user identity can access the resources, given that his/her identity exists within the target organization and has the required permissions.

For a business-facing application (that is, a **line of business** (**LOB**) application), Azure AD B2B can provide a secure identity management solution with ease, and no additional custom implementation. However, if the application needs to be client-facing, we would need a more flexible solution with additional support for registration. Azure B2C can provide the required support for individual user accounts.

Azure AD B2C

Azure AD B2C is an identity management service for consumer-facing scenarios with the option to customize and control how customers sign up, sign in, and manage their profiles when using your applications. This targets various platforms.

B2C is a modern federated identity management service where the consumer applications (that is, the relying parties) can consume multiple identity providers and verification methods.

In the B2C realm, the user flows for sign-up and sign-in are referred to as user journeys. User journeys can be customized with policies if required. The Identity Experience Framework consumes these policies to achieve the desired user flows. The Identity Experience Framework is the underlying platform that establishes multi-party trust and completes the steps in a user journey.

Similar to Azure AD itself, a tenant describes a user domain where certain relations between the users and applications can be defined. However, in B2C, the domain is customer-specific, not organization-specific. In other words, a tenant defines an access group that is governed by the policy descriptions and the linked identity providers.

In this setup, multiple applications can be given access to multiple tenants, making B2C a perfect fit for development companies with a suite of applications that they want to publish to consumers. Consumers, once they sign up using one of the linked OIDC identity providers, can get access to multiple consumer-facing applications.

Summary

In this chapter, we have browsed through the PaaS platforms, as well as the architectural approaches that are available for hosting and implementing ASP.NET Core web services. Using the flexible infrastructure offered by ASP.NET Core, it is a relief for developers to implement microservices that consume data from Cosmos DB collections. The services that contain CRUD operations on domain objects can be optimized and improved with Redis, as well as containerization, and hosted on multiple platforms and operating systems. Security, being one of our main concerns in a distributed cloud architecture, can be ensured using the available identity infrastructure and Identity as a Service offerings such as Azure Active Directory and Azure AD B2C on Azure cloud stack.

In the next chapter, we will move on to Azure Serverless, which is yet another service platform that .NET Core can prove to be vital.

10
Using .NET Core for Azure Serverless

Azure Functions are serverless compute modules that take advantage of various triggers, including HTTP requests. Using Azure Functions, developers can create business logic containers, completely isolated from the problems brought by monolithic web application paradigms and infrastructure. They can be used as simple HTTP request processing units and so-called microservices, as well as for orchestrating complex workflows. Azure Functions come in two flavors (compiled or script-based) and can be written in different languages, including C# with .NET Core modules.

In this chapter, we will incorporate Azure Functions into our infrastructure so that we can process data on different triggers. We will then integrate Azure Functions with a Logic App, which will be used as a processing unit in our setup.

The following topics will be covered throughout this chapter:

- Understanding Azure Serverless
- Implementing Azure Functions
- Creating logic applications
- Integrating Azure services to functions

Understanding Azure Serverless

In the previous chapters, we created a document structure as a repository and implemented ASP.NET Core services as containerized microservices so that we could cover our main application use cases. These use cases can be considered the primary data flow through the application, and our main concern for performance is concentrated around these data paths. Nevertheless, the secondary use cases, such as keeping track of the statuses of auctions for a user that he/she has previously got involved in, or creating a feed to inform users about new vehicles up for auction, are the features that could increase the return rate of the users and maintain the user base. Therefore, we would need a steadfast, event-driven strategy that will not interfere with the primary functionality and should be able to scale without having to interfere with the infrastructure.

Azure Functions and other Azure Serverless components are tailor-made Azure offerings for these type of event-driven scenarios where one or more Azure infrastructure services would need to be orchestrated.

Developing Azure Functions

We can use multiple development environments to develop Azure Functions:

- Using Azure portal
- Using the Azure CLI with Azure Functions Core Tools
- Using Visual Studio or Visual Studio Code
- Using other IDEs such as Eclipse or IntelliJ IDEA

We can also use multiple languages and platforms to create our functions:

- Java/Maven
- Python
- C# (.NET Core and Scripts)
- JavaScript/Node
- F# (.NET Core)

As you can see, multiple platform and development environment combinations can be used to create Azure Functions. This means that any operating system including Windows, macOS, and Linux can be used as a development station.

Implementing Azure Functions

Let's demonstrate the cross-platform development toolset using macOS by creating our example functions using the Azure function's core tools:

1. In order to install the platform runtime, we will first register the `azure/functions` repository:

   ```
   brew tap azure/functions
   ```

2. Once the functions repository has been registered, we can continue with the installation of the Azure Functions core tools:

   ```
   brew install azure-functions-core-tools
   ```

   Once the installation is complete, we can continue with the development of our sample functions:

```
● ● ●                          ■ python — -bash — 116×13
═══> Tapping azure/functions
Cloning into '/usr/local/Homebrew/Library/Taps/azure/homebrew-functions'...
remote: Enumerating objects: 7, done.
remote: Counting objects: 100% (7/7), done.
remote: Compressing objects: 100% (7/7), done.
remote: Total 7 (delta 0), reused 5 (delta 0), pack-reused 0
Unpacking objects: 100% (7/7), done.
Tapped 1 formula (34 files, 30.7KB).
Cans-MacBook-Pro:python can.bilgin$ brew install azure-functions-core-tools
═══> Installing azure-functions-core-tools from azure/functions
═══> Downloading https://functionscdn.azureedge.net/public/2.5.553/Azure.Functions.Cli.osx-x64.2.5.553.zip
######################################################################## 100.0%
🍺 /usr/local/Cellar/azure-functions-core-tools/2.5.553: 1,151 files, 466.9MB, built in 2 minutes 5 seconds
```

3. To demonstrate functions, we will create a simple calculator function (that is, $x + y = z$). This function could be initialized as follows on Python. First, we will initialize a virtual work environment:

```
Cans-MacBook-Pro:python can.bilgin$ python3 -V
Python 3.6.4
Cans-MacBook-Pro:python can.bilgin$ python3 -m venv .env
Cans-MacBook-Pro:python can.bilgin$ ls
.env
Cans-MacBook-Pro:python can.bilgin$ source .env/bin/activate
(.env) Cans-MacBook-Pro:python can.bilgin$
```

> At the time of writing, python runtime is still in public preview and versions of python newer than 3.6 are not supported.

4. Once the environment is created and activated, we can continue with the initialization of our function project:

```
(.env) Cans-MacBook-Pro:python can.bilgin$ func init
myazurefunctions
Select a worker runtime:
1. dotnet
2. node
3. python (preview)
Choose option: 3
python
Writing .funcignore
Writing .gitignore
Writing host.json
Writing local.settings.json
Writing
/Volumes/Data/book/functions/python/myazurefunctions/.vscode/extens
ions.json
```

5. Now that the project has been created, we can create a new function called add:

```
(.env) Cans-MacBook-Pro:python can.bilgin$ cd myazurefunctions/
(.env) Cans-MacBook-Pro:myazurefunctions can.bilgin$ func new
Select a template:
1. Azure Blob Storage trigger
...
9. Timer trigger
Choose option: 5
HTTP trigger
Function name: [HttpTrigger] add
Writing
/Volumes/Data/book/functions/python/myazurefunctions/add/sample.dat
Writing
/Volumes/Data/book/functions/python/myazurefunctions/add/__init__.p
y
Writing
/Volumes/Data/book/functions/python/myazurefunctions/add/function.j
son
The function "add" was created successfully from the "HTTP trigger"
template.
```

6. Now that the function has been created, you can use any editor to edit the `__init__.py` file in order to implement the function, as follows:

```
GNU nano 2.0.6                    File: addfunction/__init__.py

import logging

import azure.functions as func


def main(req: func.HttpRequest) -> func.HttpResponse:
    logging.info('Python HTTP trigger function processed a request.')

    reqX = req.params.get('x')
    reqY = req.params.get('y')

    logging.info(f"Received Parameters: {reqX} and {reqY}")

    if reqX and reqY:
        x = int(reqX)
        y = int(reqY)

        result = x + y
        logging.info(f"Result is {result}")

        return func.HttpResponse(f"Addition result is {result}")
    else:
        return func.HttpResponse(
            "Please pass query parameters for 'x' and 'y'",
            status_code=400
        )


^G Get Help    ^O WriteOut    ^R Read File   ^Y Prev Page   ^K Cut Text    ^C Cur Pos
^X Exit        ^J Justify     ^W Where Is    ^V Next Page   ^U UnCut Text  ^T To Spell
```

7. Within the project directory, executing the following command would run the local function server:

```
func host start
```

8. Now, the function server is running and executing a get query to the given port that is displayed on the Terminal window. This would trigger the HTTP request and return the result:

```
curl 'http://localhost:7071/api/add?x=5&y=8'
Addition result is 13
```

If, in the creation step, we had chosen the first option, that is, **1. dotnet**, the project would have been created with a compiled C# function template:

```
Cans-MacBook-Pro:dotnet can.bilgin$ func init myazurefunctions
Select a worker runtime:
1. dotnet
2. node
3. python (preview)
Choose option: 1
dotnet
Writing
/Volumes/Data/book/functions/dotnet/myazurefunctions/.vscode/extens
ions.json
Cans-MacBook-Pro:dotnet can.bilgin$ cd myazurefunctions
Cans-MacBook-Pro:myazurefunctions can.bilgin$ func new
Select a template:
1. QueueTrigger
2. HttpTrigger
...
12. IotHubTrigger
Choose option: 2
Function name: add
add
The function "add" was created successfully from the "HttpTrigger"
template.
Cans-MacBook-Pro:myazurefunctions can.bilgin$ ls
add.cs    host.json   local.settings.json myazurefunctions.csproj
Cans-MacBook-Pro:myazurefunctions can.bilgin$ nano add.cs
Cans-MacBook-Pro:myazurefunctions can.bilgin$ func host start
```

Our source code for the add function would look similar to the following:

```
GNU nano 2.0.6                              File: add.cs

using System;
using System.IO;
using System.Threading.Tasks;
using Microsoft.AspNetCore.Mvc;
using Microsoft.Azure.WebJobs;
using Microsoft.Azure.WebJobs.Extensions.Http;
using Microsoft.AspNetCore.Http;
using Microsoft.Extensions.Logging;
using Newtonsoft.Json;

namespace myazurefunctions
{
    public static class add
    {
        [FunctionName("add")]
        public static async Task<IActionResult> Run(
            [HttpTrigger(AuthorizationLevel.Function, "get", "post", Route = null)] HttpRequest req,
            ILogger log)
        {
            log.LogInformation("C# HTTP trigger function processed a request.");

            string reqX = req.Query["x"];
            string reqY = req.Query["y"];

            log.LogInformation($"Received Parameters: {reqX} and {reqY}");

            if(!string.IsNullOrEmpty(reqX) && !string.IsNullOrEmpty(reqY))
            {
                var x = int.Parse(reqX);
                var y = int.Parse(reqY);

                var result = x + y;

                log.LogInformation($"Result is {result}");

                return (ActionResult)new OkObjectResult($"Addition result is {result}");
            }
        }
```

Of course, while this works for demonstration purposes, development with Visual Studio (for both macOS and Windows) as well as Visual Studio Code would provide the comfort of an actual .NET development environment.

Function triggers and bindings

The function projects and the functions that we created in the previous section, in spite of the fact that they are implemented and executed on completely separate runtimes, carry function manifests that adhere to the same schema (that is, `function.json`). While the manifest for Python implementation can be found in the folder carrying the same name as the function, the dotnet version is only generated (can be found in the `bin/output/<function>` folder) at compile time from the attributes used within the implementation. Comparing the two manifests, we can immediately identify the respective sections where the input, output, and triggering mechanisms are defined for these functions:

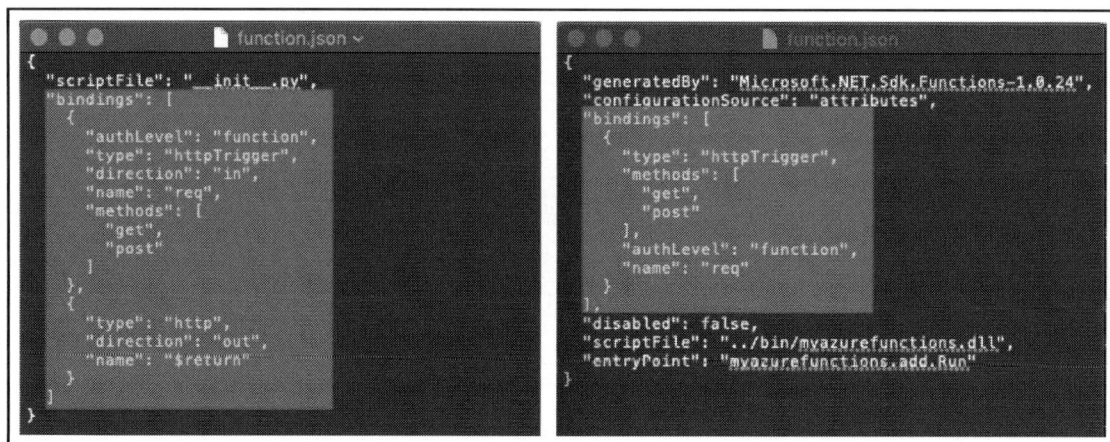

As we mentioned earlier, Azure Functions are event-triggered Azure resources. These trigger mechanisms define not only when the function is going to be executed, but also the input and output data types and/or connected services. For instance, a blob storage entry can be used as a trigger, as well as input and output. Similarly, `HttpTrigger` can define the execution method, as well as the input and out mechanisms (like in the previous examples). This way, additional service integrations can be included within the function as declarative attributes rather than functional implementations.

Some of these binding types are as follows:

Type	Trigger	Input	Output
Blob storage	✓	✓	✓
Cosmos DB	✓	✓	✓
Event grid	✓		
Event hubs	✓		✓
Http and Webhooks	✓	✓	✓
Microsoft graph OneDrive files		✓	✓
Microsoft graph Outlook email			✓
Microsoft graph events	✓	✓	✓
Queue storage	✓		✓
SendGrid			✓
Service bus	✓		✓
SignalR		✓	✓
Table storage		✓	✓
Timer	✓		
Twilio			✓

In addition to the listed items, there are other extensions available via Azure Core Tools or NuGet packages and, by default, only timer and HTTP extensions are registered in a function runtime.

Configuring functions

Azure Functions use the same configuration infrastructure as ASP.NET Core applications, hence utilizing the `Microsoft.Extensions.Configuration` module.

While the application is running on the local runtime during development, in order to read the configuration values from the `local.settings.json` file, a configuration builder needs to be created and the `AddJsonFile` extension method needs to be used. After the configuration instance is created, the configuration values, as well as the connection strings, can be accessed through the indexer property of the configuration instance.

During deployment to the Azure infrastructure, the settings file is used as a template to create the app settings that will be governed through the Azure portal, as well as the resource manager. These values can also be accessed with the same principle, but they are added as environment variables.

In order to support both scenarios, we can use the extension methods that are available during the creation of the configuration instance:

```
var config = new ConfigurationBuilder()
            .SetBasePath(context.FunctionAppDirectory)
            .AddJsonFile("local.settings.json", optional: true,
            reloadOnChange: true)
            .AddEnvironmentVariables()
            .Build();
```

Hosting functions

Azure Functions, once deployed, are hosted on the App Service infrastructure. In an App Service, as you saw in the previous examples, only the compute and other resources that are used are accumulated toward your bill. In addition, Azure Functions, in a consumption plan, are active only when they are triggered by one of the events that has been configured; hence, Azure Functions can be extremely cost-effective in mission critical integration scenarios. Function resources can also be scaled out and down, depending on the load they are handling.

The second plan that's available for functions is the premium plan. In the premium plan, you have the option to set up always running functions to avoid cold starts, as well as unlimited execution duration. Unlimited duration can come in handy on longer running processes since, by default, Azure Functions have a hard limit of five minutes, which can be extended to 10 minutes with additional configuration.

Creating our first Azure function

One of the patterns that was mentioned previously in the Azure context was the materialized view. For instance, in our initial structure, we have basic information about the auctions that are embedded within the user documents. This way, the auctions can be included as part of the user profile and users can be rated based on their involvement with successful auctions. In this setup, since the auction data on the user profile is just the denormalized data chunks from the main auction table, the services would not need to directly interact with the auctions table.

Let's take a look at the following user story and see how we can implement this solution:

> *"As a solution architect, I would like to implement an Azure Function that will update the Cosmos DB Users Collection with modified auctions data so that the Auctions API can be decoupled from the Users API."*

Our task here would be to implement an Azure function that will be triggered when an auction document is modified. The changes on this document should be propagated to the Users Collection:

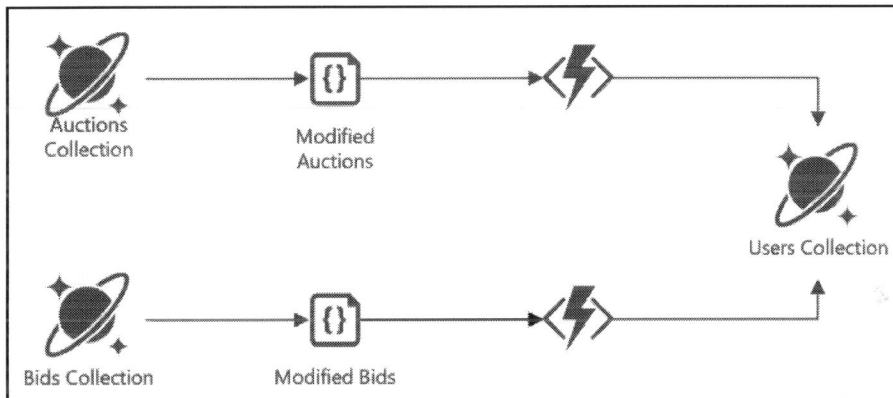

In this setup, we will start by doing the following:

1. First, we will create our **Azure Functions** project, which will be hosted as a functions app in our resource group:

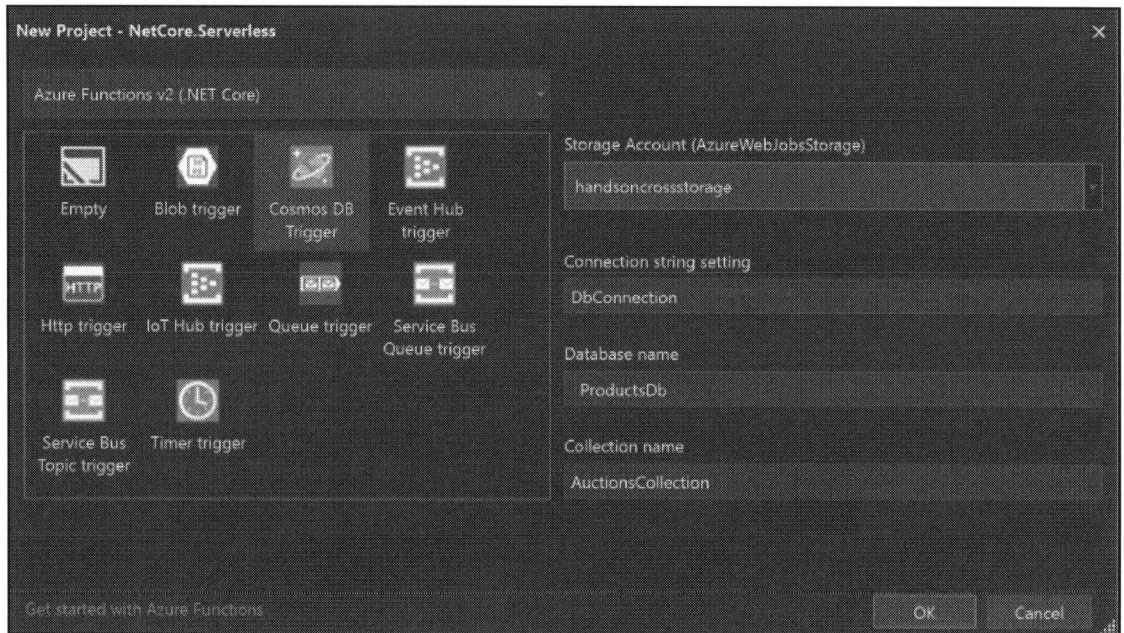

2. This will create our first function with the following declaration:

```
[FunctionName("Function1")]
public static void Run(
    [CosmosDBTrigger(
        databaseName: "ProductsDb",
        collectionName: "AuctionsCollection",
        ConnectionStringSetting = "DbConnection",
        LeaseCollectionName = "leases")]
        IReadOnlyList<Document> input, ILogger log)
```

By using CosmosDBTrigger here, we are instructing the Azure Functions runtime to create a lease so that we can connect to the Cosmos DB change feed on the given database (that is, ProductsDb) and collection (that is, AuctionsCollection) using the set connection string setting (that is, DbConnection).

3. Now, let's expand our configuration to include the given connection string setting:

```
{
  "ConnectionStrings": {
    "DbConnection":
"AccountEndpoint=https://handsoncrossplatform.documents.azure.com:4
43/;AccountKey=...;"
  }
}
```

4. Now, we will create a lease collection that will record the triggers that are registered by our Azure Functions. In order to use a single lease collection, on top of the LeaseCollectionName option, we can also add the LeasePrefix property to our declaration. This way, each lease entry will receive a prefix value, depending on the function declaration. Now, with the additional lease settings, our trigger declaration would look like this:

```
[CosmosDBTrigger(
        databaseName: "ProductsDb",
        collectionName: "AuctionsCollection",
        ConnectionStringSetting = "DbConnection",
        LeaseCollectionPrefix = "AuctionsTrigger",
        LeaseCollectionName = "LeasesCollection")]
```

5. After this, we can already run our function in debug mode and see whether our trigger is working as expected. After updating a document on the AuctionsCollection collection, you will receive the updated data almost immediately:

6. We are now receiving the modified document. If the modifications are based on the incoming data alone, we could have added an output binding with a single document or an `async` collector to modify or insert documents into a specific collection. However, we would like to update a list of auctions that the user is involved in. Therefore, we will get a Cosmos client instance using the attribute declaration:

```
[CosmosDBTrigger(
    databaseName: "ProductsDb",
    collectionName: "AuctionsCollection",
    ConnectionStringSetting = "DbConnection",
    LeaseCollectionPrefix = "AuctionsTrigger",
    LeaseCollectionName =
"LeasesCollection")]IReadOnlyList<Document> input,
[CosmosDB(
    databaseName: "ProductsDb",
    collectionName: "UsersCollection",
    ConnectionStringSetting = "DbConnection")] DocumentClient
client,
```

Now, using the client, we can execute the necessary updates on the Users Document collection.

Developing a Logic App

When tasked with implementing a Logic App, in theory, a developer would not necessarily need anything else other than a text editor since Logic Apps are an extension of ARM resource templates. The manifest for a Logic App consists of four main ingredients:

- Parameters
- Triggers
- Actions
- Outputs

Parameters, triggers, and outputs, similar to the binding concept of Azure Functions, define when and how the application is going to be executed. Actions define what the application should do.

Logic Apps can be created using an IDE with an additional schema and/or visual support, such as Visual Studio, or can be developed solely on Azure portal using the web portal.

Implementing Logic Apps

In order to create a Logic App with Visual Studio, we need to do the following:

1. We would need to use the Azure Resource Group project template and select the Logic App template from the screen that follows:

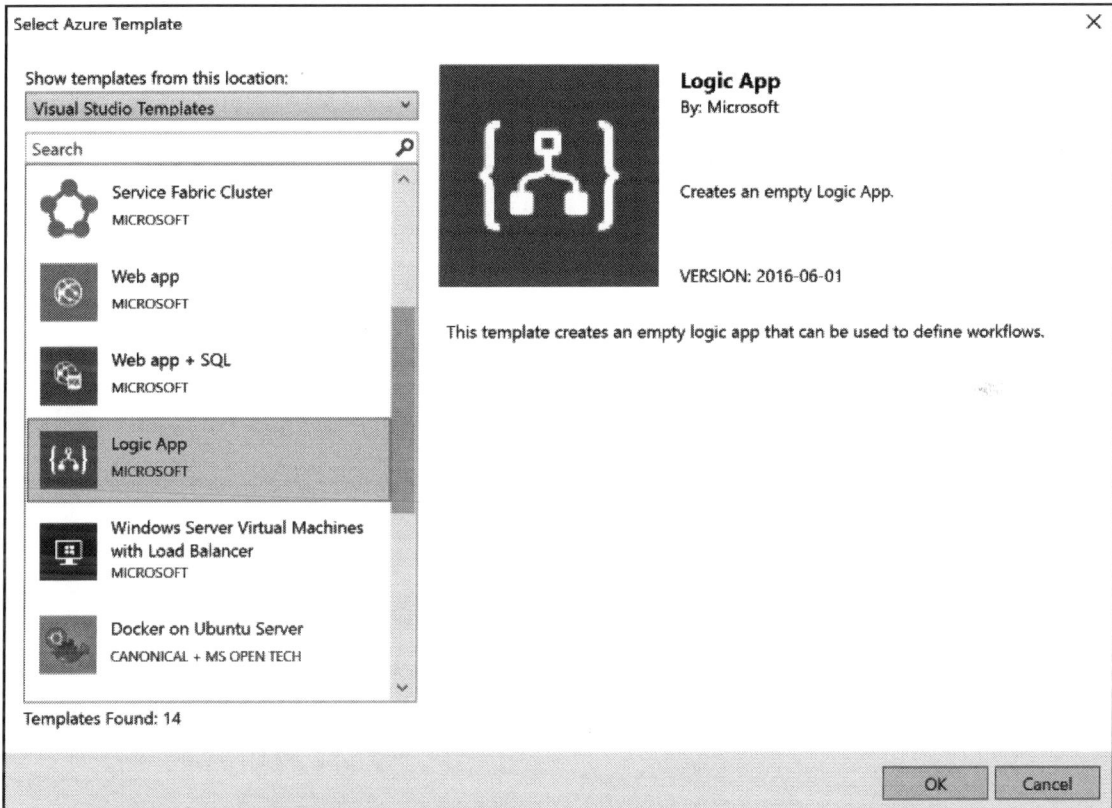

This will create a resource group manifest that contains the Logic App definition. The Logic App can now be modified using the Logic App designer within Visual Studio, given that the Azure Logic Apps Tools extension is installed (right-click on the resource group JSON file and choose **Open with Logic App Designer**).

2. The first step to implementing a Logic App is to select the trigger, which will be the initial step in our workflow. For this example, let's select **When a HTTP request is received**.

3. Now that the Logic App flow has been created, let's expand the HTTP request and paste a sample JSON payload as the body of a request we are expecting for this application trigger:

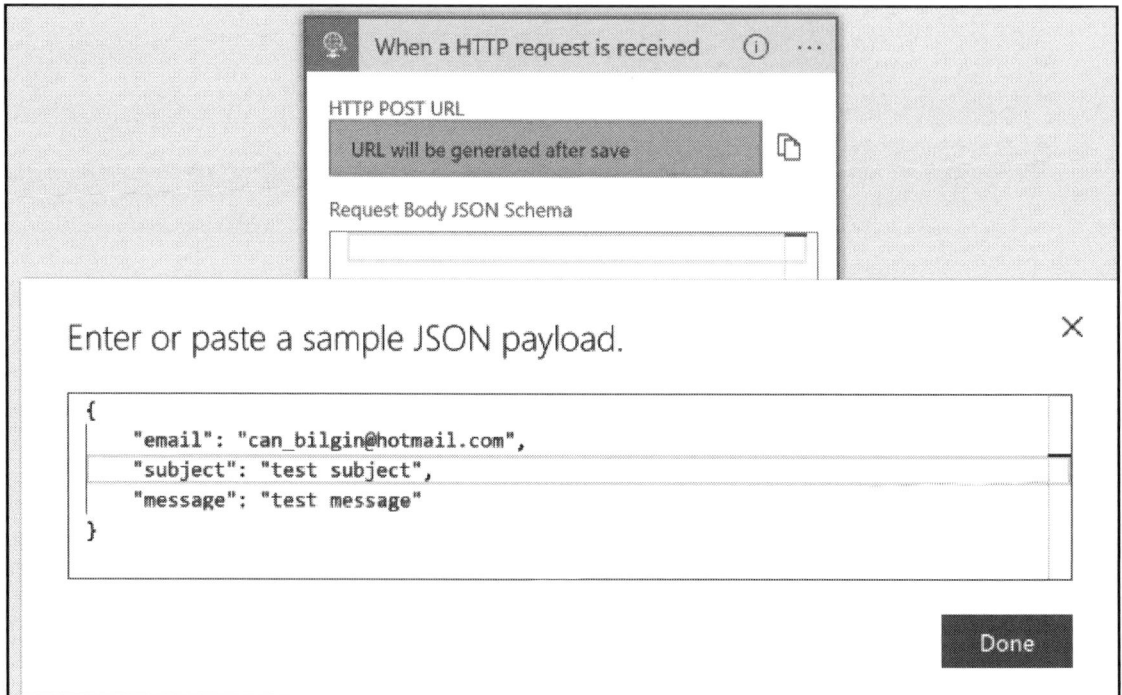

This will generate a **Request Body JSON Schema**. Now, we can send our requests, just like in the sample JSON payload.

4. Next, we will add an action to send an email (there are many email solutions; for this example, we will be using the **Send an email** action using Outlook):

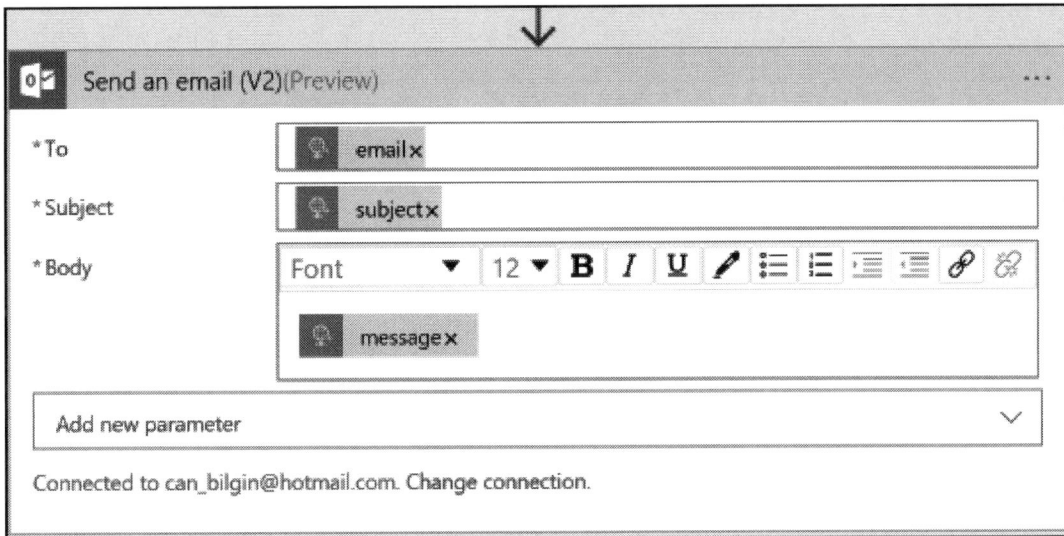

As you can see, we are in fact using the email, subject, and message parameters defined in our trigger to populate the email action.

5. Finally, we can add a **Response** action and define the response header and body. Now, our application is ready to execute:

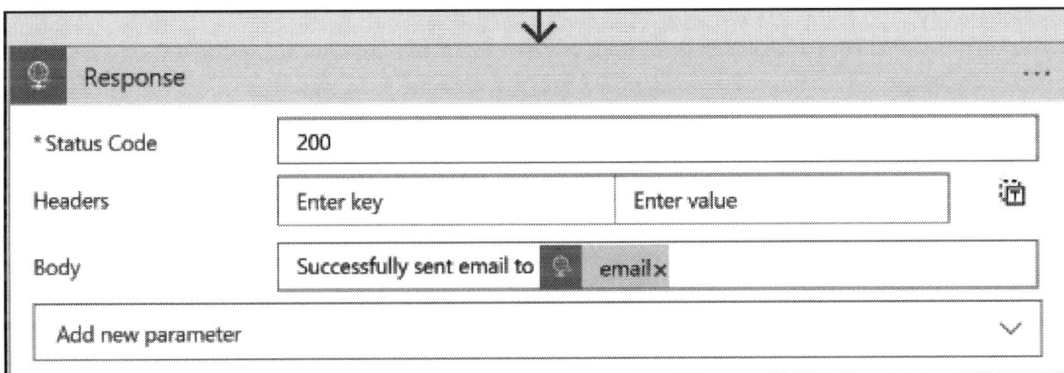

After the Logic App is deployed, you can retrieve the request URL, as well as the integrated security token from the Azure portal designer. Executing a simple POST call with the required parameters would trigger the logic application and trigger actions:

```
curl -H "Content-Type:application/json" -X POST -d
'{"email":"can.bilgin@authoritypartners.com", "title":"Test",
"subject":"Test", "message" : "Test Message"}'
```

```
"https://prod-00.northcentralus.logic.azure.com:443/workflows/5bb--
--------/triggers/manual/paths/invoke?api-
version=2016-10-01&sp=%2Ftriggers%2Fmanual%2Frun&sv=1.0&sig=eOB----
-----------"
```

```
Successfully sent the email to can.bilgin@authoritypartners.com
```

As you can see, using Logic Apps, these types of simple or even more intricate business workflows can be declaratively converted into web services, as well as executed on triggers such as queues, feeds, and web hooks. Connectors are the key components in this setup that serve these actions and the triggers that are available for logic apps.

Using connectors

In the previous example, we were using the HTTP trigger and response actions, as well as the Outlook Send Email action. These actions and triggers are packaged in so-called connectors for the Logic App infrastructure. Connectors are essentially part of a bigger SaaS ecosystem that also involves Microsoft Flow and PowerApps, as well as Logic Apps. Connectors can be described as encapsulated connectivity components for various SaaS offers (for example, email, social media, release management, HTTP requests, file transfer, and so on).

On top of the standard free set of connectors (including third-party connectors), the **Enterprise Integration Pack (EIP)**, which is a premium offering, provides building blocks for B2B enterprise integration services. These integration scenarios generally revolve around the supported industry standards, that is, **Electronic Data Interchange (EDI)** and **Enterprise Application Integration (EAI)**.

It is also possible to create custom Logic Apps connectors in order to implement additional custom use cases that cannot be realized with the available set of connectors.

If/when the actions provided through the available connectors do not satisfy the requirements, Azure Functions can be integrated as tasks into Logic Apps. This way, any custom logic can be embedded into the workflow using .NET Core and simple HTTP communication between Logic Apps and functions.

Creating our first Logic App

The main service application, so far, is built to accommodate the main application use cases and provide data for users so that they can create auctions and user profiles, as well as bid on auctions. Nevertheless, we need to find more ways to engage users by using their interests. For this type of engagement model, we can utilize various notifications channels. The most prominent of these channels is a periodic notification email setup.

The user story we will use for this example is as follows:

> *"As a product owner, I would like to send out periodic emails to registered users if there are new auctions available, depending on their previous interests, so that I can engage the users and increase the return rate."*

Before we start implementing our Logic App, in spite of the fact that it is possible to use the Cosmos DB connector to load data, let's create two more Azure Functions to load the users and latest auctions for email targets and content, respectively. Both of these functions should use `HttpTrigger` and should return JSON data as a response. Let's get started:

1. The function that will return the list of users that we will send the notifications to is as follows:

```
[FunctionName("RetrieveUsersList")]
public static async Task<IActionResult> Run(
    [HttpTrigger(AuthorizationLevel.Function, "get", "post", Route
= null)] HttpRequest req,
    ILogger log)
{
    // TODO: Retrieve users from UsersCollection
    var users = new List<User>();
    users.Add(new User{ Email = "can.bilgin@authoritypartners.com",
FirstName = "Can"});

    return (ActionResult)new OkObjectResult(users);
}
```

2. Next, we will need the data for the latest auctions:

```
[FunctionName("RetrieveLatestAuctionsList")]
public static async Task<IActionResult> Run(
    [HttpTrigger(AuthorizationLevel.Function, "get", "post", Route
= null)] HttpRequest req,
    ILogger log)
{
    // TODO: Retrieve latest auctions from AuctionsCollection
    var auctions = new List<Auction>();
    auctions.Add(new Auction
    {
        Brand = "Volvo",
        Model = "S60",
        Year = 2017,
        CurrentHighestBid = 26000,
        StartingPrice = 25000
    });

    return (ActionResult)new OkObjectResult(auctions);
}
```

Now that we have our data feeds ready, we can start implementing our Logic App.

> In this example, we have used Azure Functions to retrieve a set of DTOs in order to be cost-efficient. It it also possible to create a change feed function that will prepare the daily notification feed as soon as the data store is updated with a new user or auction/bid. This way, the Logic App could directly load the data from the daily feed document collection or table storage.

3. In our previous example, we created a Logic App using an HTTP trigger. For this example, let's start with a recurrence trigger so that we can process our data and prepare a periodic feed:

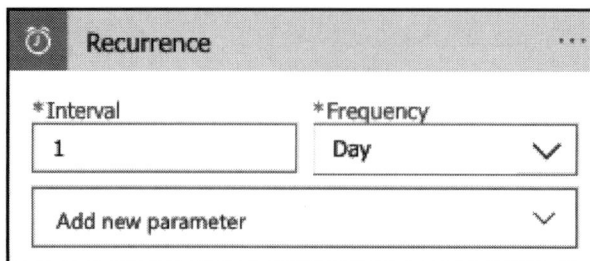

4. Next, let's retrieve the set of users using our Azure function. In order to select a function from the available actions, you should locate the **Azure Functions** action in the **Choose an action** dialog, then select the target function app that contains your functions, and finally, select the desired function:

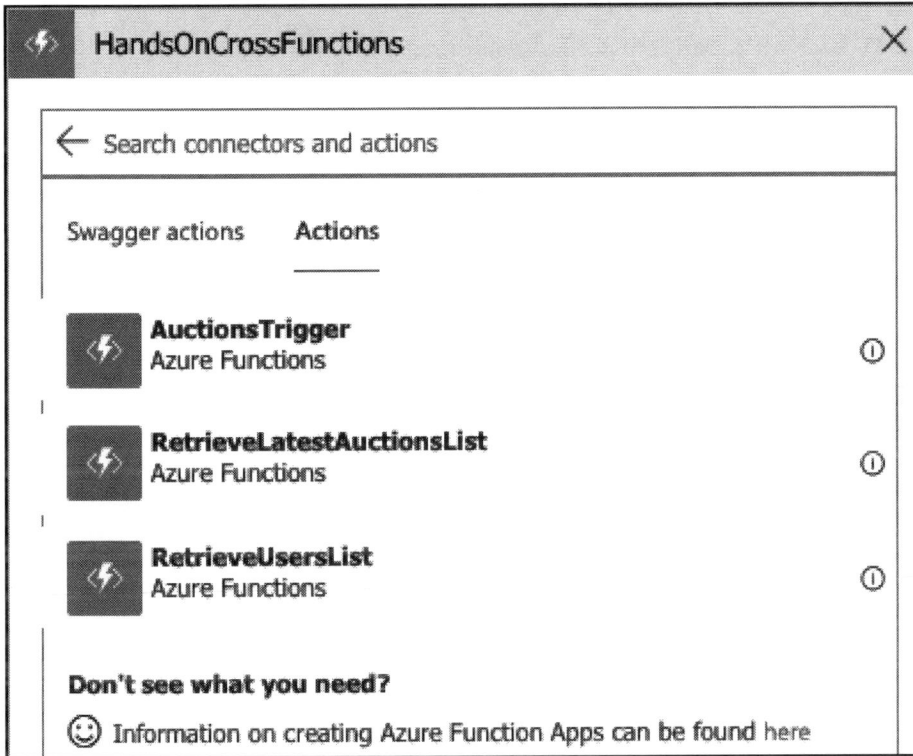

5. Once we have retrieved the results from the Azure function, it will be in JSON format. In order to ease the design and access properties of the contained data items, it would be good to include a data parse action with a predefined schema. At this point, you can have a simple run and copy the results:

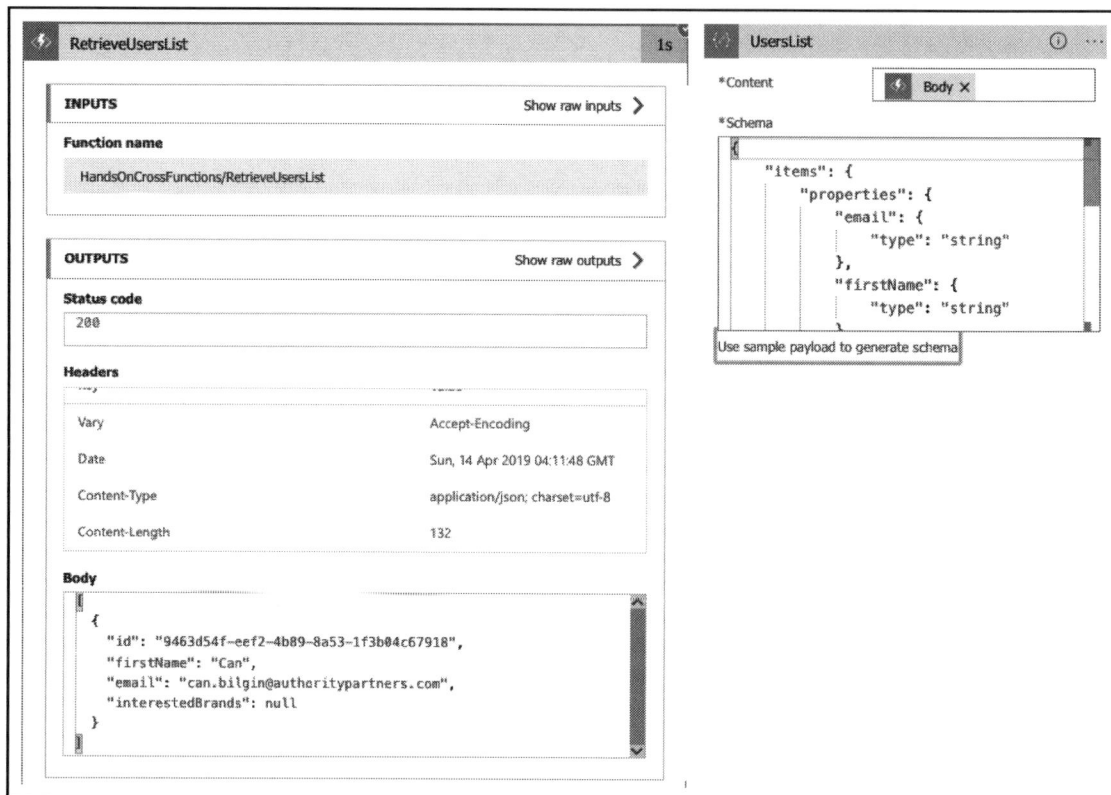

After repeating the same actions for the auctions list data, we can start building out email content. Our current workflow should look similar to the following:

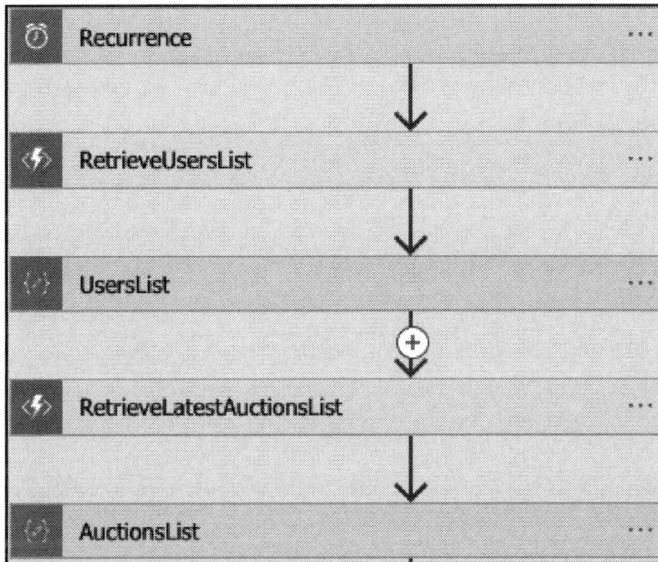

6. Before we continue with preparing the email content and sending it to each user on the list, let's structure the flow a little bit so that the data retrieval actions for users and auctions are not executed sequentially:

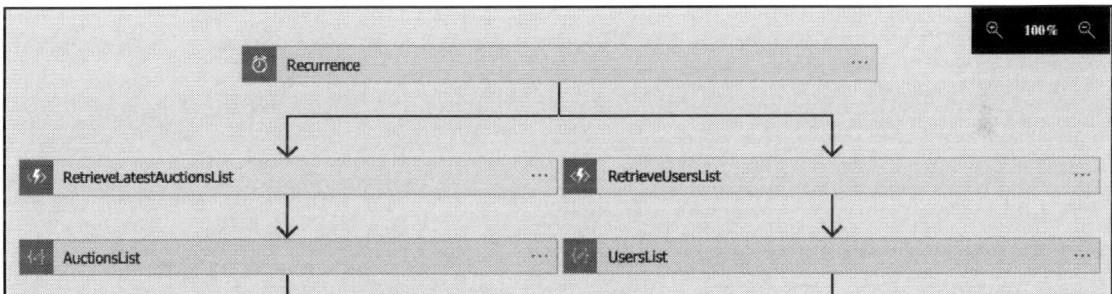

Now, we can continue with additional control statements and the preparation of the email content.

Workflow execution control

By definition, being an orchestration tool, Logic Apps utilities control statements such as foreach, switch, and conditionals in order to compose a controlled workflow using available actions. These control statements can be used as actions within the workflow using the input and output values of other actions within the Logic App context. The available set of control statements are as follows:

- **Condition**: Used to evaluate a condition and define two separate paths, depending on the result
- **Foreach**: Used to execute a path of dependent actions for each item in a sequence
- **Scope**: Used to encapsulate a block of actions
- **Switch**: Used to execute multiple separate action blocks, depending on the switch input
- **Terminate**: Terminates the current execution of the Logic App
- **Until**: Used as a while loop, where a block of actions are executed until a defined condition evaluates to true

These statements can be accessed using the **Control** action within the **Logic App Designer**:

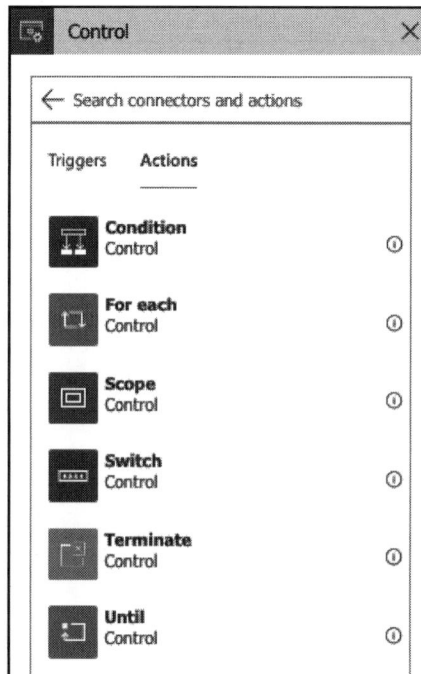

In our example, we are supposed to send an email to each user with the latest auctions list. We can achieve this with a for each action on the list of users:

As you can see, we are using the body of the **UsersList** action (that is, `body(UsersList)`, using Logic App notation), and for each item in the list, we are retrieving the email (that is, `items('For_each')['email']`) and `firstName`. In a similar manner, we can prepare the auction's email body and assign the result as the body of the subject. In addition to this simple setup, the content can be filtered according to the interest of the user using the data operations that are available:

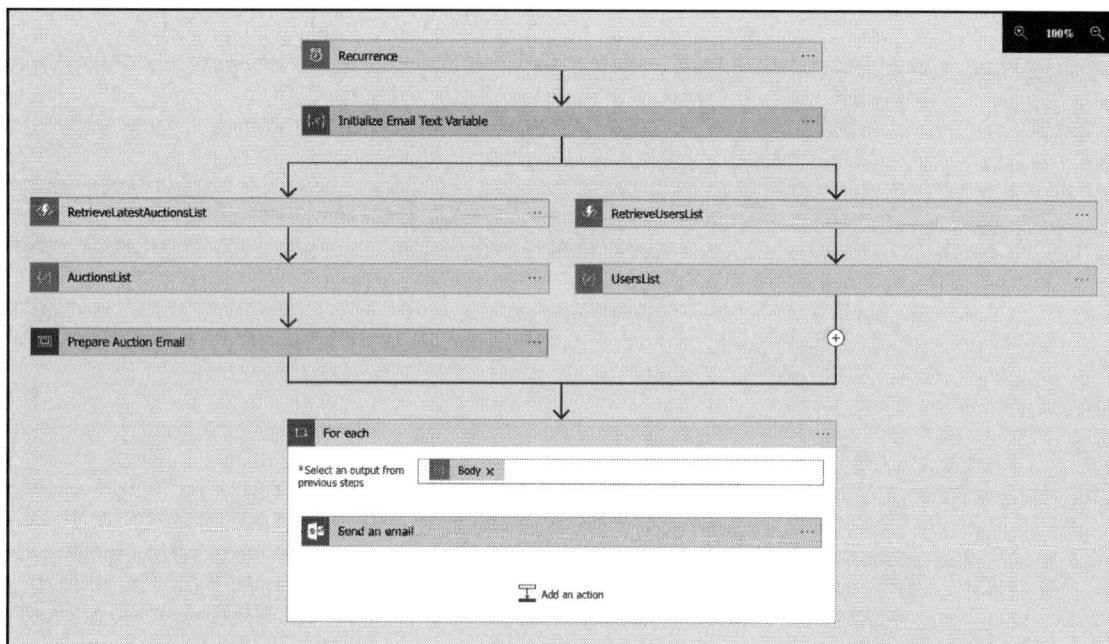

Now, we will be periodically sending auction updates to users without having to compromise or add additional complexity to our current service infrastructure.

Integration with Azure services

So far, we have only utilized Cosmos DB in the context of Logic Apps and Azure Functions among the many Azure services that we can integrate with Azure Serverless components.

As you have seen, these integrations are available through bindings for Azure Functions and through connectors for Logic Apps. Using this integrated business model, multiple architectural patterns can be composed and event-driven scenarios can be accomplished.

Let's take a deeper look at some of these integrated services.

Repository

In the Azure Serverless context, it is fair to say that almost all Azure repository models are tightly integrated with the infrastructure. Let's take a look at the following models:

- **Cosmos DB**: This has an available binding for Azure Functions. This is a connector with various available actions to execute mainstream CRUD actions, as well as advanced scenarios to create, retrieve, and execute stored procedures. Cosmos DB can also be used to trigger Azure Functions.
- **SQL server**: This is another repository service that can be integrated into the Logic Apps with the available connector, hence allowing triggers such as an item being created or modified. Logic Apps can also use the SQL and **Azure SQL Data Warehouse** connectors to execute both raw and structured queries on SQL instances. Additionally, SQL connections can be initialized within the Azure Functions using nothing but the native SQL client that's available as part of .NET Core.
- **Azure table storage**: This is another repository model that can be integrated with Azure Serverless components. Table storage tables can be used as input parameters and can be the receivers of new entities as part of an output configuration within the Azure Function infrastructure. The available connector for the Logic Apps can be utilized to execute various actions on a table storage:

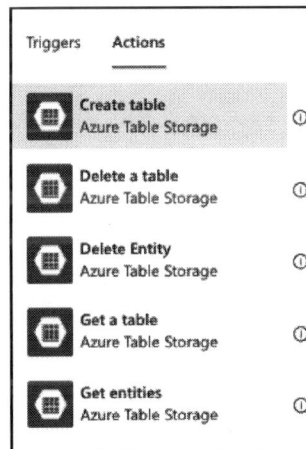

- **Azure blob storage**: This can be used as trigger for both functions and Logic Apps. The available function binding and app connector provide a variety of tasks and binding elements that can be utilized within the serverless app model.

Queue-based processing

In order to implement the aforementioned queue-based load leveling pattern, Azure distributed systems can utilize Azure queues, as well as the **Azure Service Bus**. In this way, various asynchronous processing patterns can be implemented.

Azure queues can be configured to trigger functions and Logic Apps. Both the binding and the connector have available actions so that they can listen to a specific queue. The connector contains actions for executing basic operations, including but not limited to the creation of message queues, the insertion of messages, and the retrieval of messages. An Azure message queue can also be used as an output target of an Azure function to create a message in the configured queue.

The connector and binding for the Azure Service Bus have an extended set of available actions and various triggers. Queues and topics are both available for listening for new messages as part of the trigger setup. Logic Apps can also execute almost all possible actions that a managed client can achieve through operations related to managing basic messages, dead-letter queues, locks, topics, and subscriptions.

Event aggregation

The citizen of the Azure Serverless ecosystem, **Event Grid,** is the most appropriate candidate for implementing the classic publisher/subscriber model between distributed service components, especially when Azure Functions and Logic Apps are involved. In addition to Event Grid, Event Hub is the best choice for big data pipelines that involve event streaming rather than discreet event distribution.

Event Grid aggregates the events that are collected from various so-called Event Sources, such as container registries, resource groups, service bus, and storage. Event Grid can also consume and deliver custom topics that are posted by capable components. Aggregated events are then dispersed to the registered consumers or so-called event handlers. Event handlers for Event Grid include, but are not limited to, the following:

- Azure automation
- Azure Functions
- Event Hubs

- Hybrid connections
- Logic Apps
- Microsoft flow
- Queue storage
- WebHooks

This infrastructure means that developers aren't limited by the available triggers for functions and Logic Apps as the initial point of a certain mission critical scenario. They can also create a complete event-driven subscription model.

Event hubs can be integrated as a consumer for Event Grid events and used as triggers and output for Azure Functions. A connector is available for Logic Apps with trigger and actions. The event hub, when used together with Azure Functions, can create an extremely agile scaling model for processing big data streams.

Summary

Overall, Azure Functions and Logic Apps, as part of the Azure Serverless platform, provide ad hoc, event-based solutions to fill in the gaps in any distributed cloud application. In this chapter, we have analyzed the available development options for Azure Functions. We have implemented simple Azure Functions to denormalize data on a Cosmos DB setup. Finally, we created Logic Apps by utilizing the out of the box connector tasks, as well as Azure Functions with HTTP and periodic triggers.

This chapter finalizes the Azure cloud services-related topics. In the following chapters, we will take a look at more advanced topics to improve the integration between Xamarin applications and the cloud-based service backend.

Section 4: Advanced Mobile Development

4

For developers who are not satisfied with the bare minimum, terms such as responsive, engaging, and asynchronous become the differentiating factors for a mobile application. Providing the patterns and tools to create an application that will provide a fast and fluid user interface and engage the user with a personalized approach is one of the goals of the upcoming chapters.

The following chapters will be covered in this section:

- Chapter 11, *Fluid Applications with Asynchronous Patterns*
- Chapter 12, *Managing Application Data*
- Chapter 13, *Engaging Users with Push Notifications and Graph APIs*
- Chapter 14, *Introducing Cognitive Services*

11
Fluid Applications with Asynchronous Patterns

One of the key attributes of an attractive mobile application is its responsiveness. Applications that do not interfere with the interaction of the user and, instead, maintain the rendering and execution of user gestures in a smooth manner are more desirable by users. In order to achieve fast and fluid application norms, together with performance, asynchronous execution patterns come to the rescue. When developing Xamarin applications, as well as ASP.NET Core, both the task's framework and reactive modules can help distribute the execution threads and create a smooth and uninterrupted execution flow. This chapter will go over some of the patterns associated with these modules and apply them to various sections of the application.

The following sections will walk you through some key implementation scenarios for asynchronous execution:

- Utilizing tasks and awaitables
- Asynchronous execution patterns
- Native asynchronous execution

Utilizing tasks and awaitables

User experience (UX) is a term that is used to describe the composition of UI components and how the user interacts with the UI components. In other words, UX is not only how the application is designed, but rather the impression of the user about the application. In this context, the responsiveness of the application is one of the key factors that defines the quality of the application.

In general terms, a simple interaction use case starts with user interaction. This interaction can be a tap on a certain area on the screen, a certain gesture on a canvas, or an actual user input in an editable field on the screen. Once the user interaction triggers the execution flow, the application business logic is responsible for updating the UI to notify the user about the result of their input.

As you can see, in the asynchronous version of the simple interaction model, the application starts the execution of the designated business flow and doesn't wait for it to complete. In the meantime, the user is free to interact with other sections of the UI. Once the result is available, the application UI is notified about the completion.

While this interaction model defines and satisfies simple execution scenarios (for example, validating an email field with a regular expression, or displaying a fly out to show the desired details on an item), since the interaction model as well as the business logic get more complex and additional dependencies come into the picture, such as web services, we should keep the user appraised about the work the application is doing (for example, a progress bar while downloading a remote resource). For this purpose, we can extend our interaction model to provide continuous updates to the user:

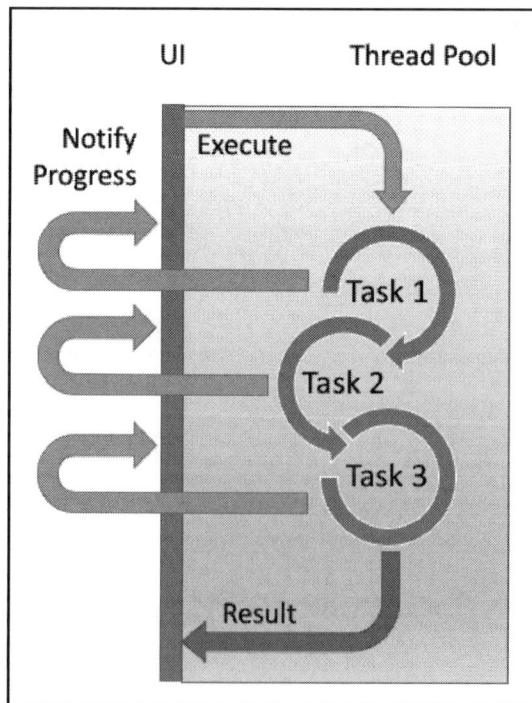

Now, the UI is continuously receiving updates from the background process. These updates can be as simple as a busy signal for a loader ring or data updates for sophisticated completion percentage components. However, this pattern raises another question as to how the application UI is going to handle multiple updates coming in from background processing. Before we can answer this question, let's take a closer look at the application UI infrastructure and task-based execution.

Task-based execution

An application UI, regardless of the platform it is implemented on, always follows a single-threaded model. Even if the underlying platform or the hardware supports multithreading, the runtime is responsible for providing a single dispatcher to render the UI in order to avoid multiple threads trying to update the same section of the screen at the same time.

In this single-threaded model, it is the application's responsibility to lay off the background processing to child threads and synchronize back to the UI thread.

The .NET Framework introduced the task-based threading model, also referred to as the **Task Asynchronous Programming** (**TAP**) model, in .NET 4.0, and since then it has become a norm for asynchronous execution – especially on mobile platforms such as Xamarin.

In simple terms, TAP provides an abstraction over the classic threading model. In this approach, the developer and, implicitly, the application, are not directly responsible for handling the thread's creation, execution, and synchronization; but simply for the creation of asynchronous work blocks (that is, tasks), allowing the underlying runtime to handle the heavy lifting. Especially considering the fact that .NET Standard is a complete abstraction for various runtimes, such as .NET Core and Mono, this abstraction allows each platform to implement the platform-appropriate handling of multithreading. This is one of the main reasons why the `Thread` class is not available in cross-platform modules, whereas platform-specific framework modules (for example, Xamarin.iOS and Xamarin.Android) provide access to the classic threading model.

A simple asynchronous block can be created using the static helper methods that are available in the `Task` class:

```
Task.Run(() =>
{
    // Run code here
})
```

In this example, we are wrapping a synchronous block of code inside a task. Now, the declaring method can return the created task block. Alternatively, if there are other asynchronous blocks, it should execute this block with an `await` keyword, thus creating an `async` method:

```
public Task SimpleyAsyncChain()
{
    return Task.Run(...);
}

public async Task MyAsyncMethod()
{
    var result = await Task.Run(...);
    await OtherAsyncMethod(result);
    // example async method
    await Task.Delay(300);
}
```

Either implementation in this example creates an asynchronous chain of methods that can be awaited on the top level. Exception handling can also be introduced using simple try/catch blocks, which is no different than using synchronous code:

```
public async Task<MyEntity> MyAsyncMethodWithExceptionHandling()
{
    MyEntity result = null;

    try
    {
        result = await Task.Run(...);
    }
    catch(Exception ex)
    {
        // TODO: Log the exception
    }

    return result;
}
```

While tasks can be executed sequentially, which is done in the `MyAsyncMethod` method, if there are no dependencies between the asynchronous blocks, they can also be executed in parallel, allowing the runtime to utilize multithreading as much as possible:

```
public async Task MyParallelAsyncMethod()
{
    var result = await Task.Run(...);
    await Task.WhenAll(OtherAsyncMethod(result), Task.Delay(300));
}
```

Using this foundation provided by the TAP model, let's take a look at the following user story:

> *"As a registered user, I would like to have a view dedicated for my profile so that I can see and verify my public information within the application."*

Perhaps the most prominent use of task-based methods is when the application needs to interact with a remote backend (for example, a RESTful web service). However, tasks are deeply integrated and the .NET Framework is the de facto way of handing multithreading. For instance, the starting point of a service proxy client would be to create a simple REST client that would execute various HTTP methods against the target API endpoint.

Before we implement our rest client, we need to define the client interface:

```
public interface IRestClient
{
    Task<TResult> GetAsync<TResult>(string resourceEndpoint, string id)
    where TResult : class;

    Task<TEntity> PostAsync<TEntity>(string resourceEndpoint, TEntity
    entity) where TEntity : class;

    Task<TEntity> PutAsync<TEntity>(string resourceEndpoint, string id,
    TEntity entity) where TEntity : class;

    Task<TResult> DeleteAsync<TResult>(string resourceEndpoint, string
    id) where TResult : class; }
```

We can extend this interface with more specialized methods, such as the GetListAsync method, which can be helpful for serializing a list of items:

```
Task<IEnumerable<TResult>> GetListAsync<TResult>(string resourceEndpoint)
where TResult : class;
```

Now, the implementation of these methods can use the simple HttpClient method to execute the remote call and some type of serialization/deserialization:

```
public async Task<TResult> GetAsync<TResult>(string resourceEndpoint,
string id)
    where TResult : class
{
    var request = new HttpRequestMessage(HttpMethod.Get, $"
    {resourceEndpoint}/{id}");
    var response = await _client.SendAsync(request);

    if (response.IsSuccessStatusCode)
    {
```

```
        var content = await response.Content.ReadAsStringAsync();
        return JsonConvert.DeserializeObject<TResult>(content);
    }

    // TODO: Throw Exception?
    return null;
}
```

Here, the client member field is initialized in the constructor of RestClient, possibly with a base URL declaration, as well as with additional HTTP handlers:

```
public RestClient(string baseUrl)
{
    // TODO: Insert the authorization handler?
    _client = new HttpClient();
    _client.BaseAddress = new Uri(baseUrl);
}
```

Using RestClient, we can then create another level of abstraction to implement the API-specific method calls that will convert the data transformation objects into the domain entities:

```
public async Task<User> GetUser(string userId)
{
    User result = null;

    // Should we initialize the client here? Is UserApi client going to
    //be singleton?
    var client = new RestClient(_configuration["serviceUrl"]);

    try
    {
        var dtoUser = await client.GetAsync<User>
        (_configuration["usersApi"], userId);
        result = User.FromDto(dtoUser);
    }
    catch (Exception ex)
    {
        // TODO:
    }

    return result;
}
```

As you can see, we have a chain of asynchronous methods finally executing a remote call. We now have to connect the user API retrieval call to our view model, which should immediately load the associated user data to be displayed on the target view. In this use case, the user interaction that triggers the business flow is possibly the user tapping on the user profile link. The application responded by navigating to the target view, which initialized the view model. The view model, in turn, requested the remote data for the user profile:

```
public async Task RetrieveUserProfile()
{
    if (string.IsNullOrEmpty(NavigationParameter))
    {
        // TODO: Error/Exception
    }

    var userId = NavigationParameter;
    var userResult = await _usersApi.GetUser(userId);

    CurrentUser = userResult;
}
```

Once the `CurrentUser` property is set, the view would be notified and updated to display the retrieved information.

This implementation would work in a simple asynchronous chain since the async/await construct provided by the language is converted into a state machine during the compilation process that ensures that the asynchronous threads yield back into the UI thread so that the view model updates can be propagated back to the UI.

If we want to make sure that the execution of the user data assignment takes place on the UI thread, we can use the `InvokeOnMainThread` method, instructing the runtime to execute a block of asynchronous code on the main UI thread:

```
Device.BeginInvokeOnMainThread (() => {
    CurrentUser = userResult;
});
```

The main thread invocation becomes an essential part of the asynchronous chain when we're dealing with multiple synchronization contexts.

Synchronization context

When dealing with asynchronous method calls using TAP, it is important to understand that `async` and `await` are language constructs provided by C# and that the actual multithreaded execution is the injected compiler-generated code that's used to replace the async/await blocks. If the compiler-generated asynchronous state machine is closely observed and the `async` method builder is analyzed, you will notice that at the start of any asynchronous await call, the current synchronization context is captured and, later on, when the asynchronous operation is completed, it is used again to execute the continuation action.

In a Xamarin application, the synchronization context – which is similar to the execution context – refers to the thread that the current asynchronous block is being called from, as well as the target thread that the current asynchronous block should yield to. If an ASP.NET core application is under the magnifying glass, the synchronization context would be referring to the current `HttpRequest` object. In certain cases, the thread pool might be acting in the synchronization context.

As we have mentioned, the previous asynchronous example for `UserProfile` would, in fact, be used in the UI thread, since the captured context at the beginning of the execution would probably be for the main UI thread. Once the retrieve operation is completed, the continuation actions (that is, the assignment of the result to the view-model) would be executed on the main thread. Nevertheless, yielding the asynchronous methods back to the UI could cause performance penalties and even deadlocks if the await chains are not handled properly. In a catastrophic scenario, the UI thread might end up waiting for an asynchronous block, which, in turn, waits for the UI thread to yield back to. In order to avoid such occurrences, it is highly advised to utilize explicit control of the captured context and yield the target context using the `ConfigureAwait` method. Moreover, especially in native mobile applications, free the UI thread from any long-running task synchronization using `ConfigureAwait(false)` (that is, do not yield back to the captured context). This ensures that the asynchronous methods are not merged back to the UI thread and that the asynchronous compositions are handled within a separate thread pool. For instance, let's say we were to add an additional asynchronous method to the `async` chain from the previous example:

```
var userResult = await _usersApi.GetUser(userId).ConfigureAwait(false);
var additionalUserData =
_usersApi.GetUserDetails(userId).ConfigureAwait(false);

CurrentUser = userResult;
```

Unlike the previous example, the last statement in this method (the continuation action) would be executed on a different thread than the UI thread. The first asynchronous call would not yield back to the UI thread because of the `ConfigureAwait` creating a secondary synchronization context. This secondary context would then be used as the captured context for the second asynchronous call, to which it would yield back to. Finally, the statement to assign the result would be executed on this secondary context. This would mean that without the `BeginInvokeOnMainThread` helper execution, the UI would likely not be updated with the incoming data.

Single execution guarantee

Another popular area of implementation for asynchronous tasks are the commands that are exposed through the view models throughout the mobile application. If the business flow that is to be executed as a response to a user input (for example, a submit button executing an update call on the user profile) depends on an asynchronous code block, then the command should be implemented in such a way that you can invoke the asynchronous function properly.

Let's demonstrate this with our existing view model. First, we would need to implement our internal execution method:

```
public async Task ExecuteUpdateUserProfile()
{
    try
    {
        await _usersApi.UpdateUser(CurrentUser);
    }
    catch (Exception ex)
    {
        // TODO:
    }
}
```

Now, let's declare our command:

```
public ICommand UpdateUserCommand
{
    get
    {
        if (_updateUserCommand == null)
        {
            _updateUserCommand = new Command(async () => await
            ExecuteUpdateUserProfile());
        }
```

```
            return _updateUserCommand;
        }
    }
```

At this point, if the command is bound to a user control such as a button, multiple taps on the button would result in multiple executions of the same command. While this might not cause any issues on the business flow (that is, the user will be updated with the current data multiple times), it could cause performance degradation and unnecessary resource consumption on the service side.

Locks and monitors as well as mutex implementations, which we are familiar with from classical threading, can also be implemented in task-based asynchronous code blocks using SemaphoreSlim. The main usage of SemaphoreSlim can be summarized as throttling one or more asynchronous blocks.

For this scenario, we can initialize a semaphore with only one available slot:

```
    private static readonly SemaphoreSlim Semaphore = new SemaphoreSlim(1);
```

In the execution block of the command method, we can check whether there is any lease on the current semaphore. If there isn't, we put a lease on one slot, and release it once the command's execution is complete:

```
    public async Task ExecuteUpdateUserProfile()
    {
        if (Semaphore.CurrentCount == 0)
        {
            return;
        }

        await Semaphore.WaitAsync().ConfigureAwait(false);

        try { ... } catch { ... }
        Semaphore.Release();
    }
```

This way, the command cannot be executed more than once at the same time, thus avoiding any data conflicts. It is important to note here that since the semaphore count is released after the command's execution, it is a must to use a try/catch block in order to avoid the semaphore being locked after an error occurs.

Logical tasks

In the retrieve example, we executed the view-model data assignment block within a
`BeginInvokeOnMainThread` block. While this actually guaranteed that the view-model
change will be propagated to the UI thread, with this type of an execution, we cannot really
say that once the asynchronous method that's awaiting the execution is complete, and that
the view-model has been updated. Moreover, the UI execution block could have used
another asynchronous code block (for example, to show a popup once the data is retrieved).
In this type of a situation, we can utilize a task completion source so that we have more
stricter control over when the asynchronous code block is truly completed:

```
public async Task RetrieveUserProfile()
{
    // Removed for brevity
    TaskCompletionSource<int> tcs = new TaskCompletionSource<int>();

    var userResult = await _usersApi.GetUser(userId);

    Device.BeginInvokeOnMainThread(async () => {
        CurrentUser = userResult;
        await ShowPopupAsync(); // async method
        tcs.SetResult(0);
    });

    await tcs.Task.ConfigureAwait(false);
}
```

In this example, we are using `TaskCompletionSource`, which represents the
asynchronous state machine and accepts a result or an exception. This state machine gets a
result only when the UI block execution is completed and the `RetrieveUserProfile`
method is finalized.

`TaskCompletionSource` can also prove useful for describing native UI flows in terms of
asynchronous blocks. For instance, the complete UI flow for a user to pick a media file from
available content providers can be described as an asynchronous block. In this case, the
completion source would be initialized once the user opens the file picker dialog, and the
result would be set once the user picks a certain file from the selected content source. The
implementation can be extended to throw an exception if the user taps on the cancel button
on a certain dialog. This way, user flows that are composed of multiple screens and
interactions can be abstracted into asynchronous methods so that they can be easily used by
the view or the view model of the application.

The command pattern

The command pattern is a derivation of the flux pattern for reactive mobile applications. In the Android world, this pattern is implemented in a similar way under the name **Model View Intent** (**MVI**), whose sole purpose is creating a unidirectional flow of data and decreasing the complexities that stem from the duplex nature of **Model–View–ViewMode** (**MVVM**).

In this pattern, each view is equipped with multiple commands that are self-contained execution blocks with references to the underlying application infrastructure (similar to a unit of work). The user interaction in this case is routed to the respective command, and the command result is propagated to the concerned controls through broadcasts (for example, using the `BroadcastReceiver` implementation).

The task infrastructure in .NET Standard and the implemented runtimes like .NET Core allow developers to implement awaitable context elements, which can easily represent commands and can be awaited using the Task syntax.

In order for a class instance to be awaitable, it should implement the `GetAwaiter` method, which, in return, is used by the .NET task infrastructure. In a command pattern implementation, we can start by creating a base abstract class that we will use for dependency injection, and also implement the `awaitable` method:

```
public abstract class BaseCommand
{
    protected BaseCommand(IConfiguration configurationInstance, IUserApi
userApi)
    {
        ConfigurationService = configurationInstance;
        UserApi = userApi;
    }

    protected IConfiguration ConfigurationService { get; private set; }

    protected IUserApi UserApi { get; private set;}

    public virtual TaskAwaiter GetAwaiter()
    {
        return InternalExecute().GetAwaiter();
    }

    protected virtual async Task InternalExecute()
    {
    }
}
```

We can also extend the command implementation to actually return a result:

```
public abstract class BaseCommand<TResult> : BaseCommand
{
    protected TResult Result { get; set; }

    public new TaskAwaiter<TResult> GetAwaiter()
    {
        return ProcessCommand().GetAwaiter();
    }

    protected override async Task InternalExecute()
    {
        Result = await ProcessCommand().ConfigureAwait(true);
        await base.InternalExecute().ConfigureAwait(true);
    }

    protected virtual async Task<TResult> ProcessCommand()
    {
        // To be implemented by the deriving classes
        return default(TResult);
    }
}
```

Now, the implementation of an actual command, for instance, to update a user profile, would look similar to the following:

```
public class UpdateUserCommand : BaseServiceCommand<User>
{
    User _userDetails;

    public UpdateProfileCommand(IConfiguration configuration, IUsersApi
    usersApi, User user):
        base(configuration, usersApi)
    {
        _userDetails = user;
    }
    protected async override Task<string> ProcessCommand()
    {
        try
        {
            Result = await _usersApi.UpdateUser(CurrentUser);
            return Result;
        }
        catch (Exception ex)
        {
            // TODO:
        }
```

```
        }
    }
```

Finally, the implemented command can be initialized and executed like so:

```
var result = await new UpdateUserCommand(configuration, usersApi, user);
```

Here, the base command can also utilize a service locator or some type of property injection so that the set of services would not need to be injected together with the command parameters. Additionally, the messenger service can also be utilized to broadcast the successful execution of the command to multiple user controls.

Creating producers/consumers with blocking collections

Thread-safe collections are an invaluable member of the asynchronous toolset in .NET Core, just like they were in the full .NET Framework. Using blocking collections, concurrent models can be implemented to provide a common ground for asynchronous tasks on multiple threads. The most prominent of these models is, without a doubt, the producer/consumer pattern implementation. In this paradigm, a method executing on a parallel thread/task will produce the data items that will be consumed by another parallel operation, called the consumer, until a bounding limit is reached or production is completed. The two methods will be sharing the same blocking collection, where the blocking collection would act as a broker between the two asynchronous blocks.

Let's illustrate this pattern with a small implementation:

1. We will start by creating the blocking collection that will be used as storage for, let's say, `Auction` items:

```
BlockingCollection<Auction> auctions = new BlockingCollection(100);
```

2. We can now add the `Auction` items to the blocking collection using a background task. Here, the `GetNewAuction` method will be retrieving/creating auction instances and pushing them down to the pipeline:

```
Task.Run(() =>
{
    while(hasMoreAuctions)
    {
        auctions.Add(GetNewAuction);
    }
```

```
    auctions.CompleteAdding();
});
```

3. Similar to the producer, we can start a separate consumer thread that will be processing the items that are delivered:

```
Task.Run(() =>
{
    while (!auctions.IsCompleted)
    {
        Process(auctions.Take());
    }
}
```

4. Taking this implementation one step further, we can use the `GetConsumingEnumerable` method to create a blocking enumerable:

```
Take.Run(() =>
{
    foreach(var auction in auctions.GetConsumingEnumerable())
    {
        Process(auction);
    }
}
```

5. Finally, by utilizing `Parallel.ForEach`, we can add even more consumers without having to go through non-trivial synchronization implementations:

```
Parallel.Foreach(auctions.GetConsumingEnumerable(), Process);
```

Now, the data that's produced by the producer will be consumed by multiple consumers until the auction's collection sends the `IsCompleted` signal, which will cause the consuming enumerable to break and continue with the rest of the code execution.

Asynchronous execution patterns

Tasks are generally used to create an easy sequential execution of asynchronous blocks. Nevertheless, in certain scenarios, waiting for a task to complete might be unnecessary or not possible. We can enumerate a couple of scenarios where awaiting a task is not possible or required:

- If we are executing the asynchronous block, similar to the update user command, we would simply bind the command to the control and execute it in a throw-and-forget manner

- If our asynchronous block needs to be executed in a constructor, we would have no easy way to await the task
- If the asynchronous code needs to be executed as part of an event handler

Multiple examples can be listed here with common concerns, such as the following:

- Method declaration should not exhibit the `async` and `void` return types
- Methods should not be forced to execute synchronously with the `Wait` method or the `Result` property
- Methods that are dependent on the result of an asynchronous block; race conditions should be avoided

These not-to-be-awaited scenarios can be circumvented with various patterns. Next, we will take a closer look at some of them.

Service initialization pattern

In the constructor scenario we described previously, let's assume that the constructor of a view-model should retrieve a certain amount of data that will then be used by the methods or commands of the same view-model. If we execute the method without awaiting the result, there is no guarantee that, when the command is executed, the `async` constructor execution would have completed.

Let's demonstrate this with an abstract example:

```
public class MyViewModel
{
    public MyViewModel()
    {
        // Can't await the method;
        MyAsyncMethod();
    }
    private async Task MyAsyncMethod()
    {
        // Load data from service to the ViewModel
    }

    public async Task ExecuteMyCommand()
    {
        // Data from the MyAsyncMethod is required
    }
}
```

When the `ExecuteMyCommand` method is called immediately after the view-model is initialized, there is a big chance that we will have a race condition and a possible bug that would elude the development for a while.

In a so-called service initialization pattern, in order to verify that the `MyAsyncMethod` execution is successfully executed, we can, in fact, assign the resultant task to a field and use this field to await the previously started task:

```
public class MyViewModel
{
    private Task _myAsyncMethodExecution = null;

    public MyViewModel()
    {
        _myAsyncMethodExecution = MyAsyncMethod();
    }

    // ...

    public async Task ExecuteMyCommand()
    {
        await _myAsyncMethodExecution;
        // Data from the MyAsyncMethod is required
    }
}
```

This way, the asynchronous race condition is avoided and the command execution will need to ensure the completion of the task reference.

Asynchronous event handling

As we mentioned earlier, if the chain of calls demand asynchronous execution, the `async` chain should in fact be propagated all the way to the top of the call hierarchy. Deviating from this setup may cause threading issues, race conditions, and possibly deadlocks. Nevertheless, it is also important that the methods should not deviate from the async Task declaration, ensuring that the `async` stack and generated results and errors are preserved.

Event handlers with asynchronous code are a good example of where we would not have much to say about the signature of a method. For instance, let's take a look at the button click handler that should execute an `awaitable` method:

```
public async void OnSubmitButtonTapped(object sender, EventArgs e)
{
    var result = await ExecuteMyCommand();
```

```
            // do additional work
    }
```

Once this event handler is subscribed to the button's clicked event, the asynchronous code would be executed properly and we would not notice any issues with it. However, the method declaration that's using void as the return type would bypass the error handling infrastructure of the runtime and, in the case of an error, regardless of the exception source, the application would crash without leaving any trace as to what has gone wrong. We should also mention that the compiler warnings associated with this type of declaration would be added to the technical debt of the project.

Here, we can, in fact, create a TAP to **Asynchronous Programming Model (APM)** conversion, which can convert the asynchronous chain into a callback method, as well as introduce an error handler. This way, the OnSubmitButtonTapped method would not need to be declared with an asynchronous signature. We can easily introduce an extension method that will execute the asynchronous task with callback functions:

```
public static class TaskExtensions
{
    public static async void WithCallback<TResult>(
        this Task<TResult> asyncMethod,
        Action<TResult> onResult = null,
        Action<Exception> onError = null)
    {
        try
        {
            var result = await asyncMethod;
            onResult?.Invoke(result);
        }
        catch (Exception ex)
        {
            onError?.Invoke(ex);
        }
    }
}
```

Another extension method can be introduced to convert tasks without any returned data:

```
public static async void WithCallback(
    this Task asyncMethod,
    Action onResult = null,
    Action<Exception> onError = null)
{
    try
    {
        await asyncMethod;
        onResult?.Invoke();
```

```
    }
    catch (Exception ex)
    {
        onError?.Invoke(ex);
    }
}
```

Now, our asynchronous event handler can be rewritten to utilize the extension method:

```
public void OnSubmitButtonTapped(object sender, EventArgs e)
{
    ExecuteMyCommand()
        .WithCallback((result) => {
            //do additional work
        });
}
```

This way, we will break the asynchronous chain gracefully without jeopardizing the task infrastructure.

The asynchronous command

In an asynchronous UI implementation, it is almost impossible to avoid command declarations and bindings that handle asynchronous tasks. The general approach here would be to create an `async` delegate and pass it as an action for the command. However, this type of promise-based execution diminishes our capacity to see through the complete life cycle of the asynchronous block. This makes it harder to create unit test for these blocks and avoid scenarios where a terminal event, such as the navigation to a different view or closing the application, can interrupt the execution.

Let's look at `UpdateUserCommand`, which we implemented previously:

```
_updateUserCommand = new Command(async () => await
ExecuteUpdateUserProfile());
```

Here, the command is simply responsible for initiating the user profile update. However, once the execution of the command is completed, there is absolutely no guarantee that the complete execution of the `ExecuteUpdateUserProfile` method took place.

In order to remedy the asynchronous execution monitoring, or lack thereof, we can implement an asynchronous command that follows the task's execution within the command itself. Let's start by declaring our asynchronous command interface:

```
public interface IAsyncCommand : ICommand
{
```

```
    Task ExecuteAsync(object parameter);
}
```

Here, we have declared the asynchronous version of the main execute method, which will be used by the actual command method. Let's implement the `AsyncCommand` class:

```
public class AsyncCommand : IAsyncCommand
{
    // ...

    public AsyncCommand(
        Func<object, Task> execute,
        Func<object, bool> canExecute = null,
        Action<Exception> onError = null)
    {
        // ...
    }

    // ...
}
```

This command will be receiving an asynchronous task and an error callback function. `async` will then use the asynchronous delegate, as follows:

```
public async Task ExecuteAsync(object parameter)
{
    if (CanExecute(parameter))
    {
        try
        {
            await _semaphore.WaitAsync();

            RaiseCanExecuteChanged();

            await _execute(parameter);
        }
        finally
        {
            _semaphore.Release();
        }
    }
    RaiseCanExecuteChanged();
}
```

Notice that we have now successfully integrated our one-time-execute fix that we previously implemented in the asynchronous command block. Each time the semaphore is leased, we will be raising an event propagating the `CanExecute` change event to the bound user control.

Finally, the actual `ICommand` interface will use the `ExecuteAsync` method using the callback conversion extension method:

```
void ICommand.Execute(object parameter)
{
    ExecuteAsync(parameter).WithCallback(null, _onError);
}
```

Now, the application unit tests can directly use the `ExecuteAsync` method, whereas the bindings would still use the `Execute` method. We can even extend this implementation further by exposing a Task-typed property, like we did in the service initialization pattern, allowing consecutive methods to check for method completion.

Native asynchronous execution

Other than the asynchronous infrastructure provided by .Net Core, Xamarin target platforms also offer some background execution procedures that might assist developers who are implementing modules to do work when the app is actually not working. In return, various business processes are executed separately from the main application UI, creating a lightweight and responsive UX.

Android services

On the Android platform, the background process can be implemented as services. Services are execution modules that can be initiated on demand or with a schedule. For instance, a started service can be initiated with an intent. This would run until it is requested to be terminated (or self-terminates). Here, it is important to note that there is no direct communication between the process that initiated the service and the service itself, once the intent is actualized.

In order to implement a simple started service, you would need to implement the `Service` class and decorate the started service's `ServiceAttribute` attribute so that it can be included in the application manifest:

```
[Service]
public class MyStartedService : Service
```

```
{
    public override IBinder OnBind(Intent intent)
    {
        return null;
    }

    public override StartCommandResult OnStartCommand(
        Intent intent, StartCommandFlags flags, int startId)
    {
        // DO Work, can reference common core modules
        return StartCommandResult.NotSticky;
    }
}
```

Once the service is created, you can initiate the service with an `Intent`, as follows:

```
var intent = new Intent (this, typeof(MyStartedService));
StartService(intent);
```

You can also use `AlarmManager` to initiate the service periodically.

Another option for service implementations is the so-called bound service. Bound services, unlike started services, keep an open channel of communication through the usage of a binder. The binder service methods can be called by the initiating process, such as an activity.

iOS backgrounding

The iOS platform also provides background mechanisms for fetching additional data from remote servers, even though the application or even the device is inactive. While it is not as reliable as the alarm manager on Android, these background tasks are highly optimized for preserving battery. These background tasks can be executed as a response to certain system events, such as geolocation updates or on certain nominal intervals. We have used the word *nominal* here since the periods a background task is executed in are not deterministic and could change over time according to the execution performance of the background task, as well as the system resources that are available.

For instance, in order to perform a background fetch, you would first need to enable the background fetch from the **Background Modes** section:

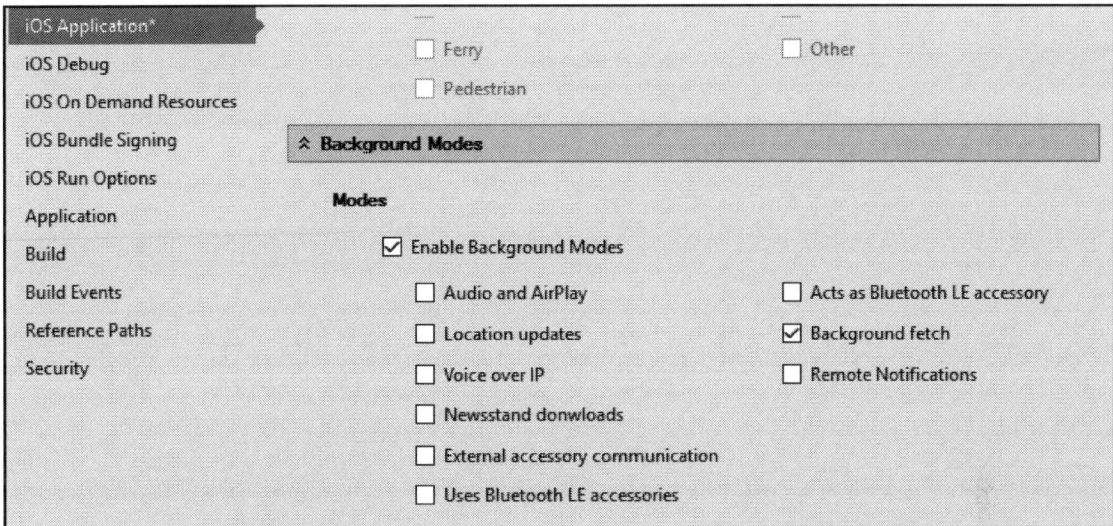

Once the background fetch is enabled, we can introduce our fetch mechanism, which will be executed periodically. The fetch mechanism would generally make a remote service call to update the data to be displayed so that once the application foregrounds, it would not need to repeat these refresh data calls. A perform fetch can be set up in `AppDelegate`:

```
public override bool FinishedLaunching(UIApplication app, NSDictionary options)
{
    global::Xamarin.Forms.Forms.Init();
    LoadApplication(new App());
    UIApplication.SharedApplication
.SetMinimumBackgroundFetchInterval(UIApplication.BackgroundFetchIntervalMinimum);

    return base.FinishedLaunching(app, options);
}
```

Now, the iOS runtime will be calling the `PerformFetch` method on regular intervals, so we can inject our retrieval code here:

```
public override void PerformFetch(
    UIApplication application, Action<UIBackgroundFetchResult>
    completionHandler)
{
```

```
    // TODO: Perform fetch

    // Return the correct status according to the fetch execution
    completionHandler(UIBackgroundFetchResult.NewData);
}
```

The returned result status is important since the runtime utilizes the result to optimize the fetch interval. The result status can be one of the following three statuses:

- `UIBackgroundFetchResult.NewData`: Called when new content has been fetched, and the application has been updated
- `UIBackgroundFetchResult.NoData`: Called when the fetch for new content went through, but no content is available
- `UIBackgroundFetchResult.Failed`: Useful for error handling; this is called when the fetch was unable to go through

In addition to the background fetch, `NSUrlSession`, coupled with the background transfer infrastructure, can provide background retrieval mechanisms that can be incorporated into the background fetch operations. This way, the application content can be kept up to date, even though it is in an active state.

Summary

In short, mobile applications should not be designed to undertake long-running tasks on the user interaction tier, but rather user asynchronous mechanisms to execute these workflows. The UI, in this case, would just be responsible for keeping the user informed about the background execution status. While in the past, background tasks were handled through classic .NET threading model, nowadays, the TAP model provides a rich set of functionality, which releases the developers from the burden of creating, managing, and synchronizing the threads and thread pools. In this chapter, we have seen that there are various patterns that would allow for the creation of background tasks that would yield back to the UI thread so that the asynchronous process results can be propagated to the UI. We also discussed different strategies for synchronous mechanisms, together with Tasks, thus avoiding deadlocks and race conditions. Additionally, we have looked into the native background procedures on iOS and Android.

Overall, asynchronous tasks, as well as background techniques, are mostly used for one common goal: to keep the data up to date within the application domain. In the next chapter, we will take a closer look at different techniques for managing the application data effectively.

Managing Application Data

12

Most mobile applications are coupled with certain datasets that are either downloaded through a service backend or bundled into the application. These datasets can vary from simple static text data that's used throughout the application to real-time updates related to a certain context. In this context, the developers are tasked with creating the optimal balance between remote interaction and data caching. Moreover, offline support is becoming the norm in the mobile application development world. In order to avoid data conflicts and synchronization issues, developers must be diligent about the procedures that are implemented according to the type of data at hand. In this chapter, we will discuss possible data synchronization and offline storage scenarios using products such as SQLite and Realm, as well as the new .NET Core modules, such as the Entity Framework Core.

The following sections will give you insights into how to manage application data:

- Improving HTTP performance with transient caching
- Persistent data cache using SQLite
- Data access patterns
- Understanding Realm

Improving HTTP performance with transient caching

In our previous examples, the client application held a direct service communication line with the service infrastructure. This way, the mobile application would load fresh data that's required to display a certain view on every view-model creation. While this provides an up-to-date context for the application, it might not be the most desirable experience for the user since, when we're dealing with mobile applications, we would need to account for bandwidth and network speed issues.

When developing a mobile application, it is a common mistake to assume that the application running on the simulator would behave the same once deployed to a physical device. In other words, it is quite naive to assume that the high-speed internet connection that is used on the development machine would be the same as the possible 3G network connection that you will have on the mobile device.

Fortunately, developers can emulate various network scenarios on device simulator/emulators for iOS and Android. On Android, the emulator features a valuable emulation option for the network type. The **Network type** option allows you to select various network types, from **Global System for Mobile Communications (GSM)** to **Long Term Evolution (LTE)**, as well as the **signal strength** (that is, poor, moderate, good, great). On iOS, the easiest way to emulate various network connections is to install the Network Link Conditioner tool, which can be found within the **Additional Tools for Xcode** developer download package. Once the package has been installed, the network connection of the host computer and implicitly the connectivity of the iOS simulated device can be adjusted.

Now that we can emulate network connectivity issues, let's take a look at ways we can improve the responsiveness of our application, even under subpar network conditions.

Client cache aside

We have already utilized server-side caching by implementing a simple cache-aside pattern using the Redis cache. However, this cache only helps us to improve the service infrastructure's performance. In order to be able to create and display view-model data with cached data, we would need to implement a caching store on the mobile application side.

The easiest way to implement this pattern would be to create simple caching store(s) for different entities that we are retrieving from the server. For instance, if we were to retrieve the details of a certain auction, we could first check that the data exists in our caching store. If the data item does not exist, we could retrieve it from the remote server and update our local store with the entity.

In order to demonstrate this scenario, let's start by creating a simple cache store interface that will exhibit only to methods for setting and retrieving a certain entity type:

```
public interface ICacheStore<TEntity>
{
    Task<TEntity> GetAsync(string id);
    Task SetAsync(TEntity entity);
}
```

This implementation assumes that the entities that it will handling will all have a string typed identifier.

Next, we will implement the constructor for our Auctions API:

```
public class AuctionsApi : IAuctionsApi
{
    private readonly IConfiguration _configuration;
    private readonly ICacheStore<Auction> _cacheStore;
    public AuctionsApi(IConfiguration configurationInstance,
ICacheStore<Auction> cacheStore)
    {
        _configuration = configurationInstance;
        _cacheStore = cacheStore;
    }
}
```

It is important to note that we are intentionally skipping the implementation of the cache store. A simple in-memory cache or a sophisticated local storage cache would both satisfy the interface requirements in this case.

Now that we have created our IAuctionsApi implementation, we can go ahead and implement the get method using the cache store as the first address to check for the target auction:

```
public async Task<Auction> GetAuction(string auctionId)
{
    // Try retrieve the auction from cache store
    Auction result = await _cacheStore.GetAsync(auctionId);
    // If the auction exists in our cache store we short-circuit
    if (result != null) { return result; }

    // ...
}
```

We have the data entity that was returned from the cache store. Now, we will implement the remote retrieval procedure in case the data item does not exist in the local store:

```
var client = new RestClient(_configuration["serviceUrl"]);

try
{
    result = await client.GetAsync<Auction>(_configuration["auctionsApi"],
auctionId);
    await _cacheStore.SetAsync(result);
}
catch (Exception ex)
```

```
{
    // TODO:
}

return result;
```

This finalizes a simple cache-aside pattern implementation. Nevertheless, our work is not finished with the caching store since this implementation assumes that, once the entity is cached in our local store, we won't need to retrieve the same entity from the remote server.

Entity tag (ETag) validation

Naturally, in most cases, our previous assumption about static data would fail us, especially when dealing with an entity like Auction, where the data entropy is relatively high. When the view-model is created, we would need to make sure that the application presents the latest version of the entity. This would not only avoid incorrect data being displayed on our view, but also avoid conflict errors (that is, referring to timestamp integrity checks on Cosmos DB).

In order to achieve this, we would need a validation procedure to verify that the entity at hand is the latest one and that we don't need to retrieve a newer version. The **Entity tag (ETag)** is part of the HTTP protocol definition. It is used as part of the available web cache validation mechanisms. Using ETag, the client application can make conditional requests that return the complete dataset if there is a more recent version of the entity that is being retrieved. Otherwise, the web server should respond with the 304 (not modified) status code. One of the ways for the client to execute the conditional requests is by using the If-None-Match header, accompanied by the existing ETag value.

Now, let's take a step back and take a look at the auctions controller that we implemented in our RESTful facade:

```
public async Task<IActionResult> Get(string key)
{
    var cosmosCollection = new
CosmosCollection<Auction>("AuctionsCollection");
    var resultantSet = await cosmosCollection.GetItemsAsync(item => item.Id
== key);
    var auction = resultantSet.FirstOrDefault();

    if(auction == null)
    {
        return NotFound();
    }
```

```
        return Ok(auction);
}
```

Since we have used the `Timestamp` field that is retrieved from Cosmos DB for validating whether the entity we are trying to push to the document collection is the latest version, it is fair to use the same field to identify the current version of a specific entity. In other words, the timestamp field, in addition to the ID of that specific entity, would define a specific version of an entity. Let's utilize the `If-None-Match` header to check whether there were any changes made on the given entity since it was loaded by the client application. First, we will check whether the client has sent the conditional header:

```
// Get the version stamp of the entity
var entityTag = string.Empty;

if (Request.Headers.ContainsKey("If-None-Match"))
{
    entityTag = Request.Headers["If-None-Match"].First();
}
```

Regardless of the entity tag value, we will retrieve the latest version of the entity from the document collection. Nevertheless, once the entity is retrieved, we will be comparing the ETag header value to the timestamp that's retrieved from the Cosmos DB collection:

```
if (int.TryParse(entityTag, out int timeStamp) && auction.TimeStamp ==
timeStamp)
{
    // There were no changes with the entity
    return StatusCode((int)HttpStatusCode.NotModified);
}
```

This completes the server-side setup. Now, we will modify our client application so that it sends the conditional retrieval header:

```
// Try retrieve the auction from cache store
Auction result = await _cacheStore.GetAsync(auctionId);

Dictionary<string, string> headers = null;

// If the auction exists we will add the If-None-Match header
if (result != null)
{
    headers = new Dictionary<string, string>();
    headers.Add(HttpRequestHeader.IfNoneMatch.ToString(),
result.Timestamp.ToString());
}

var client = new RestClient(_configuration["serviceUrl"], headers);
```

At this point, every time we are retrieving an auction entity, we would try to load it from the local cache. However, instead of short-circuiting the method call, we will be just adding the conditional retrieval header to our `RestClient`. We can, of course, further refactor this to create an `HttpHandler` that we can pass as a behavior to the `HttpClient`, or even introduce a behavior on the `RestClient` to handle the caching in a generic manner.

Additionally, we will also need to modify `RestClient` so that it can handle the `NotModified` response that will be returned by the server. Open source implementations of a caching strategy using the standard web caching control mechanisms can be another solution for these modifications.

Key/value store

Akavache is another one of the available caching solutions in the open source scene. Technically, Akavache is a .NET Standard implementation for various caching stores. In the words of the author himself, it's an asynchronous persistent key-value store. Having been implemented on the .NET Standard framework, it can be utilized on both Xamarin and .Net Core applications.

In Akavache, cache stores are implemented as blob storage on various mediums, such as in-memory, local machine, or user account. While these stores do not refer to a specific system location across platforms, each has a respective translation, depending on the target platform. For instance, the user account and secure stores refer to shared preferences on iOS, and they would be backed up to the iTunes cloud, whereas on the UWP platform, these two would refer to user settings and/or roaming user data and they would be stored on the cloud associated with the user's Microsoft account. Because of this, each platform imposes their own restrictions on these local blob storages.

Making use of these stores is also quite easy using the extensions methods available for the blob cache storage abstraction. The extension method that pertains to the retrieval of data using the cache-aside pattern is outlined as follows:

```
// Immediately return a cached version of an object if available, but
*always*
// also execute fetchFunc to retrieve the latest version of an object.
IObservable<T> GetAndFetchLatest<T>(this IBlobCache This,
    string key,
    Func<IObservable<T>> fetchFunc,
    Func<DateTimeOffset, bool> fetchPredicate = null,
    DateTimeOffset? absoluteExpiration = null,
    bool shouldInvalidateOnError = false,
    Func<T, bool> cacheValidationPredicate = null)
```

As you can see, the data is retrieved using `fetchFunc` and is put into the current `IBlobCache` object. Generally, the key that's used for local caching is the resource URL itself. Additionally, a cache validation predicate can be included to verify that the retrieved cache data is still valid (for example, not expired).

It is also important to note that Akavache makes heavy use of reactive extensions and that the return types are generally observables rather than simple tasks. Therefore, the returned data should be handled mostly with event subscriptions. The completion might be fired multiple times, depending on the status of the cached data (that is, once for the cached version of the data and once for the remote retrieval).

The transient cache can be a life-saver in low bandwidth connectivity scenarios, but, at times, you might need to store the data in a relational model, especially if the data you are retrieving does not have too much entropy. In these type of situations, you would need a little more than a key/value store.

Persistent data cache using SQLite

In the previous examples, we didn't use a relational data store for local data. In most cases, especially if we are dealing with a NoSQL database, the relational data paradigm loses its appeal and data denormalization replaces the data consistency concerns in favor of performance. Nevertheless, in certain scenarios, in order to find the optimal compromise between the two, we might need to resort to relational data mappings.

In the relational data management world, the most popular data management system for mobile applications is without a doubt SQLite. SQLite is, at its core, a relational database management system contained in a C programming library. What differentiates SQLite from other relational data management systems is the fact that the SQLite engine does not use or require a standalone process that the consuming application communicates with. The SQLite data store and the engine is, in any application scenario, an integral part of the application itself. In simple terms, SQLite is not a client-server engine, but rather an embedded data store.

SQLite has various implementations that span a wide set of platforms, such as native mobile platforms and web and desktop applications, as well as embedded systems. Even though certain browsers still do not and will probably never support SQLite, it still remains the norm as the local store for Xamarin applications and targeted mobile platforms.

SQLite.NET

SQLite.NET was one of the earliest implementations of SQLite on the **Portable Class Library** (**PCL**) platform, and it still remains one of the most popular cross-platform implementations (now targeting .NET Standard).

The implementation patterns with SQLite.NET are based on domain model entity attributes that define the entity indexes and other data columns:

```
public class Vehicle
{
    [PrimaryKey, AutoIncrement]
    public int Id { get; set; }
    [Indexed]
    public string RemoteId { get; set; }
    public string Color { get; set; }
    public string Model { get; set; }

    public int Year { get; set; }

    public string AuctionId { get; set; }
}
```

Relationships between entities can be introduced using the attributes (for example, ForeignKey, OneToMany, and ManyToOne) that are included in the SQLite.NET Extensions module, which allows developers to create the ORM model and helps with data retrieval and update processes.

After the entity model is prepared, the db connection can be created using various file path combinations:

```
// Path to the db file
var dbPath =
Path.Combine(Environment.GetFolderPath(Environment.SpecialFolder.MyDocument
s), "Auctions.db");

var db = new SQLiteAsyncConnection(dbPath);
```

With the db connection, various actions can be executed using the LINQ syntax and familiar table interaction context:

```
await db.CreateTableAsync<Vehicle>();

var query = db.Table<Vehicle>().Where(_ => _.Year == 2018);

var auctionsFor2018 = await query.ToListAsync();
```

`Sqlite.NET` also supports text-based queries without resorting to LINQ-2-Entity syntax.

`SQLCipher` is another extension that can be used to create encrypted database stores for sensitive data scenarios.

Entity Framework Core

Entity Framework Core combines the years of accumulated ORM structure with SQLite support, making it a strong candidate for local storage implementations. Similar to the classic .NET version of Entity Framework, data contexts can be created and queried using the `UseSqlite` extension with a file path for `DbContextOptionsBuilder`:

```
public class AuctionsDatabaseContext : DbContext
{
    public DbSet<Auction> Auctions { get; set; }

    protected override void OnConfiguring(DbContextOptionsBuilder
optionsBuilder)
    {
        var dbPath = string.Empty;

        switch (Device.RuntimePlatform)
        {
            case Device.Android:
                dbPath = Path.Combine(
                        Environment.GetFolderPath(
                            Environment.SpecialFolder.MyDocuments),
                        "Auctions.db");

            // removed for brevity
        }

        optionsBuilder.UserSqlite($"Filename={dbPath});
    }
}
```

Now, the `DbSet<TEntity>` declarations can be used to construct LINQ queries, which are used to retrieve data, while the context itself can be used to push updates.

In addition to the extensive relational functionality, Entity Framework Core also offers `InMemory` database support. By using an `InMemory` database instead of the application storage, developers can easily mock the local cache implementation to create unit and integration tests.

Data access patterns

So far, we have defined various data stores that will be used within the client boundaries as a secondary source of data to help with offline scenarios, as well as when network connectivity is problematic. This way, the user interface does not appear blank—or worse, goes into an infinite loading loop. Instead, a previous version of the dataset is displayed to the user immediately while remote retrieval takes place.

Before we can dive into the different architectural patterns to coordinate the local and remote data, let's try to describe the meaning of coordination by means of identifying the types of data that we have in our application. According to the life cycle and entropy of the specific data element, we can categorize the data types into the following groups:

- **Transient data**: These data elements will constantly be changing and the local storage should be continuously invalidated. When dealing with this data type, the application should first load the local cache to respond to user input as fast as possible, but each time, we should execute a remote call to update the local cache and the view-model data. Examples of this would be auctions information and bidding entries.
- **Reference data**: These data elements will change from time to time, but it is acceptable to use cache storage as a source of truth (if it exists) with a reasonable time-to-live period. User profiles and vehicle data might be good examples of this data type. When dealing with reference data, the view-model should try to load the entities from the cache. If it does not exist or the data has expired, the application should reach out to the remote server and update the local cache with the remote data.
- **Static data**: There are certain data elements in an application that will probably never change. These static data elements, such as the list of countries or internal enumerator descriptions, would, under normal circumstances, be loaded only once and served from local storage.

Other than these data types, we should also mention volatile entity types, which should not be cached by the application at all unless there is a clear indication for it. For instance, imagine a search query with an arbitrary set of query parameters on vehicles or auctions sets returning a paginated list of entries. It is quite unreasonable to even try to cache this data and server through the view-model. On the other hand, the most recent entries list that we might want to display on the main dashboard can, in fact, be categorized as transient data.

Now, let's take a look at some simple implementation patterns that might help with implementing a fluid data flow and user experience.

Implementing the repository pattern

Naturally, while talking about data stores, the foremost abundant pattern in development work is the repository pattern. In the context of local and remote data, we could create small repositories that implement the same exact repository interface and create a manager class to coordinate the repositories.

Let's start with a transient data repository, such as Auctions. For the Auctions API, let's assume that we have three methods to retrieve:

- A list of auctions
- The details of a specific auction
- A method to update a specific auction

Let's define our interface for these methods:

```
public interface IAuctionsApi
{
    Task<IEnumerable<Auction>> GetAuctions();

    Task<Auction> GetAuction(string auctionId);

    Task UpdateAuction(Auction auction);
}
```

We will now have two competing repositories that implement the same interface:

```
public class RemoteAuctionsApi : IAuctionsApi
{
    // ...
}

public class LocalAuctionsApi : IAuctionsApi
{
    // ...
}
```

Here, the simplest way to handle the data would be to implement a wrapper that will keep the local cache up to date and pass the wrapper instance, as well as the local repository instance, to the view-model so that the view-model can handle the initial cache load scenario, thus utilizing the local repository and then calling the wrapper instance.

Let's go ahead and implement the wrapper class, which we will call the manager:

```
public class ManagerAuctionsApi : IAuctionsApi
{
```

```
        private readonly IAuctionsApi _localApi;
        private readonly IAuctionsApi _remoteApi;
        // ...
        public async Task<Auction> GetAuction(string auctionId)
        {
            var auction = await _remoteApi.GetAuction(auctionId);
            await _localApi.UpdateAuction(auction);
            return auction;
        }
    }
```

The view-model load strategy will make the local API call, update the view model data, and call the remote API with a `ContinueWith` action to update the model:

```
Action<Auction> updateViewModel = (auction) => {
    Device.BeginInvokeOnMainTread(() => this.Auction = auction)
};

var localResult = await _localApi.GetAuction(auctionId);

updateViewModel(localResult);

// Not awaited
_managerApi.GetAuction(auctionId).ContinueWith(task =>
updateViewModel(task.Result));
```

The implementation for a reference repository would utilize a different strategy:

```
public async Task<User> GetUser(string userId)
{
    var user = await _localApi.GetUser(userId);
    if (user != null) { return user; }

    user = await _remoteApi.GetUser(userId);
    await _localApi.UpdateUser(user);

    return user;
}
```

A static data load would adhere to the same strategy as the reference repository, but it is also possible to initialize the static cache when the application is first run using a background task, or embed the static data with a JSON file or with a seeded SQLite database file into the application package.

Observable repository

In the previous example, the responsibility of choosing between the remote data and the local data fell onto the view model. If we could actually notify the view-model about the changing result (that is, local and remote data), we would need to pass the local repository instance to the view-model. By using reactive extensions, we could implement a notifying observable and subscribe to various events on the view-model:

- **OnNext**: We would update the data within the view model, which would be fired twice
- **OnCompleted**: We would hide any progress indicator we have been showing up until this point

Before we can implement the view-model subscriptions, let's implement the `GetAuction` method in a reactive manner:

```
public IObservable<Auction> GetAuction(string auctionId)
{
    var localAuction = _localApi.GetAuction(auctionId).ToObservable();
    var remoteAuction = _remoteApi.GetAuction(auctionId).ToObservable();
    // Don't forget to update the local cache
    remoteAuction.Subscribe(auction => _localApi.UpdateAuction(auction));

    return localAuction.Merge(remoteAuction);
}
```

Now that we have our observable result, we can implement the subscription model in our view model:

```
var auctionsObservable = _managerAuctionsApi.GetAuction(auctionId);
auctionsObservable.Subscribe(auction => updateViewModel(localResult));

await auctionsObservable();

IsBusy = false;
```

With the conversions from observable results to the task or vice-versa, it is important to be careful with cross thread issues. As a general rule of thumb, all of the code that updates elements on the UI thread should be executed with `Device.BeginInvokeOnMainTread`.

Data resolver

The aforementioned static data elements—especially the data points that are attached to the main collections with an ID reference—would probably never change, and yet they would be carried over the wire together with the main data elements. In order to decrease the payload size, we can in fact dismiss these data objects on the DTO models and represent them only by ID. However, in such an implementation, we would need to find a way to resolve these ID references into real data elements.

For instance, let's take a look at the engine entity that we are transferring as part of a vehicle description:

```
public class Engine
{
    [JsonProperty("displacement")]
    public string Displacement { get; set; }

    [JsonProperty("fuel")]
    public FuelType Fuel { get; set; }

    // ...
}
```

The same DTO object is used on the client side. Here, the `Fuel` type is returned as a complex object rather than a single identifier (the available options are diesel or gas). Assuming that the static data for the fuel type is initialized on the client application, we do not really need to retrieve from the server. Let's replace the property with an identifier (that is, of the string type).

On the client side, let's extend our model so that it has a reference ID, as well as an entity:

```
public class Engine
{
    [JsonProperty("displacement")]
    public string Displacement { get; set; }

    [JsonProperty("fuelId")]
    public string FuelId { get; set; }
    [JsonIgnore]
    public KeyIdentifier Fuel { get; set; }
}
```

With the ignore attribute, the reference ID will not be omitted during the serialization phase. We will, however, still need to resolve this entry every time an `Engine` instance is retrieved from the server.

Additionally, we have represented similar key/value pairs, such as the `FuelType` complex object, with a generic entity:

```
public class KeyIdentifier
{
    public string KeyGroup { get; set; } // for example, "FuelType"

    public string Key { get; set; } // for example, "1"

    public string Value { get; set; } // for example, "Diesel"
}
```

The easiest way for the runtime to identify the properties that will need to be resolved and the `Target` property that the data will be assigned to is to create a custom data annotation attribute:

```
[AttributeUsage(AttributeTargets.Property)]
public class ReferenceAttribute : Attribute
{
    public ReferenceAttribute(string keyType, string field)
    {
        KeyType = keyType;
        Field = field;
    }

    public string KeyType { get; set; }
    public string Field { get; set; }
}
```

Now, using the annotation we have just created, the engine DTO class would look like this:

```
public class Engine
{
    [JsonProperty("displacement")]
    public string Displacement { get; set; }

    [Reference("FuelType", nameof(Fuel))]
    [JsonProperty("fuelId")]
    public string FuelId { get; set; }
    [JsonIgnore]
    public KeyIdentifier Fuel { get; set; }
}
```

We are now declaring that the `FuelId` field is a reference to a `KeyIdentifier` object of the `FuelType` type and that the resolved data should be assigned to a property called `Fuel`.

Assuming that the static data about `FuelType` is stored in local storage and can be retrieved as a `KeyIdentifier` object, we can now implement a generic translator for the static data:

```
public async Task TranslateStaticKeys<TEntity>(TEntity entity)
{
    var entityType = typeof(Entity);

    var properties = entityType.GetRuntimeProperties();

    foreach(var property in properties)
    {
        var refAttribute =
        property.CustomAttributes.FirstOrDefault(item =>
        item.AttributeType == typeof(ReferenceAttribute));

        if(customAttribute == null) { continue; }
        var keyParameters = refAttribute.ConstructorArguments;
        // e.g FuelType => Fuel
        var keyType = keyParameters.Value.ToString();
        var propertyName = keyParameters.Value.ToString();
        targetProperty = properties.FirstOrDefault(item => item.Name ==
        propertyName);

        try
        {
            await TranslateKeyProperty(property, targetProperty,
            entity, keyType);
        }
        catch(Exception ex)
        {
            // TODO:
        }
    }
}
```

The `TranslateStaticKeys` method iterates through the properties of an entity and if it identifies a property with `ReferenceAttribute`, it invokes the `TranslateKeyProperty` method.

The actual translation method would follow a similar implementation path, hence retrieving the set of key/value pairs for the given type, resolving the key ID into `KeyIdentifier`, and finally assigning the data to the target property:

```
public async Task TranslateKeyProperty<TEntity>(
    PropertyInfo sourceProperty,
    PropertyInfo targetProperty,
```

```
        TEntity entity,
        string keyType)
    {
        IEnumerable<KeyIdentifier> keyIdentifiers = await
    _api.GetStaticValues(keyType);

        if(targetProperty != null)
        {
            var sourceValue = sourceProperty.GetValue(entity)?.ToString();
            if(!string.IsNullOrEmpty(sourceValue))
            {
                var keyIdentifier = keyIdentifiers.FirstOrDefault(item =>
                item.Id == sourceValue);
                targetProperty.SetValue(entity, keyIdentifier);
            }
        }
    }
}
```

Now, after retrieving an entity, we can simply invoke the translate method to resolve all outstanding data points:

```
var auction = await _managerAuctionsApi.Get(auctionId);
await Translator.TranslateReferences(auction.Vehicle.Engine);
```

The implementation can be extended to collections and to walk the complete object tree.

Understanding Realm

Throughout this chapter, our main goal was to create an infrastructure that will synchronize and coordinate the data flow between the local and remote storages so that a pleasant user experience can be provided to the users. The Realm platform, which is composed of database and server components, coupled with the live objects concept, is created with offline data first scenarios in mind.

The Realm database provides a relational data store that should be used as a local persistence store for various runtimes, including Xamarin target platforms. The Realm database is a feature-rich, lightweight, and highly performant implementation that's provided with native implementation on all of the supported platforms.

The Realm database is referred to as object-oriented, mainly because of the query structures. In Realm, the relationships are handled through natural class declarations, without the need for annotations or relational model descriptions. If we were to create an auction entity, a simple skeleton might look like this:

```
public class Auction : RealmObject
{
    public string Id { get; set; }
    public string Description { get; set; }
    public Vehicle Vehicle { get; set; }
}
```

Now, if we were to create a new auction, we would simply use the `CreateObject` method within a `Write` block:

```
var realm = Realm.GetInstance();

realm.Write(()=>
{
    var vehicle = realm.CreateObject<Vehicle>();
    vehicle.Make = "Volvo";
    vehicle.Model = "S60";

    var auction = realm.CreateObject<Auction>();
    auction.Id = Guid.NewGuid().ToString();
    auction.Description = "Family Car for Sale";
    auction.Vehicle = vehicle;
});
```

The Realm Object Server is the complimentary component to the Realm database, enabling simultaneous data synchronization across devices and platforms. This synchronization is propagated to the owners of the so-called live objects through standard events, which makes this platform an ideal candidate for Model-View-ViewModel-based implementations. The server is offered as a hosted solution or as a platform that can be hosted on-premise or on managed cloud environments, including Microsoft Azure, AWS, and Google Cloud.

Realm offers an extensive set of features that might help to remove the service layer interaction completely for some applications.

Summary

In this chapter, we implemented and employed different patterns and technologies to create offline capable and responsive applications. Initially, we used used the cache-aside pattern and utilized standard HTTP protocol definitions to cache and validate the cached data. We also analyzed technologies such as SQLite, Entity Framework Core, and Akavache. Finally, we briefly looked at the Realm components and how they can be used to manage cross-device and platform data. Overall, there are so many technologies available for developers that have been conceived within the community, and it is important to choose the correct patterns and technology stack for your specific use cases to achieve the best user experience and satisfaction.

In the next chapter, we will take a look at the Graph API, push notifications, and additional ways to engage the customer.

13
Engaging Users with Notifications and the Graph API

Push notifications are the primary tools for an application infrastructure to deliver a message to the user. They are used to broadcast updates to users, send notifications based on certain actions, and engage them for customer satisfaction or according to metrics. On the other hand, the Microsoft Graph API is a service that provides a unified gateway to data that's accumulated through Office 365, Windows 10, and other Microsoft services. This chapter will explain, in short, how notifications and the Graph API can be used to improve user engagement by taking advantage of push notifications. We will be creating a notification implementation for cross-platform applications using Azure Notification Hub. We will also create so-called activity entries for our application sessions so that we can create a timeline that is accessible on multiple platforms.

The following sections will drive the discussion about user engagement:

- Understanding Native Notification Services
- Configuring Azure Notification Hub
- Creating a Notification Service
- The Graph API and Project Rome

Understanding Native Notification Services

Push notifications, as it can be deduced from the name, is the generalized name for the messages that are pushed from the backend server applications to the target application user interface. Push notifications can be targeting a specific device, or they can target a group of users. They can vary from a simple notification message to a user-invisible call to the application backend on the target platform.

Notification providers

Push notifications are, in general terms, sent from the application backend service to the target devices. The notification delivery is handled by platform-specific notification providers, which are referred to as **Platform Notification Systems** (**PNS**). In order to be able to send push notifications to iOS, Android, and UWP applications, as a developer, you would need to create notification management suites/infrastructure for the following:

- **Apple Push Notification Service** (APNS)
- **Firebase Cloud Messaging** (FCM)
- **Windows Notification Service** (WNS)

Each of these notification providers implement different protocols and data schemas to send notifications to users. The main issues with this model for a cross-platform application are as follows:

- **Cross-Platform Push Notifications**: Each push service (APNS for iOS, FCM for Android, and WNS for Windows) has different protocols
- **Different Content Templates**: Notification templates, as well as the data structure, are completely different on all major platforms
- **Segmentation**: This is based on interest and location for outing only the most relevant content to each segment
- **Maintaining Accurate Device Registry**: Dynamic user base, done by adding registration upon startup, updating tags, and pruning
- **High Volume with Low-Latency:** Application requirements in high-volume with low latency hard to meet with cross platform implementation

Sending notifications with PNS

Using the native notification providers, the application backend can send notifications to target devices if the target device has already opened a notification channel. While the implementation of this process highly depends on the platform at hand, at a high level, the registration and notification flow are very similar on all of the PNSes.

The device should start this flow by registering itself with the PNS and retrieving a so-called PNS handle. This handle can be a token (for example, APNS) or a simple URI (for example, WNS). Once the PNS is retrieved, it should be delivered to the backend service as a calling card, as shown in *step 2* of the following diagram:

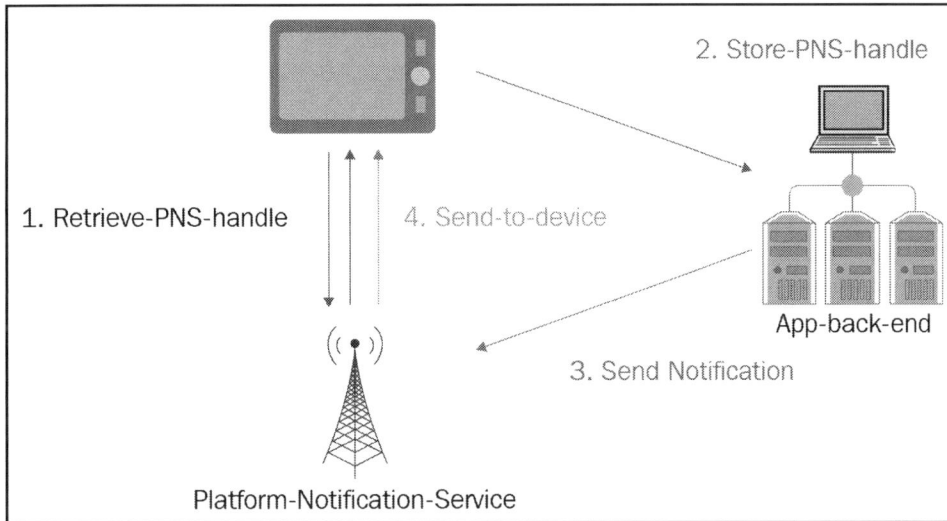

When the backend service wants to send a notification to this specific device, it contacts the PNS with the saved PNS handle for the device so that the PNS can deliver the notification.

General constraints

The **Apple Push Notification Service (APNS)** and **Firebase Cloud Messaging (FCM)** infrastructure both make use of certain manifest declarations that are very specific to the application that's installed on the target device.

APNS mandates that the application ID (and certificate) used to sign the application package have a non-wildcard bundle declaration (that is, the com.mycompany.* type of certificates cannot be used for this implementation).

The device handle (an alphanumeric token or a URL, depending on the platform) is the only piece of device information that is required and used in the notification process. In contrast, the notification service should support multiple applications and platforms.

Azure Notification Hub

In an environment where multiple platforms and multiple service providers exist, Azure Notification Hub acts as a broker between backend services that creates the notification requests and the provider services that deliver these notification requests to target devices.

Notification Hub infrastructure

Considering the release environments for an application (that is, alpha, beta, and prod) and the notification hubs, each environment should be set up as a separate hub on the Azure infrastructure. Nevertheless, notification hubs can be united with a so-called namespace so that application environments for each platform can be managed in one place.

Notification hub

Semantically, a notification hub refers to the smallest resource in the Azure Notification Hub infrastructure. It maps directly to the application running on a specific environment and holds one certificate for each **Platform Notification System** (PNS). The application can be hybrid, native, or cross-platform. Each notification hub has configuration parameters for different notification platforms to support the cross-platform messaging infrastructure:

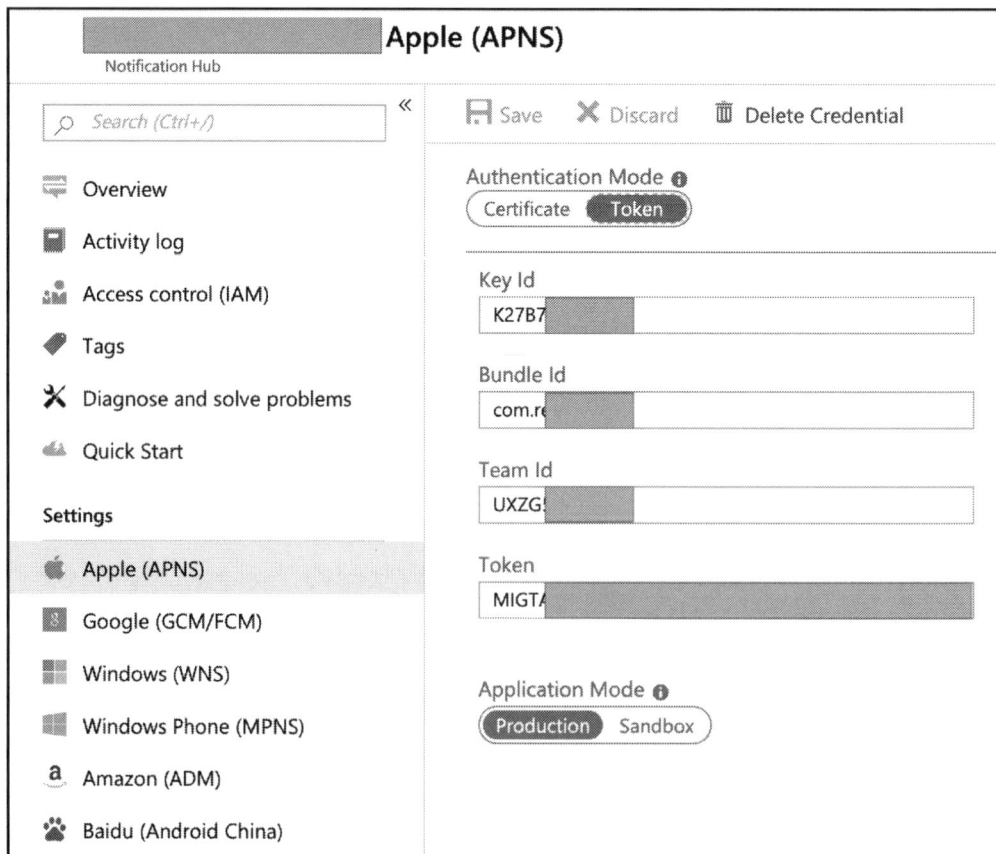

Notification namespace

A namespace, on the other hand, is a collection of hubs. Notification namespaces can be used to manage different environments, as well as create clustered of notification hubs for bigger target audiences. Using the namespace, developers or the configuration management team can track the notification hub status and use it as the main hub for dispatching notifications, as shown in the following screenshot:

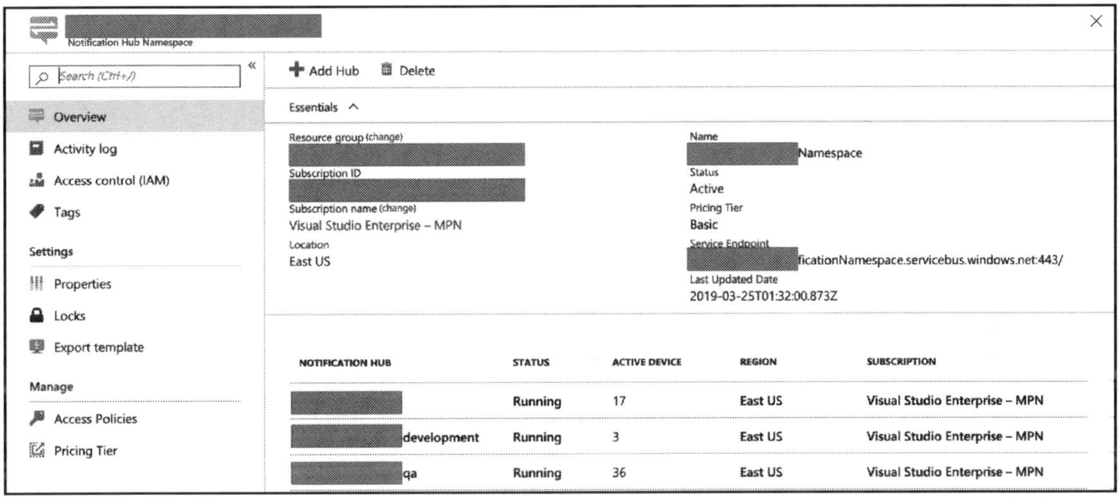

In an ideal setup, an application-specific notification service would communicate with the specific hubs, while a namespace would be used to configure and manage these hubs:

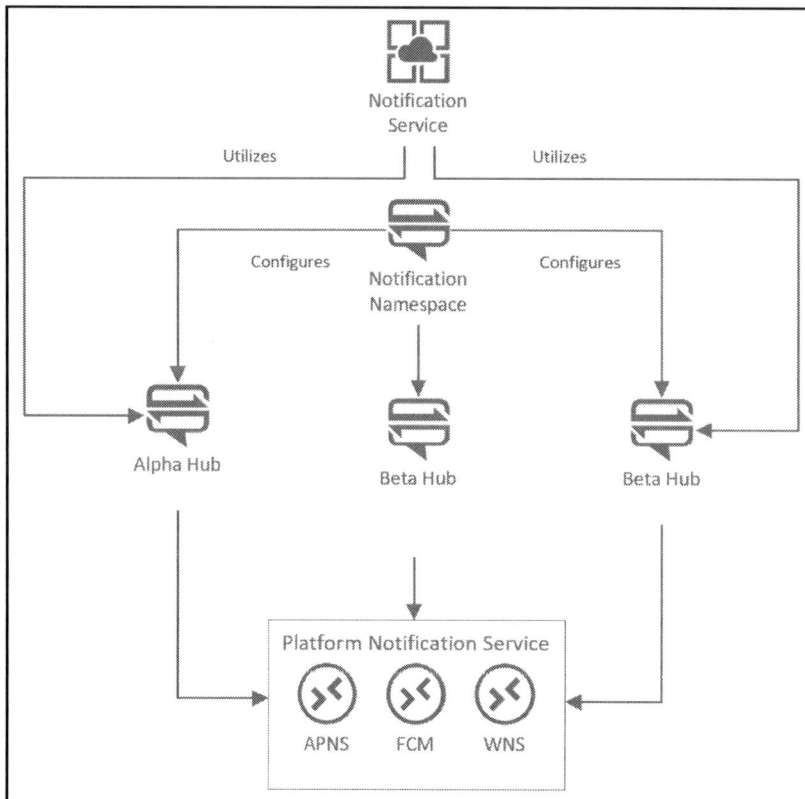

In this configuration, the application-specific notification service never directly communicates with the platform-specific notification services; instead, notifications would deliver the messages to the target platforms. In a similar approach, multiple applications that share a common target user group can also be unified under a namespace and use the same notification namespace.

In order to understand the application registration process, we should take a closer look at the notification hub registration and notification process.

Notifications using Azure Notification Hub

Similar to the PNS process, the Azure Notification Hub flow can be defined in two parts:

- Registration
- Notification

Registration

Azure Notification Hub supports two types device registration, client and backend registration:

- In the client registration scenario, the application calls the Azure Notification Hub with the device handle that's received on the native platform. In this scenario, only the device handle can be used to identify the user. In order for the backend services to send notification messages to the target device, the device should pass on the notification information to the backend service.
- In a backend registration scenario, the backend service handles registration with the Azure Notification Hub. This scenario allows the backend service to insert additional information to the registration data, such as user information (for example, user ID, email, role, and group), application platform information (for example, application runtime, release version, and release stage), and other identifiers for the application and/or user. This type of registration process is more suited for implementations where multiple applications might be receiving notifications from multiple sources so that the notification service can act as an event aggregator.

The registration of the target device with a notification provider using the notification hub is done either using data packages, called registrations, or with extended device information sets, called installations. While the registration package contains the notification handle or the notification URI for a device, installation contains additional device-specific information.

During the registration process, the client application (or the backend notification service) can also register notification templates associated with tags to personalize the notifications and create notification targets. The tags can be used later to broadcast notifications to groups of devices that belong to a single user, or groups of users with multiple devices, hence enabling a broadcast-ready infrastructure.

Notification

Once the device is registered with the backend service and the notification hub, the backend service can send notifications using the PNS handle of the device that's targeting a specific platform. With this type of notification, the backend server should identify what is the target platform, prepare a notification message using the target platform's schema, and finally send the notification message to the given handle.

Another approach for sending notifications to users is to use notification tags. A specific user tag can be used with an associated template to target multiple platforms. In this case, during registration, the device would be registered with a specific template (depending on the platform) and a user tag identifying the user (for example, `username: can.bilgin`). This way, any notification message that is sent to the notification hub with the given tag will be routed to the target devices that has this registration.

Finally, broadcast messages using various tags that define a certain interest of a user can be employed with notifications. In this notification scenario, the application interface would expose certain notification settings, and these notification settings would then be translated into registration tags. Each time the user updates these settings, the device registration with the backend service would need to be updated with the updated set of tags. For instance, in our application, if we were to include a notification setting that allows the user to receive notifications when a new auction is created for a specific make or model of a car, we could use this tag to send notifications to similar users as soon as a new entry is created in the application data store.

Creating a notification service

Depending on the granularity of the notification target group, the application design can include a notification service; otherwise, another option would be to have a setup where the notification infrastructure can be integrated into either one of the existing services. For instance, if the application does not require a user-specific messaging scenario, there is no need to track device registrations. In this case, the client-side registration mechanism can be implemented so that notification categories are created and the backend services can send the notifications to target tags.

As part of our auctions application, let's use the following user story to start the implementation of our notification service:

> *"As a product owner, I would like to have a notification service created so that I can have an open channel to my user group through which I can engage them individually or as a group to increase the return rate."*

In light of this request, we can start analyzing different use cases and implementation patterns.

Defining the requirements

Since the notification service will be supporting incoming notifications from various modules, the architectural design and implementation would need to adhere to certain guidelines, as well as satisfy registration and notification requirements:

- The notification service should be designed to support multiple notification hubs for different application versions (for example, alpha, beta, and prod)
- The notification service should be designed to receive notification requests for a specific device, specific user, or a group of users (for example, interest groups, certain roles, and people involved in a certain activity)
- Different notification events should be assignable with different notification templates
- Users should be registered with different templates according to their language preferences
- Users should be able to opt in or out of certain notifications

Overall, the notification service implementation should implement an event aggregator pattern, where the publishers are the source of the notification and are responsible for determining the notification specifications, as well as a definition for the target, while the subscribers are the native mobile applications that submit registration requests defining their addressing parameters. Once a notification that meets the criteria for a target comes down the pipeline, the notification is routed to the associated target.

Device registration

Device registration is the first use case to be implemented and tested by the development team. In this scenario, the user would open (or install) the Auctions application and authenticate through one of the available identity providers. At this stage, we have the device information as well as the user identity, which we can use for registration. Let's see how this is done:

1. The notification channel will need to be opened as soon as the user opens the application (in the case of a returning user—that is, an existing identity), or once registration/authentication is completed. On iOS, we can use the `UserNotificationCenter` module to authorize for notifications:

```
UNUserNotificationCenter.Current.RequestAuthorization(
    UNAuthorizationOptions.Alert |
    UNAuthorizationOptions.Sound |
    UNAuthorizationOptions.Sound,
    (granted, error) =>
{
    if (granted)
    {
        InvokeOnMainThread(UIApplication.SharedApplication
            .RegisterForRemoteNotifications);
    }
});
```

2. Once the registration is invoked, the override method in the `AppDelegate.cs` file can be used to retrieve the device token:

```
public override void RegisteredForRemoteNotifications(
    UIApplication application,
    NSData deviceToken)
```

3. We can now send this token using the authorization token for identifying the user to the notification service device registration controller using a registration data object:

```
public class DeviceRegistration
{
    public string RegistrationId { get; set; } // Registration Id
    public string Platform { get; set; } // wns, apns, fcm
    public string Handle { get; set; } // token or uri
    public string[] Tags { get; set; } // any additional tags
}
```

Before we begin interacting with the notification hub, we would need to install the `Microsoft.Azure.NotificationHubs` NuGet package, which will provide the integration methods and data objects. The same package can, in fact, be installed on the client side to easily create the notification channel and retrieve the required information to be sent to the backend service.

4. Once the device registration is received on our service, depending on whether it is a new registration or an update of a previous registration, we can formulate the process to clean up previous registrations and create a notification hub registration ID for the device:

```
public async Task<IActionResult> Post(DeviceRegistration device)
{
    // New registration, execute cleanup
    if (device.RegistrationId == null && device.Handle != null)
    {
        var registrations = await
        _hub.GetRegistrationsByChannelAsync(device.Handle, 100);
        foreach (var registration in registrations)
        {
            await _hub.DeleteRegistrationAsync(registration);
        }

        device.RegistrationId = await
_hub.CreateRegistrationIdAsync();        }

    // ready for registration
    // ...
}
```

5. We can now create our registration description with the appropriate (that is, platform-specific) channel information, as well as the registration ID we have just created:

```
RegistrationDescription deviceRegistration = null;

switch(device.Platform)
{
    // ...
    case "apns":
        deviceRegistration = new
        AppleRegistrationDescription(device.Handle);
        break;
    //...
```

```
}

        deviceRegistration.RegistrationId = device.RegistrationId;
```

6. We will also need the current user identity as a tag during the registration. Let's add this tag, as well as the other ones that are sent by the user:

```
deviceRegistration.Tags = new HashSet<string>(device.Tags);

// Get the user email depending on the current identity provider
deviceRegistration.Tags.Add($"username:{GetCurrentUser()}");
```

7. Finally, we can complete the registration by passing the registration data to our hub:

```
await _hub.CreateOrUpdateRegistrationAsync(deviceRegistration);
```

Here, if we were using installation-based registration rather than device registration, we would have a lot more control over the registration process. One of the biggest advantages is the fact that device installation registration offers customized template associations:

```
var deviceInstallation = new Installation();
// ... populate fields
deviceInstallation.Templates = new Dictionary<string,
InstallationTemplate>();

deviceInstallation.Templates.Add(
    "type:Welcome",
    new InstallationTemplate
    {
        Body = "{\"aps\": {\"alert\" : \"Hi ${FullName} welcome to
Auctions!\" }}"
    });
```

To take this one step further, during the registration process, the template can be loaded according to the preferred language of the user:

```
var template = new InstallationTemplate()
template.Body = "{\"aps\": {\"alert\": \"
            + GetMessage("Welcome", user.PreferredLanguage) + "\" }}";
```

This is how a user registers the device. Now, we will move on to transmitting notifications.

Transmitting notifications

Now that the user has registered a device, we can implement a send notification method on the server. We would like to support free text notifications, as well as templated messages with data. Let's begin:

1. Let's start by creating a base notification method that will define the destination and the message:

```
public class NotificationRequest
{
    public BaseNotificationMessage Message { get; set; }

    public string Destination { get; set; }
}
```

2. For the simple message scenario, the calling service module is not really aware of the target platform—it just defines a user and a message item. Therefore, we need to generate a message for all three platforms, hence covering all possible devices the user might have:

```
public class SimpleNotificationMessage : BaseNotificationMessage
{
    public string Message { get; set; }

    public IEnumerable<(string, string)> GetPlatformMessages()
    {
        yield return ("wns", @"<toast><visual><binding
        template=""ToastText01""><text id=""1"">"
        + Message + "</text></binding></visual></toast>");
        yield return ("apns", "{\"aps\":{\"alert\":\"" + Message +
        "\"}}");
        yield return ("fcm", "{\"data\":{\"message\":\"" + Message
        + "\"}}");
    }
}
```

3. For the templated version, the message looks a little simpler since we assume that the device already registered a template for the given tag and so we don't need to worry about the target platform. Now, we just need to provide the parameters that are required for the template:

```
public class TemplateNotificationMessage : BaseNotificationMessage
{
    public string TemplateTag { get; set; }
```

```
        public Dictionary<string, string> Parameters { get; set; }
}
```

> **TIP**
>
> It is generally a good idea to create a generic message template that will be
> used to send simple text messages and register this template as a simple
> message template so that we don't need to create a separate message for
> each platform and push it through all of the available device channels.

4. Now, we can process our notification request on the notification service and
   deliver it to the target user:

```
if (request.Message is SimpleNotificationMessage simpleMessage)
{
    foreach (var message in simpleMessage.GetPlatformMessages())
    {
        switch (message.Item1)
        {
            case "wns":
                await _hub.SendWindowsNativeNotificationAsync(
                    message.Item2,
                    $"username:{request.Destination}");
                break;
            case "aps":
                await _hub.SendAppleNativeNotificationAsync(
                    message.Item2,
                    $"username:{request.Destination}");
                break;
            case "fcm":
                await _hub.SendFcmNativeNotificationAsync(
                    message.Item2,
                    $"username:{request.Destination}");
                break;
        }
    }
}
else if(request.Message is TemplateNotificationMessage
templateMessage)
{
    await _hub.SendTemplateNotificationAsync(
        templateMessage.Parameters,
$"username:{request.Destination}");
}
```

We have successfully sent our message to the target user using the `username:<email>`
tag. This way, the user will be receiving the push notification on their device. It is up to the
platform-specific implementation to handle the message.

Broadcasting to multiple devices

In the previous example, we used the username tag to define a specific user as the target of the notification message. In addition to single tags, tag expressions can also be used to define a larger group of users that subscribed to a specific notification category. For instance, in order to send the notification to all of the users that accepted auction updates on new auctions and who are also interested in a specific auto manufacturer, the tag may look as follows:

```
notification:NewAuction && interested:Manufacturer:Volvo
```

Another scenario would be to send a notification about a new highest bid on an auction:

```
notification:HighestBid && auction:afabc239-a5ee-45da-9236-37abc3ca1375 &&
!username:john.smith@test.com
```

Here, we have combined three tags, the last of which is the tag that ensures that the user with the highest bid does not get the message that was sent to the rest of the users who are involved in the auction so that they can receive a more personalized *Congratulations* type of message.

Advanced scenarios

The notification messages demonstrated here are only simple models of notifications and do not require extensive implementation, neither on the client application nor on the server side.

Now, we will take a look at some more advanced usage scenarios.

Push to pull

Notification hubs customers in banking, healthcare, and government segments cannot pass sensitive data via public cloud services such as APNS, FCM, or WNS. In order to support these types of scenarios, we can utilize a scheme where the notification server is only responsible for sending a message ID and the client application only retrieves the target message using the given message ID. This pattern is generally referred to as the Push-2-Pull pattern, and it is a great way to migrate the communication channel from the notification hubs and PNSes to the common web service channel.

Rich Media for push messages

Notification messages sent by the native notifications provider (PNS) can also be intercepted and processed before they are presented to the user. On the iOS platform, the Notification Extension framework allows us to create extensions that can modify mutable notification messages.

In order to create a notification extension, you can add a new project and select one of the Notification Extensions that suits your requirements the most:

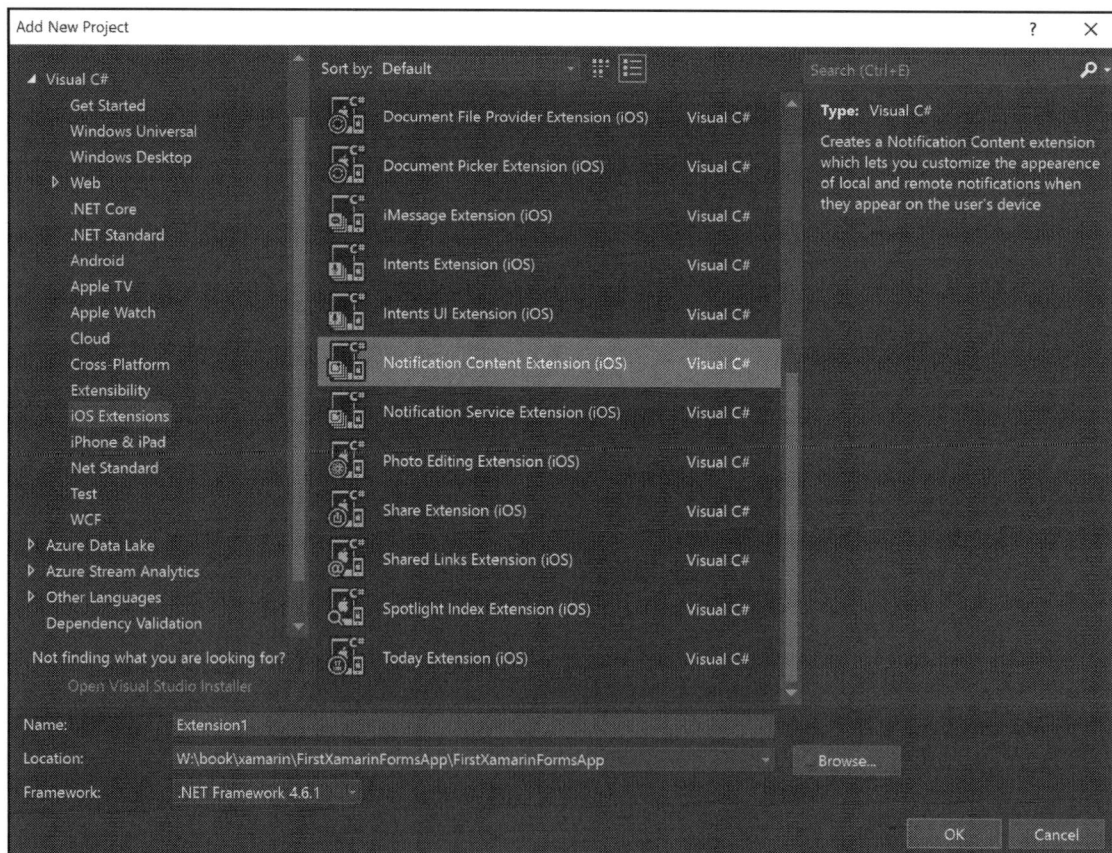

With the context extension, it is possible to create interactive notification views that can display rich media content, while the service extension can be used to intercept and process the notification payload (for example, decrypt the secure payload).

Android has similar mechanisms that allow you to modify the notification content before it reaches the user. The toast messages can also exhibit a reply control, which accepts in-place user interaction. While this engagement model increases the usability of the application, it does not contribute to the return rate.

Overall push notifications are powerful developer tools that attract users back to the application and keep them engaged, even when the application is not running.

The Graph API and Project Rome

Since we're discussing cross-platform engagement, this is a great time to talk about the Graph API and Project Rome. These interwoven infrastructure services that are available in the Microsoft Cloud infrastructure allow developers to create application experiences that span across platform and device boundaries.

The Graph API

The Graph API is a collection of Microsoft cloud services that are used to interact with the data that is collected through various platforms, including Microsoft Office 365 and Microsoft Live. The data elements in this web of data are structured around the currently signed in user. Interactions with certain applications (for example, a meeting that's been created, an email that's been sent, or a new contact being added to the company director) or devices (for example, a sign-in on a new device) are created as new nodes in the relationship graph.

These nodes can then be used to create a more immersive experience for the user, who is interacting with the application data from various sources.

For instance, simply by using the Microsoft Identity within the application, after the user authorizes to access the resources, the application downloads documents, creates thumbnails, retrieves destination users, and creates a complete email before sending it to a group of users. Other than Microsoft-based applications, third-party applications can also create graph data to increase engagement with the user.

Project Rome

Project Rome, which is built upon the premises of the Graph API, can be defined as a device runtime for connecting and integrating Windows-based and cross-platform devices to the Project Rome infrastructure services. This runtime is the bridge between the infrastructure services in the Microsoft Cloud and the APIs that are delivered as a programming model for Windows, Android, iOS, and Microsoft Graph, hence enabling client and cloud apps to build experiences using Project Rome's capabilities.

Project Rome exposes several key APIs, most of which can be used in cross-platform application implementations. The current API set is composed of the following:

- Device relay
- User activities
- Notifications
- Remote Sessions
- Nearby sharing

Some of these features are available only for the Windows platform, but all of them require a Microsoft identity to be present within the application domain. One important note is that all of these features can be consumed through the Graph API using the REST interface.

Let's take a closer look at these features and work out different use cases.

Device relay

The device relay is a set of modules that allows device to device communication and handover functionality. Semantically, the feature set resembles the previously available application services and URI launch functionalities on the UWP platform. However, with the device relay, an application can essentially communicate with another application on another device or even on another platform (for example, using the SDK, it is possible to launch an application on a Windows device from an Android phone).

User activities

User activities is one of the closest integrations with the Graph API. Using the user activities and the activity feeds, it is possible for an application to create a history feed with relevant actions when the application was last used. These feeds are fed into the Graph API and synchronized across devices. For instance, using the activity feeds, we can create a history of the auctions/vehicles that the user viewed in a certain session. Then, once these items are synchronized through the same live ID over to a Windows device, the user can easily click on an item from the feed and get back to their previous session. If the Windows device does not have the application installed, the operating system will advise the user to install the same application on the Windows platform. This way, the penetration of your application will increase on multiple platforms. In simple terms, this feature is solely dedicated to the continuity of user experience across devices and platforms.

A user activity might contain deep links, visualizations, and metadata about the activity that is created as an entry in activity history.

Notifications

Project Rome notifications, also known as Microsoft Graph Notifications, are another way of notifying users. What differentiates Graph Notifications from other platform-specific notification providers is the fact that Graph Notifications target users that are specifically agnostic to the platform they are using. In other words, the destination of the Graph Notifications is not devices, but users. In addition to the user-based notification model, the notification state within Graph Notification infrastructure is synchronized across devices so that the application itself does not need to anything more to set a certain message as dismissed or similar to reflect the user interaction on multiple devices.

Remote Sessions

The Remote Sessions API is a Windows-only API, and it allows devices to create shared sessions, join the sessions, and create an interactive messaging platform between different users. The created sessions can be used to send messages across devices, as well as keep shared session data within the joint session.

Nearby sharing

Nearby sharing allows apps to send files or websites to nearby devices using Bluetooth or Wi-Fi. This API can be used on Windows as well as native Android runtimes. During a share operation, nearby sharing functionality is also intelligent enough to pick up on the quickest path between the two devices by selecting either a Bluetooth or network connection. The discovery function that is part of the sharing module allows the application to discover the possible targets for the share operation via Bluetooth.

Summary

In this chapter, we have taken a look at ways to improve user engagement using push notifications and Microsoft Graph API implementations. Keeping user engagement alive is the key factor to maintaining the return rate of your application. Push notifications is an excellent tool that you can use engage your users, even when your application is not active. Azure Notification Namespaces and Hubs make this engagement a lot easier to implement by creating an abstraction layer between the PNSes and the target device runtimes. On top of the push notifications, we analyzed various APIs that are available on the Graph API through Project Rome and RESTful APIs that are readily available.

We have successfully engaged our user, so in the next chapter, we will try to surprise them with machine learning and cognitive services gimmicks to create not only a responsive and fluid application user interface, but also an intelligent one.

Introducing Cognitive Services 14

Azure Cognitive Services are a set of machine learning algorithms that are used to solve most common AI programming tasks. Azure Cognitive Services streamlines complex algorithms and turns them into trivial tasks, such as executing a simple remote query. It is composed of five distinct categories of algorithms, namely, vision, speech, language, knowledge, and search. In this chapter, we will be adding speech recognition to our application using the speech API.

The following sections will walk you through how you can introduce cognitive services into your projects:

- Understanding Cognitive Services
- Speech APIs
- Computer vision
- Search APIs

Understanding Cognitive Services

AI and machine learning have always been attractive topics for computer scientists and software developers. Nevertheless, these topics require a deep understanding of complex subjects, such as neural networks, convolutions, ane clustering. In addition to the complexities of these machine learning concepts, the applications of these concepts can become even more complex and time-consuming. For developers that do not posses the skillset or the time to introduce complex machine learning algorithms to their projects, Azure Cognitive Services is a reasonable alternative.

Azure Cognitive Services is a set of APIs that can help developers to create intelligent applications with minimal knowledge of artificial intelligence and machine learning. These APIs can be used as part of the Azure cloud infrastructure and are grouped under five categories:

- **Vision APIs**: The Vision API consists of several APIs that deal with image processing and pattern recognition. Computer vision, face detection, and the ink recognizer are only a few of the APIs that are available in this category.
- **Speech APIs**: The Speech API services deal with audio and natural language processing. Speech Services is the main service bundle that deals with natural language processing and synthesized speech. The speaker recognition API, on the other hand, is the vocal recognition service offered within this category.
- **Language APIs**: Natural language processing and translation services offered under this category can provide valuable assets for processing lexical data.
- **Search APIs**: The Search API bundle exposes a wide spectrum of API endpoints that expose various features from Bing search.
- **Decision APIs**: The Decision API services are most apparent in the implementation of pattern recognition. The personalizer API can be of great use for identifying user patterns and anticipating user behavior to customize application content.

As a complete package, Azure Cognitive Services can be introduced to your Azure subscription via the Azure Portal:

Cognitive Services
Microsoft

Cognitive Services
Microsoft

Create   ♡ Save for later

Cognitive Services is a product bundle that enables customers to access multiple services with a single API key.

Product features:

Access to Vision, Language, and Search services using a single API

Quickly connect services together to achieve more insights into your content

Easily integrate with other services like Azure Search

Legal Notice

Microsoft will use data you send to Bing Search Services to improve Microsoft products and services. Where you send personal data to this service, you are responsible for obtaining sufficient consent from the data subjects. The Data Protection Terms in the Online Services Terms do not apply to Bing Search Services.

Please refer to the Online Services Terms for details. Microsoft offers policy controls that may be used to disable new deployments.

Similarly, standalone services can also be created within a resource group instead of us having to select the complete set of services.

At the time of writing, there is no complete SDK available for all of the cognitive services. Hence, once the services are created within a resource group, the services can be accessed through the service APIs. The more straightforward way of authentication and/or authorization is to use the `Ocp-Apim-Subscription-Key` header to define the specific API subscription that is desired. Additionally, access tokens that are trusted by Azure Active Director can also be used to execute RESTful calls.

Speech APIs

The Speech API is the unified version of the speech-to-text and text-to-speech services, and is the successor of the Bing Speech API. In addition to recognizer and synthesizer, speech translation provides real-time, multi-language translation capabilities.

Speech to text

Azure Speech Services provides a very competent alternative to the native speech recognition features of mobile platforms. Azure Speech Services is a proven technology that is also used for Cortana and supports multiple languages. Speech to text uses the universal language model that was trained by Microsoft-owned data, but it can also be trained further for custom acoustics, language, and pronunciation.

Speech services are available through various SDKs and the RESTful API. Unfortunately, the Xamarin SDK is currently not available (only native versions are available for Java and Objective-C). Nevertheless, .NET Core SDK is ready to use and can be used with UWP applications.

Some of the key features of this transcription service are as follows:

- **Short utterance**: It can transcribe a short utterance (that is, less than 15 seconds). This functionality is available through the SDK and the REST API.
- **Continuous transcription**: This feature provides a transcription of longer utterances, as well as continuous streaming audio. It is only available through the SDK.
- **Intent recognition**: Using the language services provided by LUIS, the speech SDK can identity intents and entities. The REST API does not have direct integration with this feature.
- **Batch transcription**: The REST API can process multiple audio files asynchronously.

Here, the most used speech to text features are the short utterance and the batch transcription since these two features provide an easy way to convert short user commands into application commands.

In order to implement speech recognition, we would need to use the cognitive services REST—more specifically, the speech services. Let's begin with the implementation.

The implementation process starts by authenticating with the identity provider using the subscription key:

```
public AuthenticationService(string apiKey)
{
    subscriptionKey = apiKey;
    httpClient = new HttpClient();
    httpClient.DefaultRequestHeaders.Add(
        "Ocp-Apim-Subscription-Key", apiKey);
}

public async Task<string> FetchTokenAsync(string fetchUri)
{
    UriBuilder uriBuilder = new UriBuilder(fetchUri);
    uriBuilder.Path += "/issueToken";
    var result = await
        httpClient.PostAsync(uriBuilder.Uri.AbsoluteUri, null);
    return await result.Content.ReadAsStringAsync();
}
```

Once the authentication token is received, it should be included in the consequent requests as an authorization token with bearer schema. The recognition API is exposed though the following endpoint:

```
https://speech.platform.bing.com/speech/recognition/
```

Now, we can execute the recognition request with a POST method call with a steam content type:

```
pulic async Task<string> SendRequestAsync(
    Stream fileStream, string contentType)
{
    if (httpClient == null)
    {
        httpClient = new HttpClient();
    }

    httpClient.DefaultRequestHeaders.Authorization = new
        AuthenticationHeaderValue("Bearer", bearerToken);

    var content = new StreamContent(fileStream);
    content.Headers.TryAddWithoutValidation(
        "Content-Type", contentType);

    var response = await httpClient.PostAsync(url, content);
    return await response.Content.ReadAsStringAsync();
}
```

The language that's used during the recognition can be adjusted with the language parameter.

Once the recognized text is returned by the recognition API, our application can start processing it. The most convenient way to parse the recognized text into lexical components and execute commands would be to use the language API.

Language Understanding Service

Audio processing, combined with Language APIs, can provide even more structured data that our application can process. The **Language Understanding Service (LUIS)** allows applications to understand what a person really wants. In other words, using LUIS, your application can receive audio input in complete sentences and convert it into commands and execution procedures.

In order to be able to utilize LUIS, we need to create an application registration on the LUIS portal (that is, `https://www.luis.ai`):

Create new app

Name (Required)

> Auctions

Culture (Required)

> English   ▾

** Culture is the language that your app understands and speaks, not the interface language.

Description

> Vehicle Auctioning Application|

Done   Cancel

After the application is created, we will need to define the schema that the application will adhere to in order to identify the commands requested by the users similar to the example schema provided by the LUIS portal:

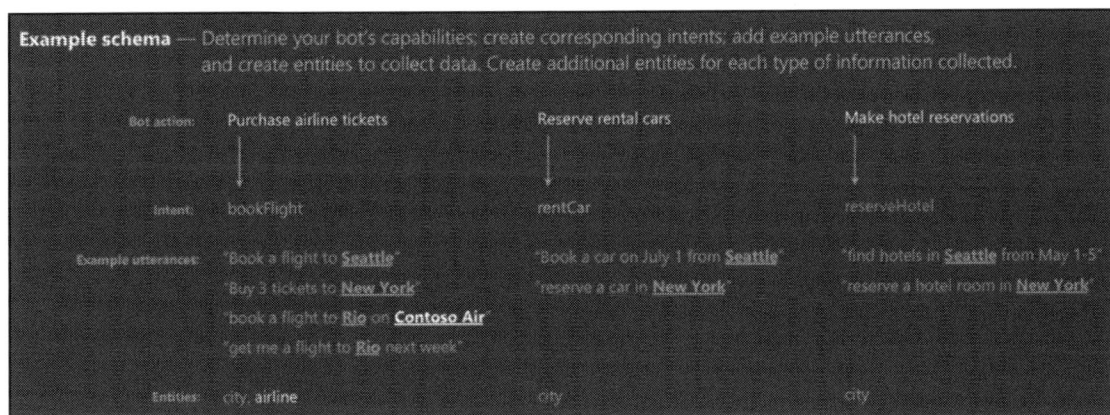

In order to be able to execute queries using the application ID and the private key, we would need to train and publish the application. This can be done on the application dashboard:

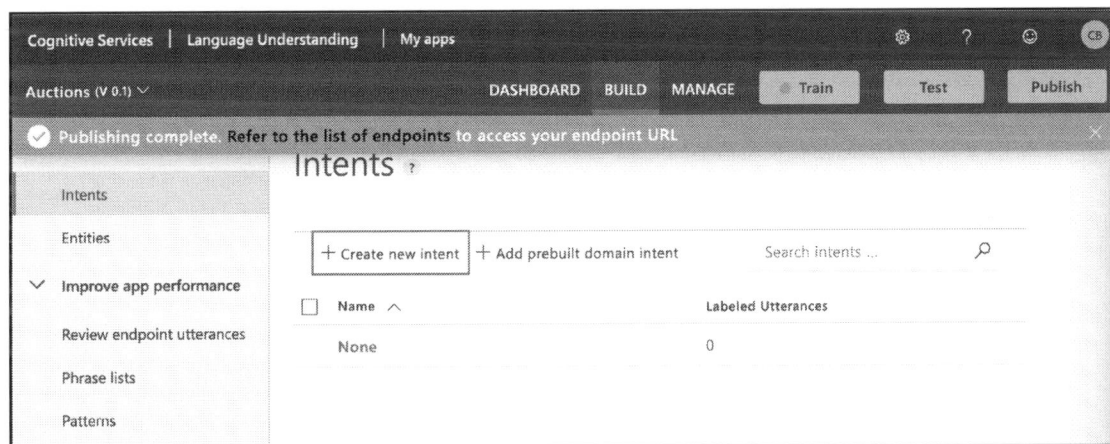

The REST queries can be executed on the developer portal, as well as by using an HTTP request tool such as CURL or postman. The URL we will be using for LUIS applications with the application ID and private key is as follows:

```
https://westus.API.cognitive.microsoft.com/luis/v2.0/apps/84d54230-f80d-41b
3-862b-343183f36411?q=I am looking for a red volvo s60 from 2018
```

Without any schema definition about intents and entities, the result to any `query` will be empty:

```
{
    "query": "I am looking for a red volvo s60 from 2018",
    "topScoringIntent": {
        "intent": "None",
        "score": 0.823669851
    },
    "entities": []
}
```

Now, let's define our main entity in this sentence (that is, **volvo**, which is a vehicle brand):

We are using the list type to create a predefined list of values for auto manufacturers. These data points will be used to determine the make of the vehicle that the users will be uttering during the voice search. Other entity type options are as follows:

- **Simple**: A simple entity describes a single concept. For example, if the user's intent is `GetWeather`, you can use `City` as a simple entity to capture the city for the weather report.
- **Composite**: You can use a composite entity to represent an object that has parts. The composite entity is made up of entities that form the whole. For example, a composite entity called `TicketsOrder` in a travel app can be composed of three child entities that describe attributes of the tickets to order, that is, `Number`, `PassengerCategory`, and `TravelClass`.
- **Regex**: A regex entity is an entity that matches based on the regular expression defined. Regex entities are not machine learned entities.
- **Pattern**: A pattern entity is a variable-length placeholder that's used only in a pattern's template utterance to mark where the entity begins and ends.

We will now move on and create an intent for the verb **look**:

As you can see from the preceding screenshot, the sentences started to make much more sense to LUIS.

After adding color as a list entity and training the application, the results for the same sentence are much more reliable:

Training the LUIS application with additional entities and other intents would give better matches and classified data that can be used as requests in our search queries.

Computer vision

Computer vision is another intelligent services that allows applications to identify various entities. Applications of this feature can vary from simple image recognition on a picture to face recognition with training data:

| Objects | [{ "rectangle": { "x": 542, "y": 115, "w": 1397, "h": 765 }, "object": "car", "parent": { "object": "Land vehicle", "parent": { "object": "Vehicle", "confidence": 0.914 }, "confidence": 0.914 }, "confidence": 0.909 }] |
| Tags | [{ "name": "outdoor", "confidence": 0.998841 }, { "name": "car", "confidence": 0.998357356 }, { "name": "road", "confidence": 0.9949555 }, { "name": "tree", "confidence": 0.9877573 }, { "name": "blue", "confidence": 0.9813743 }, { "name": "grass", "confidence": 0.968876243 }, { "name": "vehicle", "confidence": 0.9148866 }, { "name": "land vehicle", "confidence": 0.8881643 }, { "name": "transport", "confidence": 0.7687348 }, { "name": "full-size car", "confidence": 0.7539416 }, { "name": "wheel", "confidence": 0.708732 }, { "name": "ford", "confidence": 0.608252943 }, { "name": "subaru", "confidence": 0.518539 }, { "name": "mid-size car", "confidence": 0.506383657 }] |

The item identifications can be improved with domain-specific models and additional tagging.

The implementation of this API is done through the REST API, as shown in the following code:

```
static async Task MakeAnalysisRequest(string imageFilePath)
{
    try
    {
        HttpClient client = new HttpClient();
        // Request headers.
        client.DefaultRequestHeaders.Add(
            "Ocp-Apim-Subscription-Key", subscriptionKey);

        string requestParameters =
        "visualFeatures=Categories,Description,Color";
        string uri = uriBase + "?" + requestParameters;

        HttpResponseMessage response;
        byte[] byteData = GetImageAsByteArray(imageFilePath);

        using (ByteArrayContent content = new
        ByteArrayContent(byteData))
        {
            content.Headers.ContentType;
            new MediaTypeHeaderValue("application/octet-stream");
            // Asynchronously call the REST API method.
            response = await client.PostAsync(uri, content);
        }
        // Asynchronously get the JSON response.
        string contentString = await
```

```
            response.Content.ReadAsStringAsync();
            // Display the JSON response.
            Console.WriteLine("\nResponse:\n\n{0}\n",
            JToken.Parse(contentString).ToString());
        }
        catch (Exception e)
        {
            Console.WriteLine("\n" + e.Message);
        }
    }
}
```

Search API

The Bing Search API, as its name suggests, is the exposed service version of Bing search. Using the Search API, you can create customized/tailored search queries and serve content to your users through your application. These search queries are highly customizable, just like the result set.

The Search API offers a rich set of features that help you to integrate web content into mobile apps. Each of these features can abide the custom configuration you have set up, hence defining the search scope and restricting the results to certain criteria.

> In order to execute the queries outlined in the following topics, you would need to retrieve a subscription key either from your paid subscription or the trial account that can be created. The subscription key should be added using the `Ocp-Apim-Subscription-Key` header. The API management console and documentation that is available to developers on an API basis can be used to executed these queries.

Search query completion

The most attractive of these features for any developer looking to create a search dialog is the custom real-time search suggestions. We do so, using the suggestions API, possibly combining it with a custom configuration. In order to utilize custom suggestions, you would need to subscribe to the `BingCustomSearch` API. For autocomplete, you can use the REST API with the following endpoint:

```
https://api.cognitive.microsoft.com/bingcustomsearch/v7.0/suggestions/searc
h?customconfig=0&q=<myqueryparameter>
```

The response will display the result set with additional metadata, such as the display name and search type, as well as the original query string.

Web Search API

The custom search API is technically the service oriented implementation of Bing search. In addition to the complete search result metadata pertaining to the result type (for example, news, web page, and image), the result set is also categorized according to the available content-specific search result types.

The resultant metadata can be further improved by using the textDecorations query parameter. This way, the result content would have highlighted sections using specific unicode characters (that is, *U+E000* marks the beginning of the query term and *U+E001* marks the end of the query term).

Looking at the data model, we can understand the data structure and how this data can be used within the application. In the following example, you can see that the webPages property is populated with web page results:

```
"webPages": {
    "webSearchUrl": "https:\/\/www.bing.com\/search?q=volvo+s60",
    "totalEstimatedMatches": 1700000,
    "value": [{
        "id":
"https:\/\/api.cognitive.microsoft.com\/api\/v7\/#WebPages.0",
        "name": "2019 All-New S60 Luxury Sport Sedan | Volvo Car USA",
        "url": "https:\/\/www.volvocars.com\/us\/cars\/new-models\/s60",
        "isFamilyFriendly": true,
        "displayUrl": "https:\/\/www.volvo cars.com\/us\/cars\/new-
models\/s60 ",
        "snippet": "The S60  offers a choice of Drive Modes, so you can
tailor your driving experience to your mood. There are three default
settings for the engine, automatic gearbox, steering, brakes and stability
control system that put the focus on driving, comfort or fuel economy and
emissions.",
        "dateLastCrawled": "2019-05-12T02:30:00.0000000Z",
        "language": "en",
        "isNavigational": false
    },
    ...
    ]
```

The result is returned as a response to the following query:

```
https://api.cognitive.microsoft.com/bing/v7.0/search?q=volvo
s60&count=10&offset=0&mkt=en-us&safesearch=Moderate
```

Specialized search APIs are available through dedicated API endpoints, and each of these APIs have more content-specific search query parameters.

Image search

Image search is another feature that can help developers with search queries that target the image content. Image search is also available through the Bing Image Search API, and the result sets are comparable. However, the Bing Image Search API also offers parameters that can narrow down the image set with additional criteria, such as aspect, color, height, width, and size.

The Bing Image Search API is available through the following URL:

```
https://api.cognitive.microsoft.com/bing/v7.0/images/search
```

Using the Image Search API, the application can generate image content on the fly, creating a more pleasant and rich experience for the user. If you are planning to use these images as part of the user content, it is important to execute the search with the licensing options so that the returned content does not violate any restrictions. The available licensing options and associated explanation from the official provider are Any

- **Any**
- **Public**
- **Share**
- **ShareCommercially**
- **ModifyCommercially**
- **All**

In order to demonstrate the search results for images, we can execute a quick query by utilizing the query and color parameters:

```
https://api.cognitive.microsoft.com/bing/v7.0/images/search?q=volvo s60
2019&count=10&offset=0&mkt=en-us&safeSearch=Moderate&color=blue
```

As you can see, the market parameter in this query is set to en-us, which will return results that are more relevant to this market.

Other

In addition to web pages and images, Bing can execute search queries on additional content types, which can help to create connected applications and provide rich media content. The following additional specialized APIs are available through the Bing API services:

- **Entity search**: The Bing Entity Search API returns search results containing entities, which can be people, places, and so on. Depending on the query, the API will return one or more entities that satisfy the search query. The search query can include noteworthy individuals, local businesses, landmarks, destinations, and more.
- **News search**: The Bing News Search API lets you find news stories similar to Bing.com/news. The API returns news articles from either multiple sources or specific domains. You can search across categories to get trending articles, top stories, and headlines.
- **Video search**: The Bing Video Search API lets you find videos across the web. Here, you can get trending videos, related content, and thumbnail previews.
- **Visual search**: Here, you can upload an image or use a URL for the image, to get insightful information about it, such as visually similar products, images, and related searches.
- **Local business search**: The Bing Local Business Search API lets your applications find contact and location information about local businesses based on search queries.

These content types are available through the custom search API results, as well as the specialized endpoints.

Summary

Cognitive Services can provide a wide variety of tools that add intelligence to your application and create a more intuitive and easy to interact with the user experience. As we have seen in this chapter, services such as speech recognition and language processing allow the users to interact with your application in a more natural way. Using computer vision, applications can anticipate user needs and act on them without the user interacting with the application in conventional terms. In addition, the Search API makes it easier to bring in additional content and provide a richer experience.

This chapter finalizes our application development process. Now, we will continue with **application lifecycle management (ALM)** with Azure DevOps in order to create CI/CD pipelines for .NET Core and Xamarin applications.

Section 5: Application Life Cycle Management

5

Azure DevOps and Visual Studio App Center are the two pillars of application life cycle management when we talk about .NET Core and Xamarin. Azure DevOps, previously known as Visual Studio Online or Team Services, provide the complete suite for implementing DevOps principles, whereas App Center acts as a command center for mobile application development, testing, and deployment. Using these tools, developers and operations teams can implement robust and productive delivery pipelines that can take the application source from the repository to production environments.

The following chapters will be covered in this section:

- Chapter 15, *Azure DevOps and Visual Studio App Center*
- Chapter 16, *Application Telemetry with Application Insights*
- Chapter 17, *Automated Testing*
- Chapter 18, *Deploying Azure Modules*
- Chapter 19, *CI/CD with Azure DevOps*

15
Azure DevOps and Visual Studio App Center

Visual Studio App Center is an all-in-one service provided by Microsoft, and is mainly used by mobile application developers. Both Xamarin platforms as well as UWP applications are among the supported platforms. The primary goal of App Center is to create an automated Build-Test-Distribute pipeline for mobile projects. App Center is also invaluable for iOS and Android developers since it is the only platform that offers a unified beta distribution for both target runtimes that support telemetry collection and crash analytics. Using Azure DevOps (previously known as Visual Studio Team Service) and App Center, developers can set up a completely automated pipeline for Xamarin applications that will connect the source repository to the final store submission.

This chapter will demonstrate the fundamental features of Azure DevOps and App Center, and how to create an efficient application development pipeline suited to individual developers, as well as development teams.

The following topics will be covered in this chapter:

- Using Azure DevOps and Git
- Creating Xamarin application packages
- App Center for Xamarin
- Distribution with AppCenter
- App Center telemetry and diagnostics

Using Azure DevOps and Git

The first and foremost crucial module of Azure DevOps that is utilized by developers is the available source control options. Developers can choose to use either **Team Foundation Version Control** (**TFVC**) or Git to manage the source code (or even both at the same time). Nevertheless, with the increasing popularity of decentralized source control management, because of the flexibility and integration that the development toolset offers, Git is the more favorable choice to many. Git is natively integrated with both Visual Studio and Visual Studio for Mac.

Creating a Git repository with Azure DevOps

Multiple Git repositories can be hosted under the same project collection in Azure DevOps, depending on the project structure that is required. Each of these repositories can be managed with different security and branch policies.

In order to create a Git repository, we will use the Repos section of Azure DevOps. Once a DevOps project is created, an empty Git repository is created for you that needs to be initialized. The other options here would be to import an existing Git repository (not necessarily from another Azure DevOps project or organization) or push an existing local repository from your workstation:

MyFirstXamarinForms is empty. Add some code!

∧ **Clone to your computer**

HTTPS SSH https://xplatformdev@dev.azure.com/xplatformdev/MyFirstXamar...   ⧉   **OR**   ⧉ **Clone in VS Code** ∨

Generate Git credentials

ⓘ Having problems authenticating in Git? Be sure to get the latest version of Git for Windows or our plugins for IntelliJ, Eclipse, Android Studio or macOS & Linux terminal.

∧ **or push an existing repository from command line**

HTTPS SSH

```
git remote add origin https://xplatformdev@dev.azure.com/xplatformdev/MyFirstXamarinForms/_git/MyFirstXamarinForms
git push -u origin --all
```

∧ **or import a repository**

Import

∧ **or initialize with a README or gitignore**

☑ Add a README   Add a .gitignore: **None** ∨   **Initialize**

> The repository type choice used to be an initial decision while creating the Visual Studio Team Services project that couldn't be changed. However, with Azure DevOps, it is possible to create Git repositories even after the initial project's creation, side by side with a TFVC repository.

It's important to note that the clone option offers the generation of Git credentials in order to authenticate with this Git instance. The main reason for this is that Azure DevOps utilizes federated live identity authentication (possibility two factor), which Git doesn't support by itself. Thus, the users of this repository would need to generate a **Personal Access Token (PAT)** and use it as a password. The Git for Windows plugin automatically handles this authentication issue by creating the PAT automatically (and renewing it once it is expired). PAT is currently the only solution to be able to use Git with Visual Studio for Mac.

In the initialize option, we can also select the `.gitignore` file (similar to TFVC's `.tfignore` file) type to be created so that undesired user data from the project folders isn't uploaded to the source repository.

Branching strategy

Using Git and the Git flow methodology/pattern, a development team can base their local and remote branches on two main branches: development and master.

The development branch is used as the default branch (also known as the trunk) and represents the next release source code until the set of features (branches) are completed and signed off by the development team. At this point, a release branch is created, and the final stabilization phase on the next release package uses the release branch as the base for all hot-fix branches for development. Each pull request from a hot-fix branch that changes the release version will need to be merged back into the development branch.

A general flow of the feature, development, and release branches is shown in the following diagram:

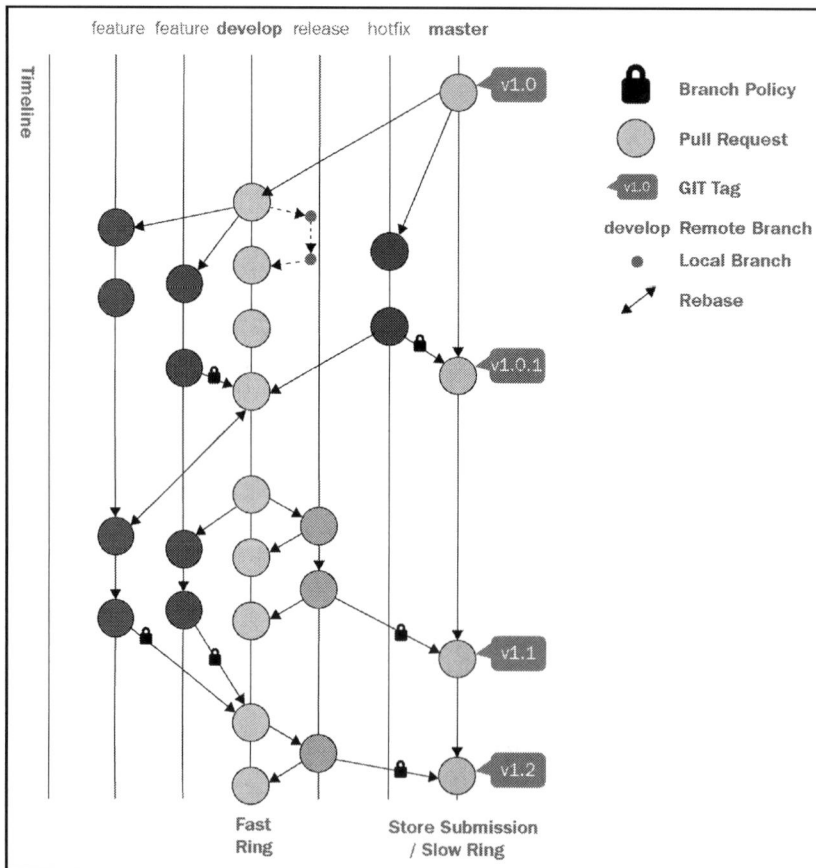

This general flow of the branches is as follows:

1. To safeguard the development branch as well as release branches, developers, whether working on local or feature branches, will need to create a pull request that will be verified per the branch policies.

   > Pull requests promote the peer review process, as well as additional static analysis that needs to be executed in order to contribute to the development or release branches.

2. In this setup, the development branch will have a continuous integration build and deployment to the fast ring of the AppCenter distribution so that the dev and QA team can verify the features that are merged into the branch immediately.

3. Once the current release branch is ready for regression, it can be merged into the master branch. This merge is generally verified with automated UI tests (that is, automated regression). The master branch is used as the source repository for the Visual Studio App Center slow ring deployments (that is, the staging environment).

4. Feature branches that go stale (see the outer feature branch) span across multiple releases and need to be rebased so that the development branch history is added to the feature branch. This allows the development team to have cleaner metadata about the commits.

5. The hot-fix branch is used to rectify either failed store submissions or regression bugs that are reported on release branches. Hot-fix branches can be tested by automated UI tests (fully automated regression) and manually using slow ring releases.

6. Once the master branch is ready for release, a new tag is created as part of a code freeze, and a manually triggered build will prepare the submission package for iOS and Android versions of the application.

This methodology can also be modified to use the release branches for staging and store deployments, rather than using the master branch. This approach provides a little more flexibility for the development team and is a little easier than managing a single release branch (that is, the master branch).

Managing development branches

During the development phase, it is important—especially if you are working with a bigger team—to keep a clean history on the development and feature branches.

In an agile-managed project life cycle, the commits on a branch (either local or remote) state that tasks belong to a user story or a bug, while the branch itself may correlate with the user story or bug. Bigger, shared branches among team members can also represent a full feature. In this context, in order to decrease the amount of commits while still keeping the source code safe, instead of creating a new commit each time a change set push occurs, you can make use of the amend commit feature:

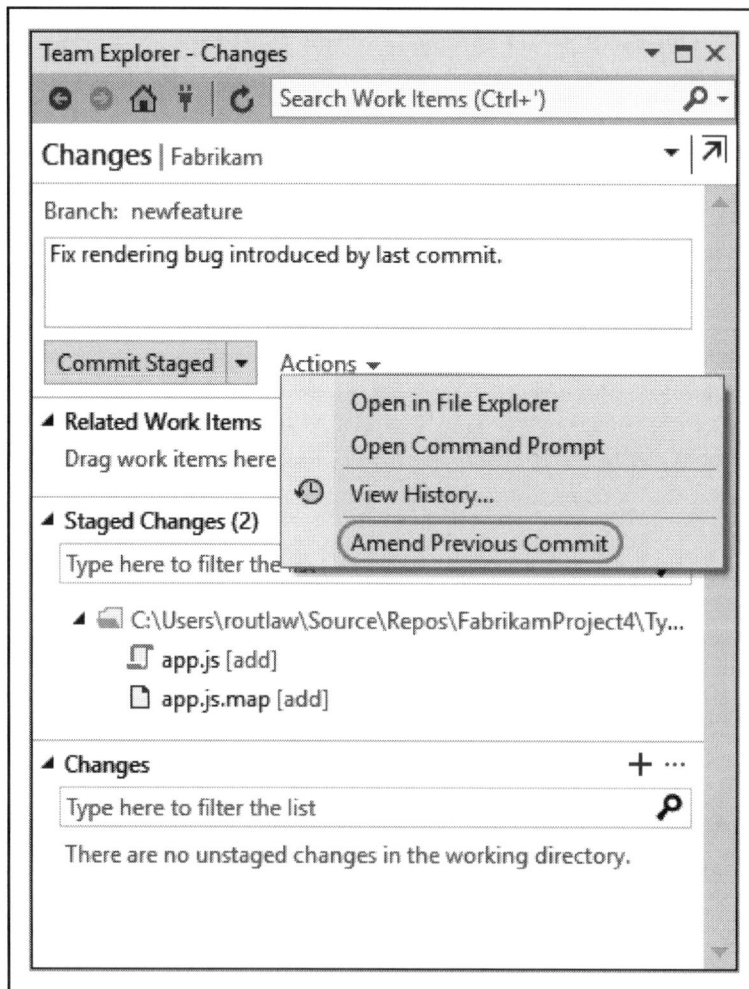

```
Team Explorer - Changes                         ▼ □ ✕

 ⟲ ⊝ ⌂ ⚑ | ⟳   Search Work Items (Ctrl+')          ⌕ ▾

Changes | Fabrikam                                ▼ | ↗

Branch: newfeature

 Fix rendering bug introduced by last commit.



 [ Commit Staged  | ▾ ]  Actions ▾
                          ┌────────────────────────────────┐
 ▲ Related Work Items     │   Open in File Explorer         │
   Drag work items here   │   Open Command Prompt           │
                          │  ⟲ View History...              │
 ▲ Staged Changes (2)     │  (  Amend Previous Commit  )    │
   Type here to filter the└────────────────────────────────┘

     ▲ 📁 C:\Users\routlaw\Source\Repos\FabrikamProject4\Ty...
          🗋 app.js [add]
          🗋 app.js.map [add]

 ▲ Changes                                         + ⋯
   Type here to filter the list                    ⌕
   There are no unstaged changes in the working directory.

```

Once the commit is amended with the changes, the local commit would need to be merged to the remote version. Here, the key to avoiding the merge commit between the remote and local branch (since the remote branch has the older version of the commit, hence a different commit) would be to use the **--force** or **--force-with-lease** options with the push command. This way, the local commit (the amended one) will overwrite the remote version:

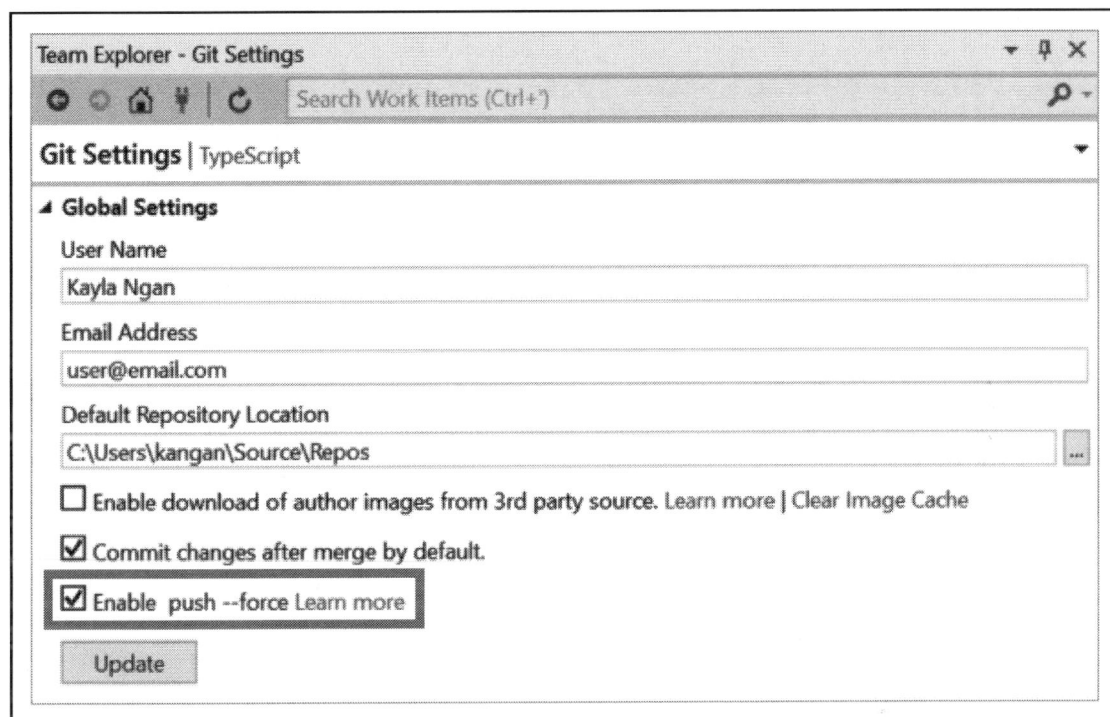

It is important to note that it is highly discouraged to amend commits and overwrite remote branches when multiple developers are working on the same branch. This can create inconsistent history for a branch on local repositories of involved parties. In order to avoid such scenarios, the feature branch should be branched out and rebased onto the latest version of the feature branch once it's ready to be merged.

Let's assume that you create a local branch from the remote feature branch and pushed several commits. Meanwhile, your teammates pushed several updates to the feature branch:

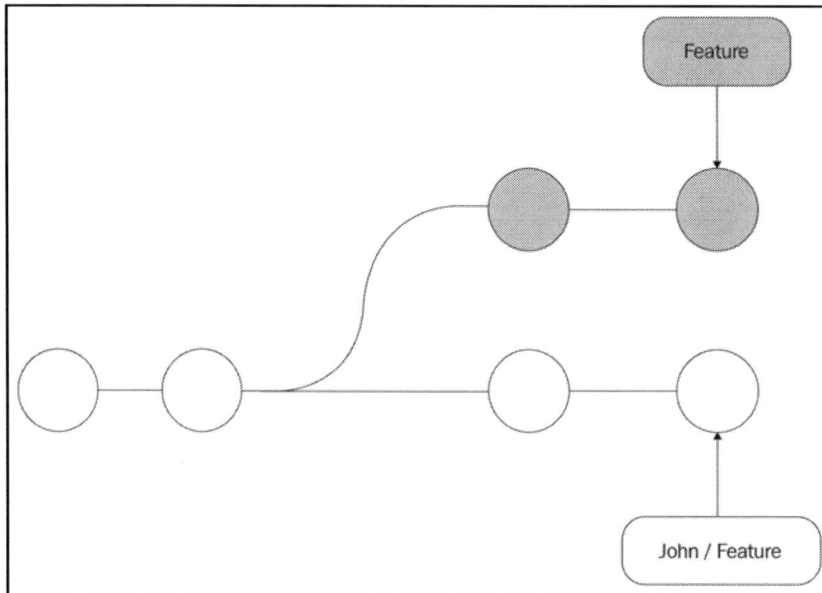

With a conventional sync (pull and push), there will be a merge commit created, and the history for the feature branch would look similar to the following:

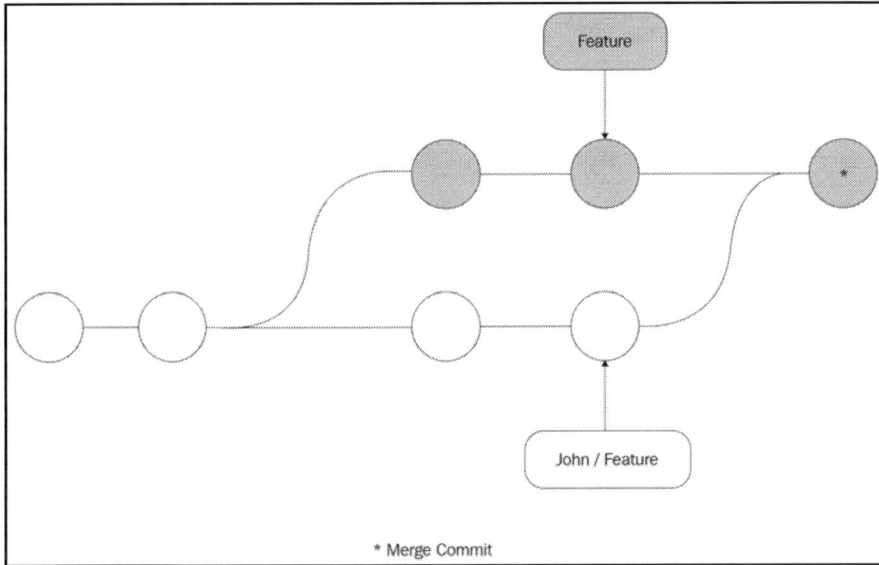

* Merge Commit

However, if the local branch is rebased onto the latest version of the remote branch prior to the push, the history would look like this:

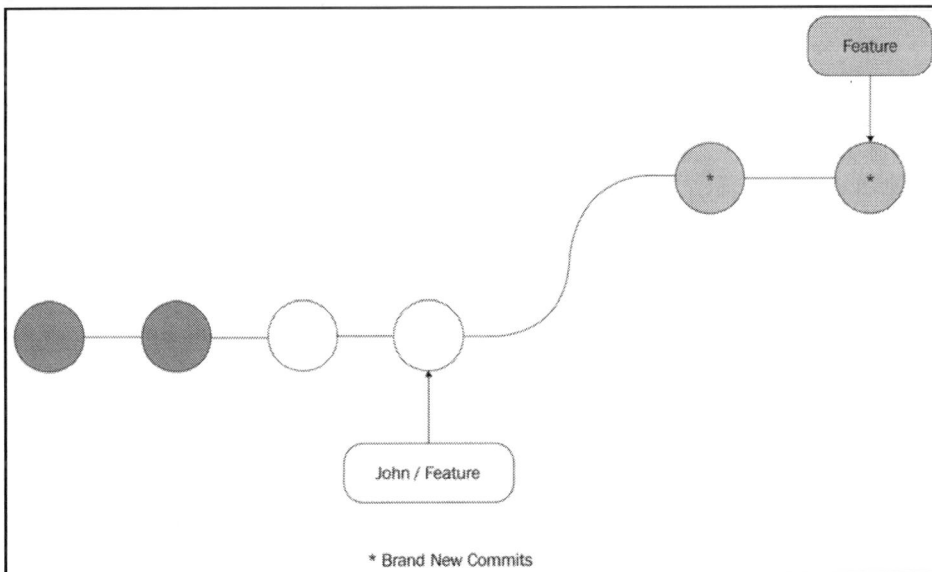

* Brand New Commits

A rebase strategy should be employed before you create a pull request to the development or the release branches so that the clean history of the branch can be preserved, thus avoiding complex merge conflicts.

Creating Xamarin application packages

Once our application is ready to be tested on real devices, we can start preparing the pipeline so that we can compile and package the application to be deployed to our alpha and beta environments on App Center. Azure DevOps provides out-of-the-box templates for both Xamarin.Android and Xamarin.iOS applications. These pipelines can be extended to include additional testing and static analysis.

Using Xamarin build templates

Creating a build and release template for Xamarin applications is as trivial as using the Xamarin template for iOS and Android. Additionally, the UWP (if your application supports this platform) template can also be used to create a UWP build:

After the pipeline is created, we will need to make several small adjustments to both platforms to be able to prepare the application so that we can put it on real devices.

Xamarin.Android build

Let's have a look at the steps to build the Android project:

1. We have to identify the correct Android project to be built using the wildcard designation and the target configuration:

Xamarin.Android ⓘ   ᱻ Link settings   🗋 View YAML   ✕ Remove

Version   1.*   ⌄

Display name *

> Build Xamarin.Android project

Project *   ⓘ

> **/*Client.Droid*.csproj   ...

Target   ⓘ

Output directory   ⓘ

> $(build.binariesdirectory)/$(BuildConfiguration)

Configuration   ⓘ

> $(BuildConfiguration)

☑ Create app package   ⓘ

☐ Clean   ⓘ

> Note that the configuration and output directory are using pipeline variables. These parameters can be defined in the variables section of the pipeline configuration page.

2. Select a keystore file so that you can sign the application package. Unsigned application packages cannot be run on real Android devices. Keystore is a store that contains certificate(s) that will be used for singing the application package. If you are using Visual Studio for development (on Mac or Windows), the easiest way to generate an ad hoc distribution certificate would be to use the Archive Manager:

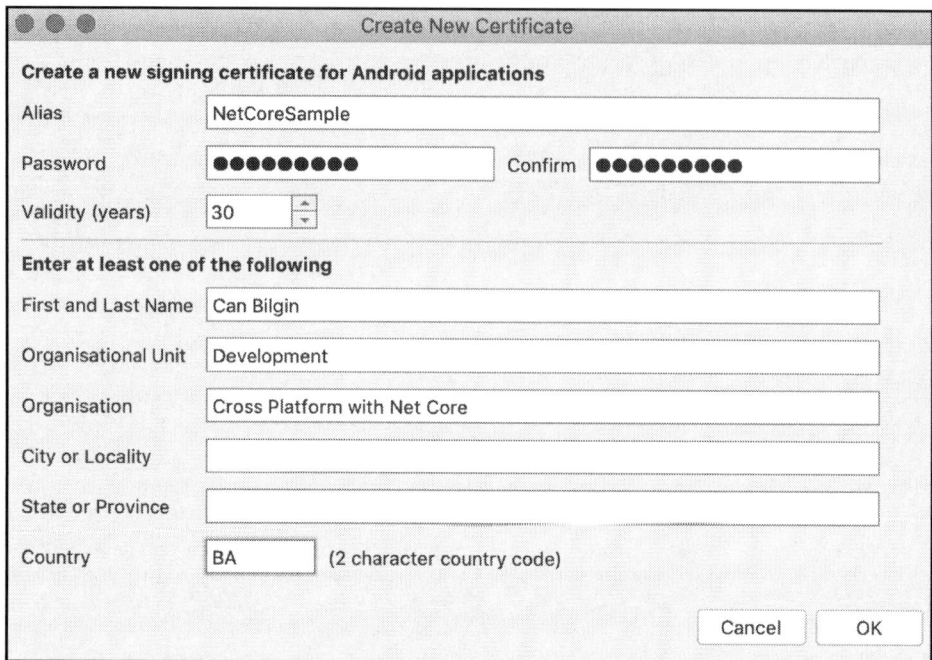

3. Once the store is created, the keystore file can be found in the following folder on Mac:

```
~/Library/Developer/Xamarin/Keystore/{alias}/{alias}.keystore
```

It can be found in the following folder on Windows:

```
C:\Users\{Username}\AppData\Local\Xamarin\Mono for
Android\Keystore\{alias}\{alias}.keystore
```

4. You can now use the .keystore file to complete the signing step of our Android build pipeline:

Note that the `.keystore` file is used by the pipeline as a secure file. In a similar fashion, the keystore password can (should) be stored as a secure variable string.

5. The pipeline, as it is right now, is ready to compile the Android version of the application in order to create an APK package.

Next, we will need to prepare a similar pipeline for the iOS platform.

Xamarin.iOS pipeline

Similar to its Android counterpart, the Xamarin.iOS template creates the full pipeline to compile the iOS project. We will need to modify the parameters for the created tasks so that we can successfully prepare the application package. Let's get started:

1. Before we start the pipeline configuration, head over to the Apple Developer site to generate a distribution certificate, Application ID, and an ah hoc provisioning profile:

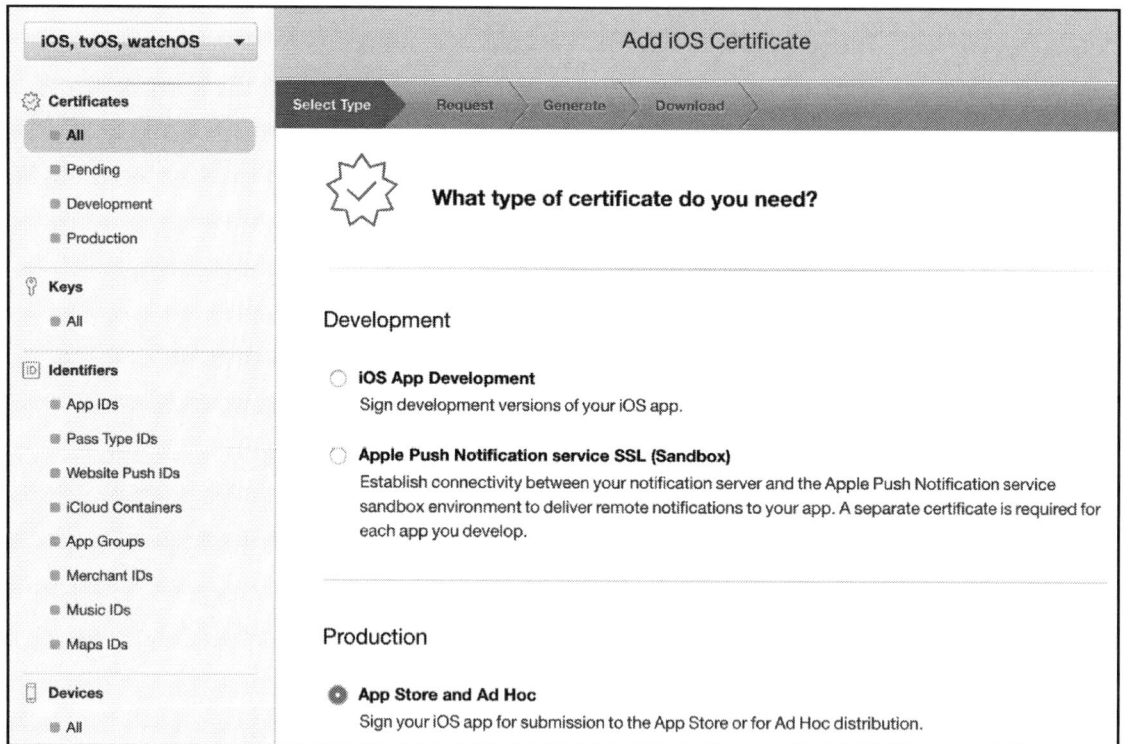

2. Once the distribution certificate option is selected, the developer site will take you through the steps of generating a CSR and generating the signing certificate.
3. After the certificate is created, you will need to download and install the certificate so that you can export it as a public/private key pair (.p12). You will need to use the following steps in order to do this:
   1. Open the Keychain Access tool.
   2. Identify the distribution certificate that we have downloaded and installed.

3. Expand the certificate revealing the private key.
4. Select both the public certificate and private key so that you can use the **Export** option.

4. Once we have the distribution certificate, we will need an app ID to be able to generate a provisioning profile. When generating the app ID, the important decision is to decide whether to use a wildcard certificate (this might be a good option to use with multiple applications in the prerelease versions) or a full resource identifier.

5. The final step in the application provisioning process is to create the application provisioning profile for ad hoc distribution. The ad hoc distribution is the most appropriate distribution option for prerelease distribution through App Center.

6. With the **p12** certificate export and the mobile provisioning profile that we have generated and downloaded from the Apple developer site, we can head over to Azure DevOps and modify the **Install an Apple Certificate** and **Install an Apple Provisioning Profile** tasks:

7. Finally, it is important to make sure that the application package is created for real devices, and not a simulator. This configuration will need to be corrected in the Xamarin.iOS build tasks:

Here, the **Signing identity** and **Provisioning profile UUID** can be left blank as these elements will be installed by the pipeline. If multiple profiles or certificates exist in the pipeline, you will need to define the specific one to use.

> The Apple ad hoc distribution profile requires the UUID of the devices that are allowed to use the distributed version of the application. In simple terms, any device involved in using and testing this version of the application should be registered in this provisioning profile, and the application should be signed with it.

Environment-specific configurations

Native applications differ from web applications from a configuration perspective since the application CI pipeline should embed the configuration parameters into the application package. While the configuration parameters for different environments can be managed with various techniques, such as separate JSON files, compile constants, and so on, the common denominator in these implementations is that each of them uses conditional compilation or compilation constants to determine which configuration parameters are to be included in the application package. In other words, without recompiling the application, it isn't possible to change the environment-specific configurations for an application.

In order to create multiple distribution rings that are pointing to different service endpoints, the application will need to have different single pipelines with multiple configurations to build, or we would need to create multiple pipelines to build the application for a specific platform and configuration.

Creating and utilizing artifacts

In order to increase cross project reusability, the package management extension of Azure DevOps can be employed. UI components across Xamarin projects, as well as DTO models shared between Azure projects and the client application, can be merged into NuGet packages with their own life cycle: develop-merge-compile-deploy.

Storing these modules in a separate project/solution in a separate Git repository within the same Azure DevOps project would make the integration into previously created builds easier.

Once the NuGet project is ready to be compiled and packaged with a defined .nuspec file, a separate DevOps pipeline can be setup to create this package and push it into an internal feed within the same team project.

A sample NuGet build pipeline would look similar to the following:

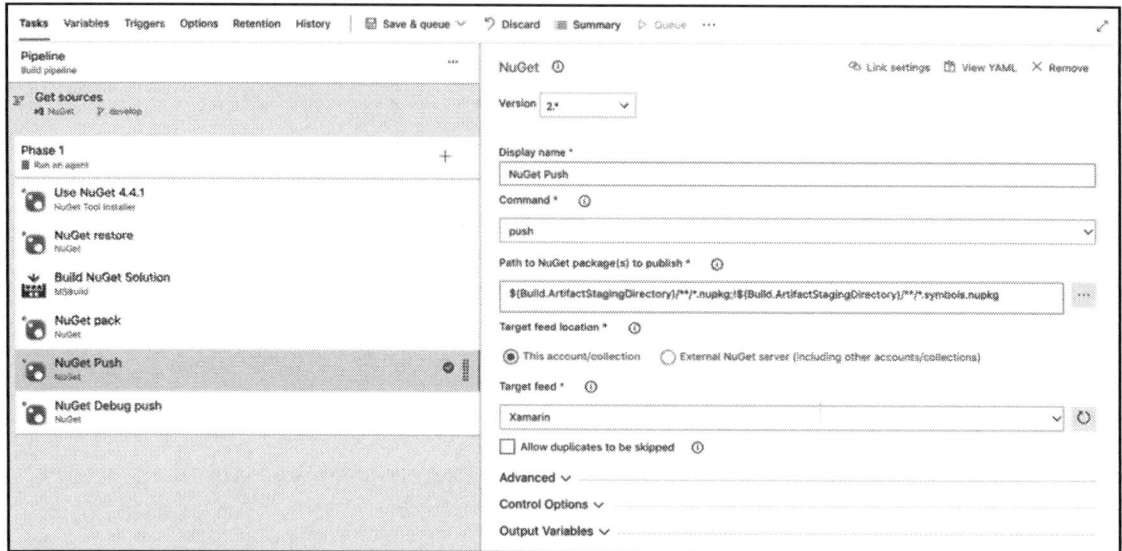

In order to include this feed in the Xamarin iOS and Android pipelines, the NuGet Restore step would need to be configured to include the internal feed, as well as the `Nuget.org` source. Additionally, `Nuget.org` can be set as an upstream source for the internal feed so that the public packages can be cached in the internal feed.

App Center for Xamarin

Visual Studio App Center, which expands on its predecessor, HockeyApp, and its feature set, is a mobile application life cycle management platform that's used to easily build, test, distribute, and collect telemetry from iOS, Android, Windows, and macOS applications. Its intrinsic integration with various repository options and build capabilities can even be used to migrate the development and release pipeline from Azure DevOps. Visual Studio App Center, just like Azure DevOps, follows a freemium subscription model, where the developers can access the majority of the functionality with a free subscription and would have to have a paid subscription for quota enhancements on certain features.

Integration with the source repository and builds

Even though we have already set up our source repository on Azure DevOps and associated build pipelines, AppCenter can be used for the same purpose.

For instance, if we were to set up the iOS build pipeline, we would follow these steps:

1. We would start by creating an application within our organization. An application on App Center also represents a distribution ring:

Add new app   ✕

ⓘ This new app will not appear in HockeyApp

App name: Icon:

NetCoreSample   N

Description: Owner:

Awesome Xamarin.iOS Application   👤 Can Bilgin

OS: ◉ iOS
              ○ Android
              ○ Windows
              ○ macOS Preview

Platform: ○ Objective-C / Swift
              ○ React Native
              ○ Cordova Preview
              ◉ Xamarin
              ○ Unity

Using a different platform? Let us know.   Add new app

2. After the application is created, in order to create a build, we will need to connect the App Center application to the target repository. Just like Azure DevOps pipelines, you are free to choose between **Azure DevOps**, **GitHub**, and **Bitbucket**:

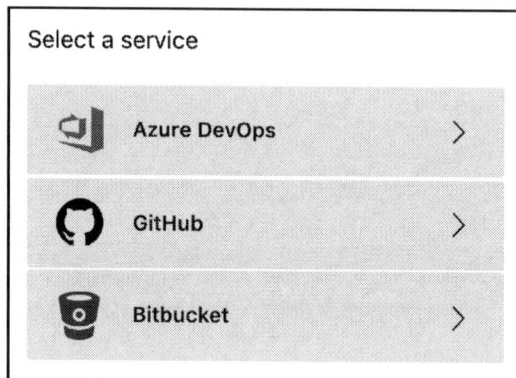

3. Once the repository is connected, we can start creating a build that will retrieve the branch content from the source repository and compile our iOS package:

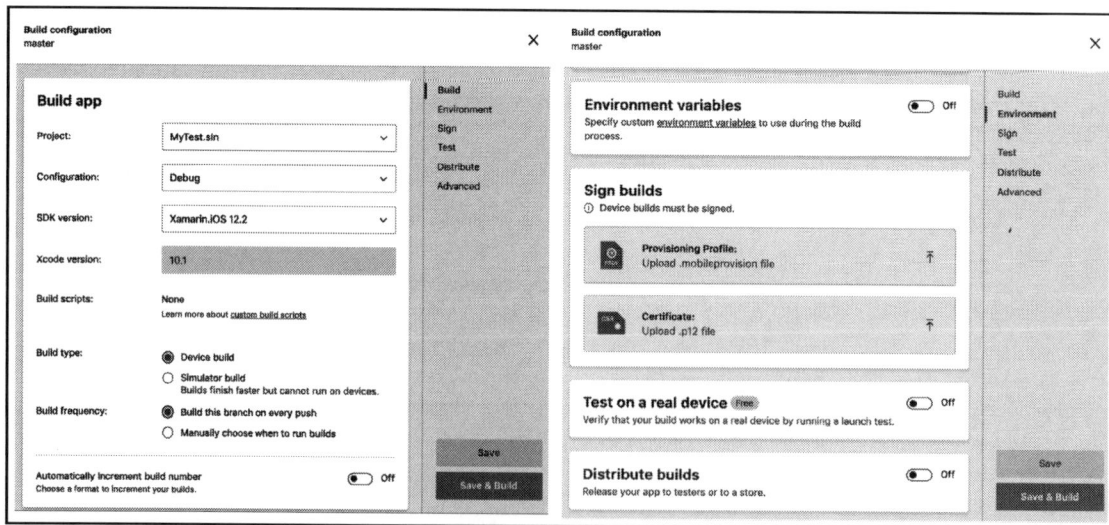

4. Once the build is set up, it can be built either manually or configured to be a CI build that will be triggered every time there is a push on the source branch (in this case, the master branch).

Setting up distribution rings

We have been mentioning distribution rings in regards to App Center since we started setting up the ALM pipeline for our applications. As we have already seen, a distribution ring refers to an app that's created on your personal or organization account. This ring represents an environment-specific (for example, Dev) compilation or a certain platform (for example, iOS) of our application.

An AppCenter application is represented through what is called an application slug. An application slug can be extracted from the URL of the app center page:

- https://appcenter.ms/users/{username}/apps/{application}:

  ```
  App Slug: {username}/{application}
  ```

- https://appcenter.ms/orgs/{orgname}/apps/{application}:

  ```
  App Slug: {orgname}/{application}
  ```

If we go back to our Azure DevOps pipeline, we can use this value to set up the deployment to AppCenter. However, before we can do this, we will need to create an App Center service connection with an API token that you can retrieve from App Center so that Azure DevOps can authorize with App Center and push application packages:

Add Visual Studio App Center service connection ✕

Connection name	App Center Connection
Server URL	https://api.mobile.azure.com/v0.1
API Token	044f5bf860b96e2cc4431746029611ce2f259786 ⓘ

☑ Allow all pipelines to use this connection.

OK   **Close**

Let's complete the rest of the configuration parameters:

Repeating the same steps for the Android version using another App Center application would complete the initial CI setup for our application. This build can be, similar to the App Center build, set up to be triggered with each merge to the develop or master branches.

It is extremely convenient to store a markdown sheet (for example, `ReleaseNotes.md`) within the solution folder (that is, in the repository) to record the changes to the application. In each pull request, when developers enter the updates to this file, the release notes about the changes being deployed can easily be pushed to the alpha and beta distribution channels.

Distribution with AppCenter

Aside from the build, test, and telemetry collection features of App Center, the main feature of App Center is to manage the distribution of prerelease applications, as well as automate submissions to public and private App Stores.

AppCenter releases

Once the application package is pushed from the build pipeline to AppCenter, an application release is created. This release represents a version of the application package. This package can be distributed to a distribution group within the current distribution ring or an external distribution target:

When a release is created, this release is accessible by the collaborators group (that is, developers who have management access to AppCenter).

AppCenter distribution groups

Distribution groups are group of developers and testers that an application release (environment and platform-specific versions of the app) can be distributed to:

New distribution group ✕

Group name:

Alpha Android Testers

Allow public access   ⬤◯ Off

Allow anyone to download

Who would you like to invite to the group?

🧑₊ Add testers or groups...

john.smith@test.com
john.smith@test.com   ⊗

jane.doe@test.com
jane.doe@test.com   ⊗

Distribution groups are extremely valuable since they provide additional staging of different distribution rings. For instance, once a release version is pushed to App Center from Azure DevOps, the first distribution group can verify the application before allowing the second distribution group access to this new release. This way, the automated releases from various pipelines can be delivered to certain target groups on both alpha and beta channels.

Additionally, if you navigate to the collaborators group details on an iOS application ring, you can identify which devices are currently included in the provisioning profile:

Devices			Export UDIDs
Model	Tester	OS	
Apple iPhone X	Can Bilgin can_bilgin@hotmail.com	iOS 12.0.1	⋮

In order to register devices, App Center offers the automatic provisioning of devices for iOS releases. In this setup, each time a new device is registered within a distribution group (given that automatic provisioning is configured), App Center will update the provisioning profile on the Apple Developer Portal and resign the release package with the new provisioning profile.

App Center distribution to production

Once the application is certified on lower rings (that is, alpha and beta), the App Center release can be pushed to the production stage. The production stage can be the target public App Store (for example, iTunes Store, Google Play Store, and so on), or the application can be published to users using Mobile Device Management or Mobile Application Management (for example, Microsoft Intune).

In order to set up the iTunes Store as the target store, you will need to add an Apple Developer account to App Center. Similarly, if your target is going to be InTune, an administrator account should be added to App Center integration:

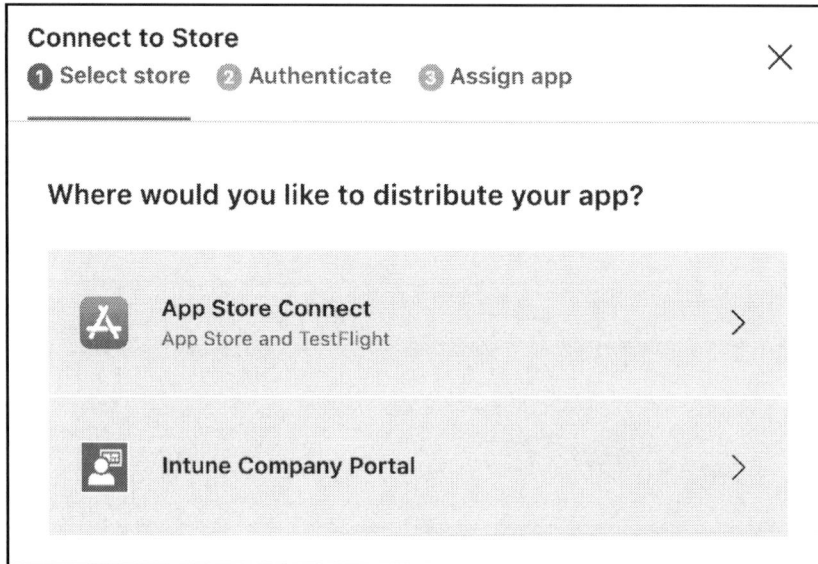

It is important to note that App Store Connect submissions will not bypass the Apple store verification of the application package—it is simply a hand-over process that is normally handled through Xcode.

App Center telemetry and diagnostics

App Center offers advanced telemetry and diagnostic options. In order to start using these monitoring features, App Center SDK needs to be installed on the application and initialized for all target platforms. Follow these steps to learn how:

1. NuGet package installation can be done from the public NuGet store. Using the package manager context:

```
PM> Install-Package Microsoft.AppCenter.Analytics
PM> Install-Package Microsoft.AppCenter.Crashes
```

2. In this case, we are creating a Xamarin.Forms application, so the initialization does not need to be platform-specific:

```
AppCenter.Start("ios={AppSecret};android={AppSecret};uwp={AppSecret
}", typeof(Analytics), typeof(Crashes));
```

3. Once the App Center SDK is initialized, default telemetry information as well as crash tracking is enabled for the application. The telemetry information can be extended with custom metrics and event telemetry using the available functionality within the SDK:

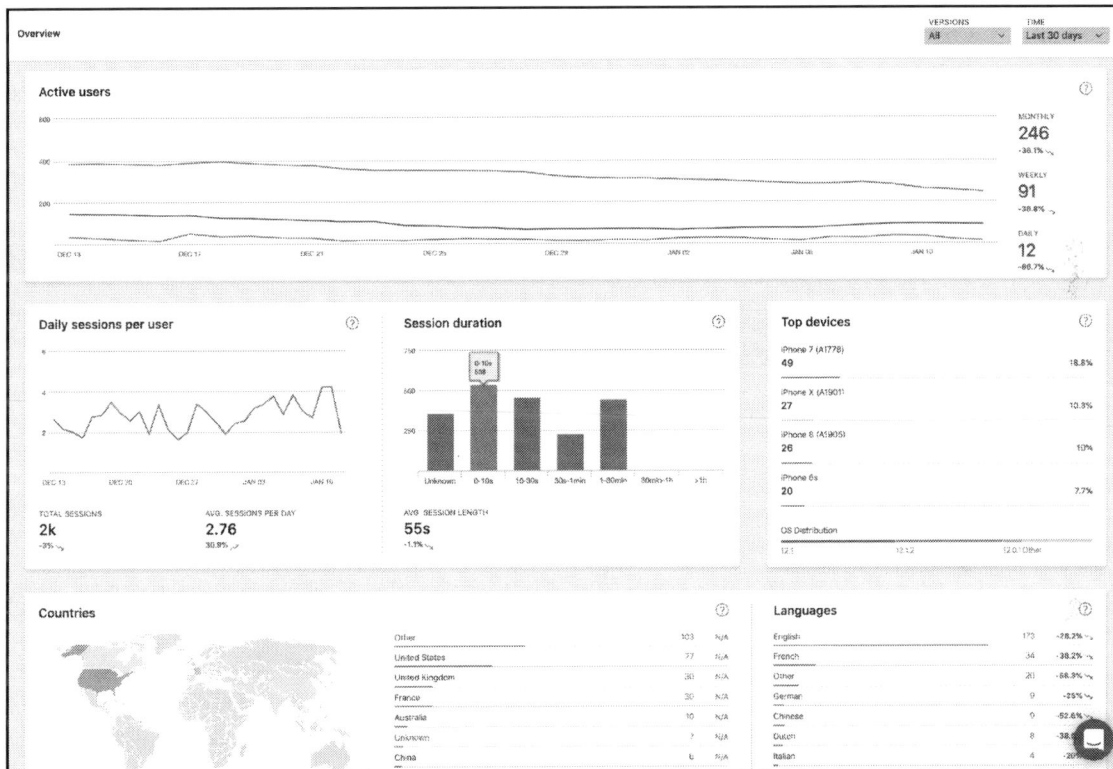

On top of the telemetry information, App Center allows you to track two types of error information: app crashes and errors. App crash information is logged together with the telemetry events that lead to the crash of the application, allowing the developers to easily troubleshoot problems.

Moreover, telemetry information can be pushed to Application Insights so that it can be analyzed on the Azure Portal.

Summary

In this chapter, we set up the initial build pipeline, which will be expanded on in the upcoming chapters. We have also discussed the available ALM features on Azure DevOps and Visual Studio App Center, and how to effectively use these two platforms together. Depending on the team's size and the application type, many different configurations can be implemented on these platforms, providing the developers with an automated and easy-to-use development pipeline.

In the next chapter, we will see how we can monitor our mobile application, as well as the Azure resources that are used with Azure Application Insights.

16
Application Telemetry with Application Insights

Agile application life cycle management dictates that after an application's release, performance and user feedback should be introduced back into the development cycle. Feedback information can provide vital clues to how to improve your application from both the business and technical perspectives. Application Insights can be a great candidate for collecting telemetry data from Xamarin applications that use an Azure-hosted web service infrastructure because of its intrinsic integration with Azure modules, as well as its continuous export functionality for App Center telemetry.

In this chapter, we will cover the following topics:

- Collecting insights for Xamarin applications
- Collecting telemetry data for Azure Service
- Analyzing data

Collecting insights for Xamarin applications

As we have previously discussed and set up, application telemetry within Xamarin applications is collected with the App Center SDK. This application data, while providing crucial information about the usage patterns of the application, cannot be further analyzed in App Center. We first need to export the App Center standard as well as the custom telemetry to an Azure Application Insights resource so that further analysis can be executed with a query language.

Telemetry data model

By using the App Center SDK, telemetry information can be collected along with events, which can contain additional information about a specific user action or application execution pattern. These additional data points, also known as dimensions, are generally used to give the user a quick snapshot of the data that is used to execute the function that triggered the telemetry event. In simple terms, a telemetry event can be described as the event name and the additional dimensions for this event. Let's begin by creating our telemetry model:

1. First, we will create our telemetry event, as follows:

```
public abstract class TelemetryEvent
{
    public TelemetryEvent()
    {
        Properties = new Dictionary<string, string>();
    }

    public TelemetryEvent(string eventName) : this()
    {
        Name = eventName;
    }

    public string Name { get; private set; }

    public virtual Dictionary<string, string> Properties { get;
    set; }
}
```

> **TIP**
>
> Any custom event will contain standard tracking metadata, such as the OS version, the request's geographical region, the device model, application version, and so on. These properties don't need to be logged as additional dimensions. Properties should be used for event-specific data.

2. Now, we will implement our telemetry writer using the App Center SDK:

```
public class AppCenterTelemetryWriter
{
    public void Initialize()
    {
        AppCenter.Start("{AppSecret}", typeof(Analytics),
        typeof(Crashes));
        AppCenter.SetEnabledAsync(true).ConfigureAwait(false);
        Analytics.SetEnabledAsync(true).ConfigureAwait(false);
```

```
            Crashes.SetEnabledAsync(true).ConfigureAwait(false);
        }

        public void TrackEvent(TelemetryEvent event)
        {
            Analytics.TrackEvent(event.Name, event.Properties);
        }
    }
```

3. In order to expand our event definition to include an `Exception` property, we
   can add an additional event for errors. In other words, we can track the handled
   exceptions within the application:

```
public void TrackEvent(TelemetryEvent @event)
{
    if(@event.Exception != null)
    {
        TrackError(@event);
        return;
    }

    Analytics.TrackEvent(@event.Name, @event.Properties);
}

public void TrackError(TelemetryEvent @event)
{
    Crashes.TrackError(@event.Exception, @event.Properties);
}
```

4. Now, we can start creating our custom events and start logging telemetry data.
   Our initial event might be the login event, which is generally the starting location
   for a user session:

```
    public class LoginEvent: TelemetryEvent
    {
        public LoginEvent()  : base("Login")
        {
            Properties.Add(nameof(Result), string.Empty);
        }

        public string Result
        {
            get
            {
                return Properties[nameof(Result)];
            }
```

```
        set
        {
                Properties[nameof(Result)] = value;
        }
    }
}
```

5. After logging in, the user will navigate to the main dashboard, so we will log our dashboard appearing event in a similar fashion:

```
protected override void OnAppearing()
{
    base.OnAppearing();

    App.Telemetry.TrackEvent(new HomePageEvent()
    {
        LoadedItems = ViewModel.Items.Count.ToString()
    });
}
```

6. Additionally, we will have to define a navigation to the details view, where various actions can be executed by the user:

```
protected override void OnAppearing()
{
    base.OnAppearing();

    App.Telemetry.TrackEvent(new DetailsPageEvent()
    {
        SelectedItem = viewModel.Title
    });
}
```

There are certain limitations on event structures that the App Center can track. The maximum number of custom event names cannot exceed 200. Additionally, the length of event names is limited to 256 characters, whereas property names cannot exceed 125 characters.

7. The resultant metadata can now be visualized on App Center Dashboards:

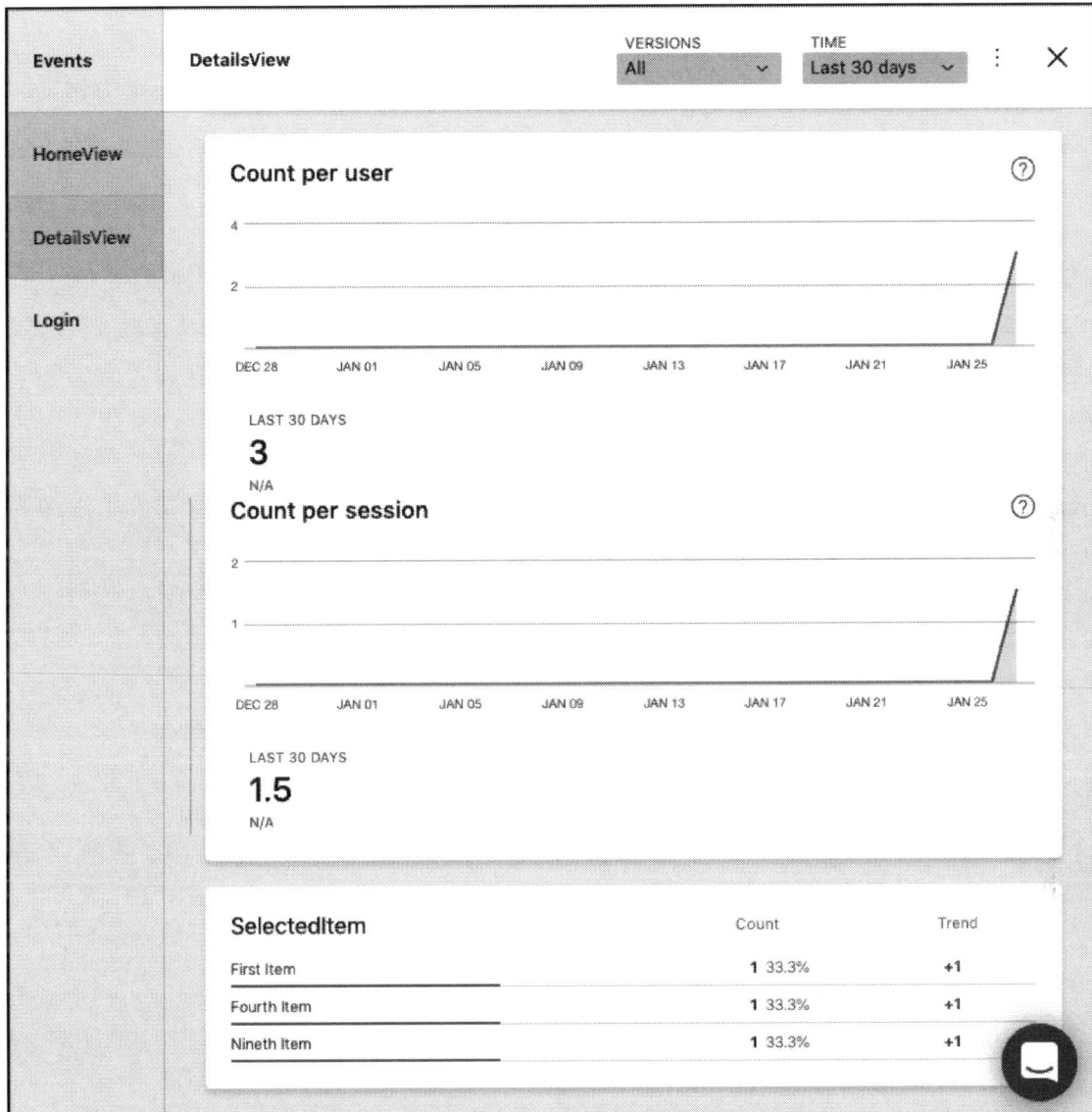

Advanced application telemetry

In the previous examples, we used a static accessor for our telemetry writer instance. However, this implementation can cause serious architectural problems, the most important of which is that we would be coupling our application class with a concrete implementation of the App Center SDK.

In order to remedy architectural issues that may arise, let's create a proxy telemetry container that will divert telemetry requests to target telemetry writers. To do so, follow the following steps:

1. We will start by creating an `ITelemetryWriter` interface to abstract our App Center telemetry handler:

```
public interface ITelemetryWriter
{
    string Name { get; }

    void TrackEvent(TelemetryEvent @event);

    void TrackError(TelemetryEvent @event);
}
```

2. Now, we will create our proxy container:

```
public class AppTelemetryRouter : ITelemetryWriter
{
    // Removed for Brevity

    public static AppTelemetryRouter Instance
    {
        get
        {
            if (_instance == null)
            {
                _instance = new AppTelemetryRouter();
            }

            return _instance;
        }
    }

    public void RegisterWriter(ITelemetryWriter telemetryWriter)
    {
        if (_telemetryWriters.Any(tw=>tw.Name ==
        telemetryWriter.Name))
        {
```

```
                throw new InvalidOperationException($"Already
                registered Telemetry Writer for
                {telemetryWriter.Name}");
            }

            _telemetryWriters.Add(telemetryWriter);
        }

        public void RemoveWriter(string name)
        {
            if(_telemetryWriters.Any(tw => tw.Name == name))
            {
                var removalItems = _telemetryWriters.First(tw =>
                tw.Name == name);

                _telemetryWriters.Remove(removalItems);
            }
        }

        public void TrackEvent(TelemetryEvent @event)
        {
            _telemetryWriters.ForEach(tw => tw.TrackEvent(@event));
        }

        public void TrackError(TelemetryEvent @event)
        {
            _telemetryWriters.ForEach(tw => tw.TrackError(@event));
        }
    }
```

It is also important to note that the container should be created on a cross-platform project where the view models are defined, since most of the diagnostic telemetry will in fact be collected on the view models rather than the views themselves. Another advantage of this implementation is the fact that we can now define multiple telemetry writers for different platforms, such as Firebase, Flurry Analytics, and so on.

While the abstract telemetry event class provides the basic data, it does not provide any metrics about the execution time. For instance, if we are executing a remote service call, or a long-running operation, the execution time can be a valuable dimension.

3. In order to collect this specific metric, let's create an additional telemetry object:

```
public abstract class ChronoTelemetryEvent : TelemetryEvent
{
    public ChronoTelemetryEvent()
```

```
    {
        Properties.Add(nameof(Elapsed), 0.ToString());
    }

    public double Elapsed
    {
        get
        {
            return double.Parse(Properties[nameof(Elapsed)]);
        }

        set
        {
            Properties[nameof(Elapsed)] = value.ToString();
        }
    }
}
```

4. Now, let's create our tracker object, which will track the execution time for events that require the execution time metric:

```
public class TelemetryTracker<TEvent> : IDisposable
    where TEvent : ChronoTelemetryEvent
{
    private readonly DateTime _executionStart = DateTime.Now;

    public TelemetryTracker(TEvent @event)
    {
        Event = @event;
    }

    public TEvent Event { get; }

    public void Dispose()
    {
        var executionTime = DateTime.Now - _executionStart;
        Event.Elapsed = executionTime.TotalMilliseconds;

        // The submission of the event can as well be moved out of
        //the tracker
        AppTelemetryRouter.Instance?.TrackEvent(Event);
    }
}
```

5. Now, using our tracker object, we can collect valuable information about time-sensitive operations within the application:

```
public async Task LoadProducts()
{
    using (var telemetry = new
TelemetryTracker<ProductsRequestEvent>(new ProductsRequestEvent()))
    {
        try
        {
            var result = await _serviceClient.RetrieveProducts();

            Items = new ObservableCollection<ItemViewModel>(
            result.Select(item => ItemViewModel.FromDto(item)));
        }
        catch (Exception ex)
        {
            telemetry.Event.Exception = ex;
        }
    }
}
```

6. The results of the service call are measured with the `Elapsed` metric, which can be observed on the LogFlow on App Center:

8092425a	STARTSESSION - 7b106777-f07c-4312-9b53-5f2930c101db	14:11:03
8092425a	STARTSERVICE	14:11:03
8092425a	EVENT - Login - {"result":"Successfully Logged In!"}	14:11:08
8092425a	EVENT - HomeView - {"loadedItems":"16"}	14:11:09
8092425a	**EVENT - ProductsServiceRequest - {"elapsed":"659.583"}**	**14:11:12**
8092425a	EVENT - DetailsView - {"selectedItem":"First Item"}	14:11:57
8092425a	EVENT - HomeView - {"loadedItems":"16"}	14:11:59
8092425a	**EVENT - ProductsServiceRequest - {"elapsed":"633.312"}**	**14:12:04**

7. Similarly, tracker objects can be created on the `OnAppearing` events of certain views and disposed of with the `OnDisappearing` event so that we can track how much time the user spent on a certain view.

Exporting App Center telemetry data to Azure

At this point, the application is collecting telemetry data and pushing it to App Center. However, as we discussed earlier, you won't be able to analyze this data – especially the custom event dimensions – any further.

In order to do this, we will need to create a new Application Insights resource and set up a continuous export so that App Center telemetry can be exported as Application Insights data. Let's begin:

1. We will start this process by creating the Application Insights resource:

Notice that, for the **Application Type**, we have selected **App Center application**. As a resource group, we will be using the same resource group as the previously created Azure services so that the complete Azure infrastructure can be deployed together.

2. Once the resource is created, go to the **Overview** section to find the instrumentation key. This key is the only requirement for setting up the continuous export process:

3. On App Center, in order to set up the export, you will need to navigate to the **Settings** section and select **Export and Application Insights** on the data export window.

   In this view, you can use the **Set up standard export** option, which is used when an Azure subscription is configured to be used with App Center. Selecting the standard export will require Admin access to the Azure subscription and will create a new resource. You can also select **Customize** and paste in the instrumentation key that we have from the Azure Portal.

4. After this setup, the event telemetry will be pushed periodically to Application Insights, which can be analyzed within the Azure Portal either using the standard analysis sections or using the query language:

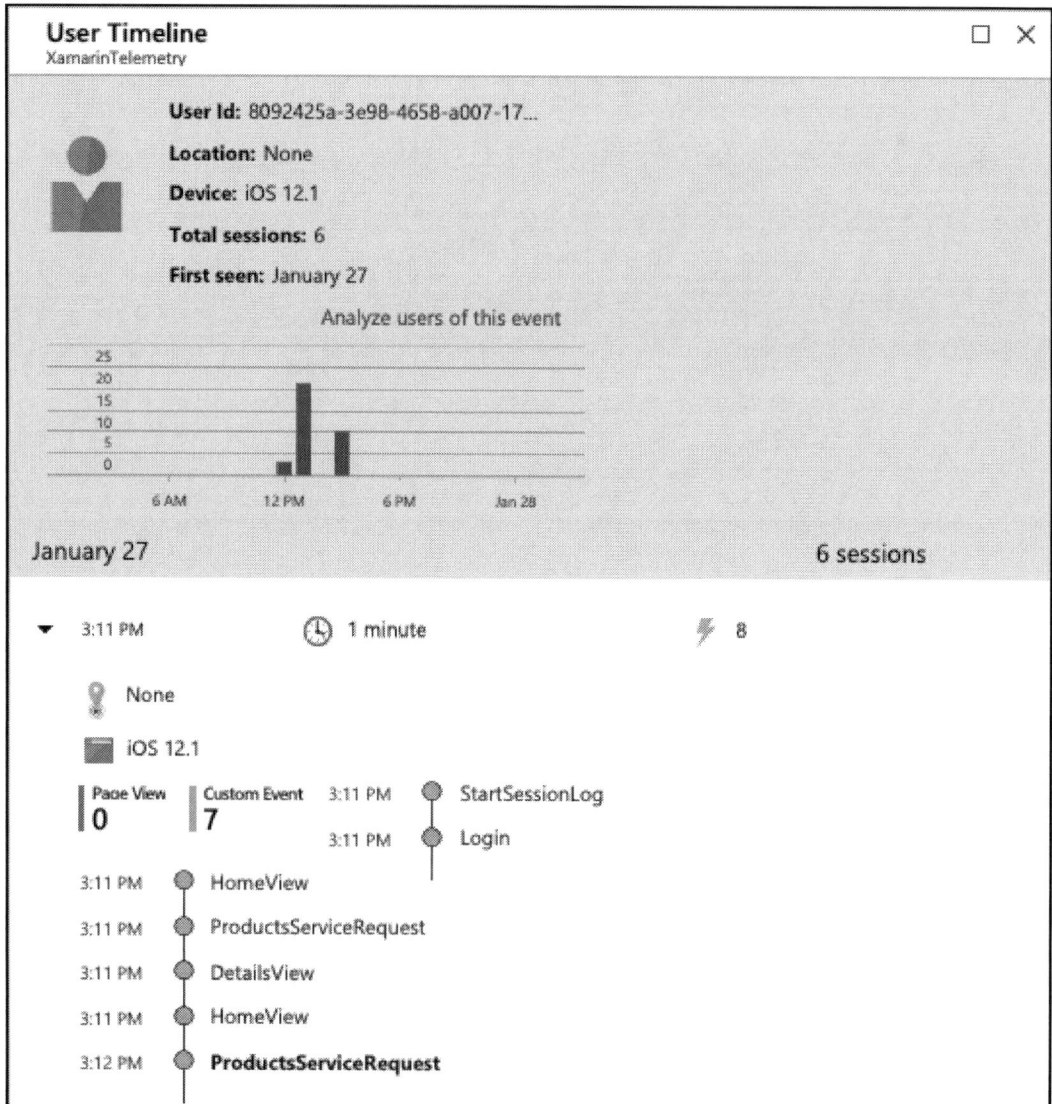

Now that the telemetry data from our application has been collected and exported to Application Insights, we can start creating the Application Insights infrastructure for our remaining modules on Azure.

Collecting telemetry data for Azure Service

Application Insights is a little more tightly integrated into Azure-based services than mobile applications. In addition to various standard, out-of-the-box, telemetries that can be collected for services such as Azure App Services and serverless components such as Azure Functions, custom telemetry, trace, and metric collection implementations are possible.

Application Insights data model

Application Insights telemetry collection can be grouped into three major groups: trace, event, and metric.

- **Trace**: Trace can be recognized as the simplest form of telemetry. Trace elements generally give a nominal description of an event and are used as a diagnostic log, similar to other flat file diagnostic log implementations. In addition to the main telemetry message, a severity level and additional properties can be defined. Trace message size limits are much larger than other telemetry types, and provide a convenient way of providing large amounts of diagnostic data.
- **Events**: Application Insights events are very similar to App Center telemetry items and are treated the same way once the App Center data is exported to Application Insights. In addition to the nominal dimensions, additional metrics data can be sent to Application Insights. Once the data is collected, the `customProperties` collection provides access to the descriptive dimensions, while the `customMeasurements` dictionary is used to access metrics.
- **Metrics**: They are generally pre-aggregated scalar measurements. If you're dealing with custom metrics, it is the application's responsibility to keep them up-to-date. The Application Insights client provides standardized access methods for both standard and custom metrics.

In addition to event-specific telemetry data types, operation-specific data types can also be traced and tracked with Application Insights, such as requests, exceptions, and dependencies. These are classed as more generalized, macro-level telemetry data, which can provide valuable information as to the health of the application infrastructure. Request, exception, and dependency data is generally tracked by default, but custom/manual implementation is also possible.

Collecting telemetry data with ASP.NET Core

Application Insights can easily be initialized using Visual Studio for any ASP.NET Core web application. In our case, we will be configuring the web API layer to use Application Insights. Let's see how we can do this:

1. If you right-click on the web application project, select **Add** | **Application Insights Telemetry**, and then click the **Get Started** link, Visual Studio automatically loads the resources that your account(s) are associated with, allowing you to choose a subscription, as well as the resource/resource group pair.

2. The **Configure Settings...** option on this page should be used to assign an already existing resource group; otherwise the application insights instance will be created on a default Application Insights resource group:

3. After the **Application Insights** configuration is complete, we can see how the telemetry data is collected without deploying the web API to the Azure App Service resource.

4. In order to see the collected telemetry data, you can start a debugging session. By using the **View** | **Other Windows** | **Application Insights** option, you can open the live Application Insights data that is collected within the debug session:

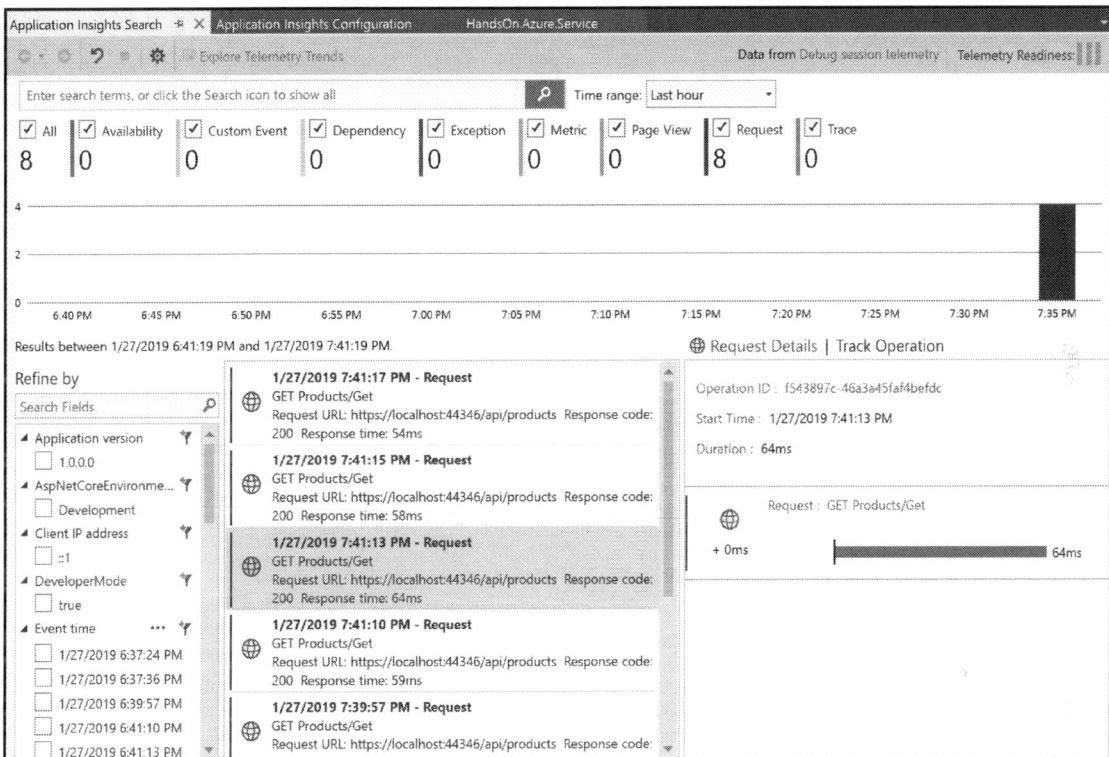

It is important to note that the data is set to use **Debug session telemetry**. The Application Insights Search toolset can also be used to read remote telemetry and execute quick search queries on live data.

Another useful tool window is Application Insights Trends, which is a quick reporting tool for various application telemetry data types, such as requests, page views, exceptions, events, and dependencies.

The same telemetry set, even if this is a debugging session, should already be available on Azure Portal as well. In other words, Application Insights does not require an Azure deployment for a server resource to be profiled and its telemetry tracked. If you navigate to the Application Insights overview page, you will notice the incoming data and collected telemetry data about the requests.

5. Additionally, the live metrics screen can provide information about the server's performance and aggregated metrics data. In order to to use profiling and performance data, the Application Insights SDK should be updated to a version higher than 2.2.0:

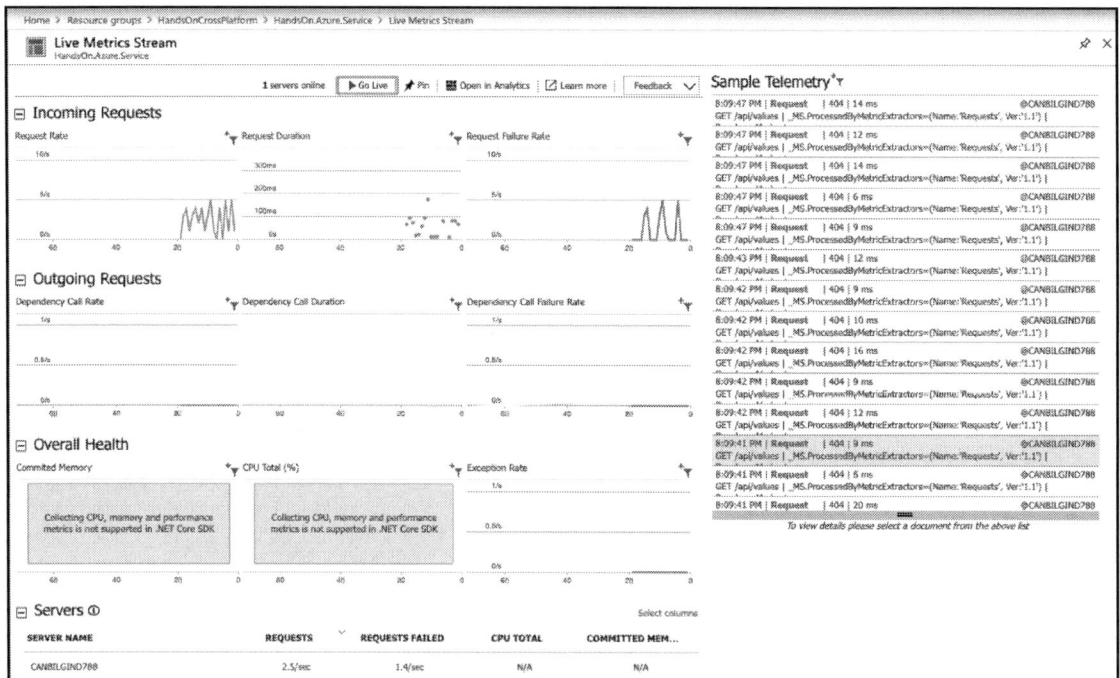

6. Now that the Application Insights infrastructure is set up, we can start creating custom telemetry and trace data. First, let's create a new operation context for the product retrieval API operation and include some additional telemetry data:

```
using (var operationContext =
_telemetryClient.StartOperation<RequestTelemetry>("getProducts"))
{
    var result = Enumerable.Empty<Product>();
    _telemetryClient.TrackTrace("Creating Document Client",
    SeverityLevel.Information);
```

```
using (var document = GetDocumentClient())
{
    try
    {
        _telemetryClient.TrackTrace("Retrieving Products",
        SeverityLevel.Information);
        result = await document.Retrieve();
    }
    catch (Exception ex)
    {
        _telemetryClient.TrackException(ex);
        operationContext.Telemetry.ResponseCode = "500";
        throw ex;
    }
}

operationContext.Telemetry.ResponseCode = "200";
return Ok(result);
}
```

7. Now, the resultant telemetry collection that contains trace entries is automatically grouped to the operation context, thus providing more meaningful information:

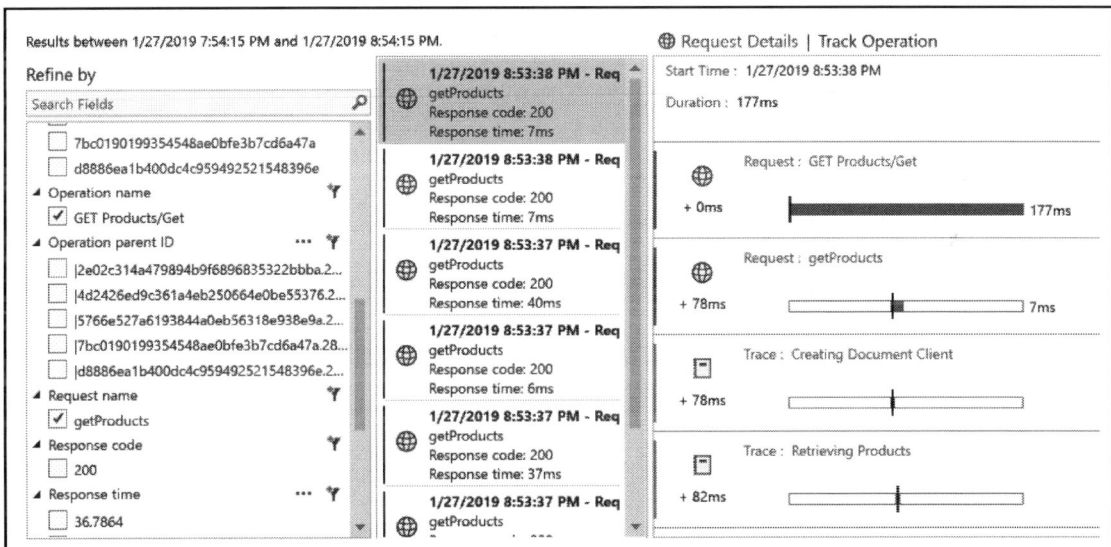

8. We can further granulize this telemetry data by separating the dependency telemetry. For instance, in this implementation we call the data provider client to load all the products. Using the telemetry client can create a dependency telemetry for this request:

```
using (var document = GetDocumentClient())
{
    var callStartTime = DateTimeOffset.UtcNow;

    try
    {
        _telemetryClient.TrackTrace("Retrieving Products",
        SeverityLevel.Information);
        result = await document.Retrieve();
    }
    finally
    {
        var elapsed = DateTimeOffset.UtcNow - callStartTime;
        _telemetryClient.TrackDependency(
            "DataSource", "ProductsDB", "Retrieve", callStartTime,
            elapsed, result.Any());
    }
}
```

9. Now, the application telemetry is tracked separately for the document source. This dependency is even created on the application map on Azure Portal:

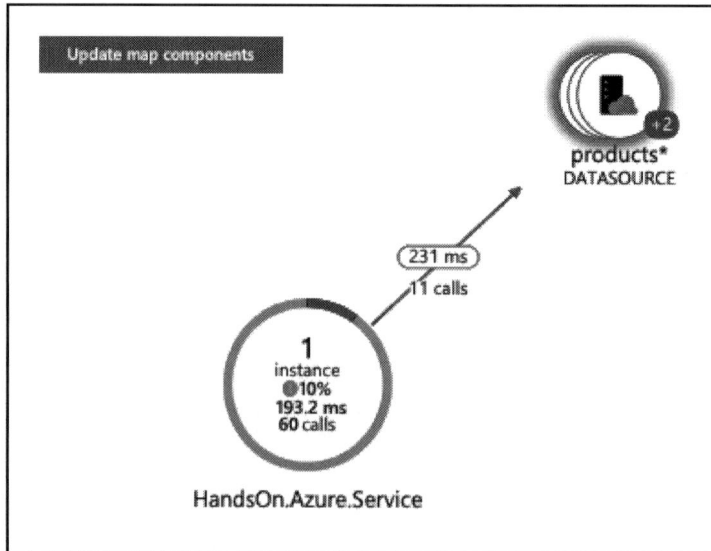

Under normal circumstances, dependency calls to resources such as Cosmos DB and SQL are automatically detected and tracked separately. The preceding implementation suits external dependencies or legacy systems.

Collecting telemetry with Azure Functions

When we talk about custom traces, collecting Application Insights telemetry data from Azure Functions is no different from using any other .NET application. When we created our Azure Functions, we injected a `TraceWriter` instance into our methods. `TraceWriter` logs are the main source of diagnostic telemetry and is collected within an Azure Function. These log entries can be filtered according to the log level using the `host.json` settings:

```json
{
  "logger": {
    "categoryFilter": {
      "defaultLevel": "Information",
      "categoryLevels": {
        "Host.Results": "Error",
        "Function": "Error",
        "Host.Aggregator": "Information"
      }
    },
    "aggregator": {
      "batchSize": 1000,
      "flushTimeout": "00:00:30"
    }
  },
  "applicationInsights": {
    "sampling": {
      "isEnabled": true,
      "maxTelemetryItemsPerSecond" : 5
    }
  }
}
```

The function section in the category levels refers to the traces that are collected within the function. `Host.Results` are automatically collected request/result telemetry data pairs, whereas aggregator data is full of aggregated metrics that the Functions host collects by default, either every 30 seconds or 1,000 results. This is then used to calculate the aggregated metrics, such as count, success rate, and so on.

In addition to the basic telemetry implementation, you can also modify default telemetry data using the Application Insights telemetry client. In this context, the telemetry client is used to modify the operation context rather than creating a new `TrackRequest`.

As you can see, the telemetry client acts more like middleware rather than the source of truth in this implementation, simply modifying the existing operation context and creating additional event data.

Analyzing data

Now that we have set up Application Insights telemetry collection on both the server side and the application side, we can try and make sense of this data.

While the Azure Portal provides quick insights into application telemetry data, if we want to really dive into application data, the Application Insights portal should be used for analysis. In the Application Insights portal, data can be analyzed using the query language. The query language, also known as the Kusto language, provides advanced read-only querying features that can help organize data from multiple sources and render valuable insights into the performance and usage patterns of your application.

For instance, let's take a look at the following simple query, which is executed on our Xamarin telemetry data:

1. We are returning the first 50 custom events that are exported from AppCenter:

```
customEvents
| limit 50
```

2. These telemetry entries contain general telemetry-related data in the root:

2019-01-27T14:11:12.681	ProductsServiceRequest	customEvent	{"WrapperRuntimeVersion":"11.2.0","WrapperSdkVersion":"1.13.0","In...

timestamp [UTC]	2019-01-27T14:11:12.681Z
name	ProductsServiceRequest
itemType	customEvent
customDimensions	{"WrapperRuntimeVersion":"11.2.0","WrapperSdkVersion":"1.13.0","IngressTimestamp":"2019-01-27T14:11:15.7610000Z"
operation_Id	0f6745b0-fa9c-4995-9a1f-f43b8009d577
operation_ParentId	0f6745b0-fa9c-4995-9a1f-f43b8009d577
session_Id	7b106777-f07c-4312-9b53-5f2930c101db
user_Id	8092425a-3e98-4658-a007-176b36504c7a
application_Version	1.0
client_Type	x86_64
client_Model	Apple
client_OS	iOS 12.1
client_IP	0.0.0.0
client_CountryOrRegion	None
appId	8f1b0388-ee97-4a76-b972-cedc497720a3
appName	XamarinTelemetry
iKey	7a584488-0b9e-42ac-8105-34e5f29e36f4
sdkVersion	appcenter.ios:1.13.0
itemId	6b7430f4-2247-11e9-93ab-ef6cd06591d6
itemCount	1

[405]

Whereas the `customDimensions` object provides more Xamarin-specific data:

customDimensions	{"WrapperRuntimeVersion":"11.2.0","WrapperSdkVersion":"1.13.0","IngressTimestamp":"2019-01-27T14:11:15.7610000Z
AppBuild	1.0
AppId	f49e730f-32de-449e-a516-5113d6d81b08
AppNamespace	com.companyname.FirstXamarinFormsApplication
CarrierCountry	None
CarrierName	None
CountryCode	None
EventId	c5129672-16ea-4377-acb2-71b467ec8ce4
IngressTimestamp	2019-01-27T14:11:15.7610000Z
Locale	en_TR
MessageType	EventLog
OsApiLevel	None
OsBuild	18C54
OsName	iOS
OsVersion	12.1
> Properties	{"elapsed":"659.583"}
ScreenSize	2436x1125
SdkName	appcenter.ios
SdkVersion	1.13.0
TimeZoneOffset	PT1H
UserId	

Finally, the `Properties` node of `customDimensions` provides the actual custom telemetry data that was sent using the AppCenter telemetry client.

3. Before we can dig into event-specific data, we need to filter telemetry events using the telemetry event name:

```
customEvents
| where name == "ProductsServiceRequest"
```

4. Then, we order the table by timestamp:

```
| order by timestamp desc nulls last
```

5. Finally, by deserializing the properties data into a dynamic field, we can use this data in our queries:

```
| extend Properties =
todynamic(tostring(customDimensions.Properties))
```

6. In order to flatten the table structure, we can assign the properties data into their own fields:

```
| extend Duration = todouble(Properties.elapsed), OperatingSystem =
customDimensions.OsName
```

7. Finally, we will project the data into a new table so that we can present it in a simpler structure:

```
| project OperatingSystem, Duration, timestamp
```

8. Now, the data from our telemetry events is structured and can be presented in reports and troubleshooting:

OperatingSystem	Duration	timestamp [UTC]
iOS	633.348	2019-01-27T15:23:21.593
OperatingSystem iOS		
Duration 633.348		
timestamp [UTC] 2019-01-27T15:23:21.593Z		
iOS	633.311	2019-01-27T15:23:18.144
iOS	640.723	2019-01-27T15:23:12.933
iOS	633.312	2019-01-27T14:12:04.082
iOS	659.583	2019-01-27T14:11:12.681

9. To take this one step further, we can also draw a chart using the final table:

```
| render timechart
```

10. This would draw a line chart with the duration values. You can also sort data by using the `summarize` function with numerous aggregate functions, such as by grouping events into hourly bins and drawing a time-based bar chart:

```
customEvents
| order by timestamp desc nulls last
| extend Properties =
todynamic(tostring(customDimensions.Properties))
| summarize event_count=count() by bin(timestamp, 1h)
| render barchart
```

Application Insights data and available query operators and methods provide countless ways for developers to act proactively on the application telemetry by gathering invaluable application data from staging environments or production and feeding it back to the application life cycle.

Summary

Application telemetry data that is collected from both the server- and the client-side can provide information that's required to improve and mould your application according to user needs. In a way, by collecting application telemetry data from various modules of the application on live environments, live application testing is executed with actual user data. This telemetry data can provide insights into the application that no other automated testing can provide. Regardless of the reality and the synthetic nature of the data, unit tests, as well as automated UI tests, should still be part of the application life cycle.

In the next chapter, we will look into various ways of testing and how we can include these tests in the development pipeline.

17
Automated Testing

Unit and coded UI tests are generally perceived by most developers as the most monotonous part of the application project life cycle. However, improving the unit test code coverage and creating automated UI tests can help to save an extensive amount of developer hours that would otherwise be spent on maintenance and regression. Especially on application projects with a longer life cycle, the stability of the project directly correlates with the level of test automation. This chapter will discuss how to create unit and coded UI tests and the architectural patterns that revolve around them. Data-driven unit tests, mocks, and Xamarin UI tests are some of the concepts that will be discussed.

The following topics will walk you through the implementation of an automatically verified application development pipeline:

- Maintaining application integrity with tests
- Maintaining cross-module integrity with integration tests
- Automated UI tests

Maintaining application integrity with tests

Regardless of the development or runtime platform, unit tests are an integral part of the development pipeline. In fact, nowadays, **Test-Driven Development** (TDD) is the most prominent development methodology and is the choice of any agile development team. In this paradigm, developers are responsible, even before the first line of actual business logic implementation is written, for creating unit tests that are appropriate for the current unit that is under development.

Arrange, Act, and Assert

Without further ado, let's take a look at the first view model in our application and implement some unit tests for it. The products view model is a simple view model that, on initialization, loads the products data using the available service client. It exposes two properties, namely, the `Items` collection and the `ItemTapped` command. Using this information, we can identify the units.

The units of the application, can be identified by implementing simple stubs as shown in the following code:

```
public class ListItemViewModel : BaseBindableObject
{
    public ListItemViewModel(IApiClient apiClient, INavigationService
    navigationService)
    {
        //...Load products and initialize ItemTapped command
    }

    public ObservableCollection<ItemViewModel> Items { get; set; }

    public ICommand ItemTapped { get; }

    internal async Task LoadProducts()
    {
        // ...
        var result = await _serviceClient.RetrieveProductsAsync();
        // ...
    }

    internal async Task NavigateToItem(ItemViewModel viewModel)
    {
        // ...
    }
}
```

Our initial unit test will set up the mock for `apiClient`, construct the view model, verify that `RetrieveProductsAsync` on the service client is called, and verify that the `ItemTapped` command is initialized properly. An additional check can be done to see whether the `PropertyChanged` event has been triggered on the `Items` property. In the context of unit testing, these three steps of a simple unit test are generally called the triple-A or AAA, Arrange, Act, Assert:

1. In the `Arrange` section, we will prepare a set of results data and return the data with a mock client:

   ```
   #region Arrange

   var expectedResults = new List<Product>();
   expectedResults.Add(new Product { Title = "testProduct",
   Description = "testDescription" });

   // Using the mock setup for the IApiClient
   _apiClientMock.Setup(client =>
   client.RetrieveProductsAsync()).ReturnsAsync(expectedResults);

   #endregion
   ```

2. Now, let's `Act` by constructing the view model:

   ```
   #region Act

   var listViewModel = new ListItemViewModel(_apiClientMock.Object);

   #endregion
   ```

3. Finally, we will execute the assertions on the view model target:

   ```
   #region Assert

   // Just checking the resultant count as an example
   // Foreach with checking each expected product has a
   // matching domain entity would improve the robustness of the test.
   listViewModel.Items.Should().HaveCount(expectedResults.Count());
   listViewModel.ItemTapped.Should().NotBeNull()
       .And.Subject.Should().BeOfType<Command<ItemViewModel>>();
   _apiClientMock.Verify(client => client.RetrieveProductsAsync());

   #endregion
   ```

4. With this simple unit test implementation, we have already reached ~80% of unit
test code coverage:

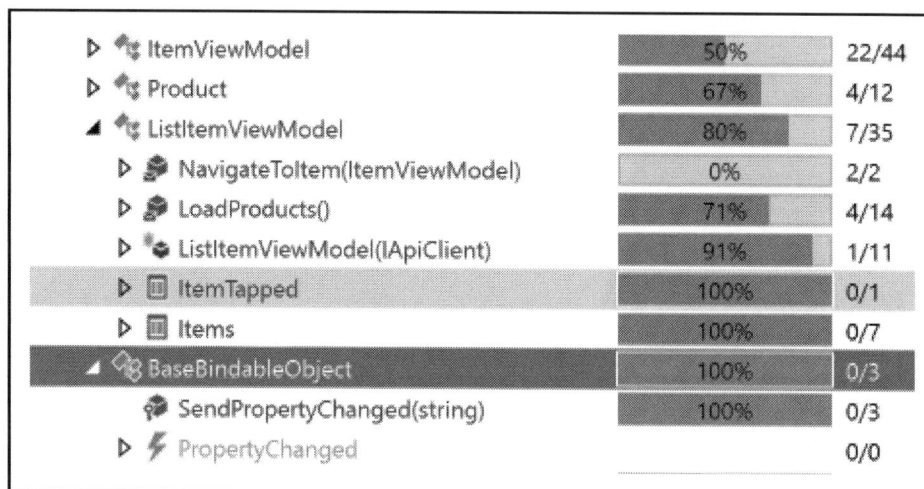

▷ ItemViewModel	50%	22/44
▷ Product	67%	4/12
◢ ListItemViewModel	80%	7/35
▷ NavigateToItem(ItemViewModel)	0%	2/2
▷ LoadProducts()	71%	4/14
▷ ListItemViewModel(IApiClient)	91%	1/11
▷ ItemTapped	100%	0/1
▷ Items	100%	0/7
◢ BaseBindableObject	100%	0/3
SendPropertyChanged(string)	100%	0/3
▷ PropertyChanged		0/0

> The xUnit.Net framework was used to implement this unit test, or so-
> called fact. Additionally, the FluentAssertions and Moq frameworks were
> utilized in order to ease the implementation and assertions. The feature
> sets of these frameworks are beyond the scope of this book.

5. The implementation is good enough for checking the initialization of the
constructor. The constructor implementation we are testing looks similar to this:

```
public ListItemViewModel(IApiClient apiClient)
{
    _serviceClient = apiClient;
    ItemTapped = new Command<ItemViewModel>(async _ => await
    NavigateToItem(_));
    if (_serviceClient != null)
    {
        LoadProducts().ConfigureAwait(false);
    }
}
```

However, notice that the LoadProducts method is in fact called without await,
and it does not merge back into the initial synchronization context. In a multithreaded environment, when executing multiple unit tests in parallel, it might
happen that the constructor is executed; however, before the asynchronous task
can complete, the assertions start. This can be worked around with a poor man's
thread synchronization—Task.Delay or Thread.Sleep.

6. However, this implementation is nothing more than a temporary workaround. Since we cannot and should not really wait for the task to complete in this scenario within the constructor, we can utilize the service initialization pattern:

```
public ListItemViewModel(IApiClient apiClient)
{
    _serviceClient = apiClient;
    ItemTapped = new Command<ItemViewModel>(async _ => await
    NavigateToItem(_));
    if (_serviceClient != null)
    {
        (Initialized = LoadProducts()).ConfigureAwait(false);
    }
}

internal Task Initialized { get; set; }
```

7. And now our Act implementation would look similar to this:

```
#region Act

var listViewModel = new ListItemViewModel(_apiClientMock.Object);
await listViewModel.Initialized;

#endregion
```

Note that we were not able to verify the PropertyChanged event trigger on the view model level for the Items property. The main reason for this is the fact that the ListItemsViewModel instance immediately executes the LoadProducts method, and before we even have a chance to subscribe to the target event, the execution is finalized. This can also be remedied with a circuit flag within the mock object we have implemented, releasing the task once the monitor is attached to the view model.

8. In order to execute these unit tests, as well as the IDE extensions, you can use the dotnet console command:

```
dotnet test --collect "Code Coverage"
```

9. This command will execute the available unit tests and generate a coverage file that can be viewed in Visual Studio:

```
W:\book\xamarin\DesignChapter\FirstXamarinFormsApplication.Client.Tests>dotnet test --collect "Code Coverage"
W:\book\xamarin\DesignChapter\FirstXamarinFormsApplication.Client.Tests\FirstXamarinFormsApplication.Client.Tests.csproj
 : warning NU1603: Castle.Core 4.0.0 depends on System.ComponentModel.TypeConverter (>= 4.0.1) but System.ComponentModel
.TypeConverter 4.0.1 was not found. An approximate best match of System.ComponentModel.TypeConverter 4.1.0 was resolved.

Build started, please wait...
W:\book\xamarin\DesignChapter\FirstXamarinFormsApplication.Client.Tests\FirstXamarinFormsApplication.Client.Tests.csproj
 : warning NU1603: Castle.Core 4.0.0 depends on System.ComponentModel.TypeConverter (>= 4.0.1) but System.ComponentModel
.TypeConverter 4.0.1 was not found. An approximate best match of System.ComponentModel.TypeConverter 4.1.0 was resolved.

Build completed.

Test run for W:\book\xamarin\DesignChapter\FirstXamarinFormsApplication.Client.Tests\bin\Debug\netcoreapp2.1\FirstXamari
nFormsApplication.Client.Tests.dll(.NETCoreApp,Version=v2.1)
Microsoft (R) Test Execution Command Line Tool Version 15.8.0
Copyright (c) Microsoft Corporation.  All rights reserved.

Starting test execution, please wait...

Attachments:
  W:\book\xamarin\DesignChapter\FirstXamarinFormsApplication.Client.Tests\TestResults\e7bf0a95-5564-41c6-b919-9967f32397
59\can.bilgin_CANBILGIND788_2019-02-11.19_51_39.coverage

Total tests: 3. Passed: 3. Failed: 0. Skipped: 0.
Test Run Successful.
Test execution time: 3.8336 Seconds
```

Creating unit tests with mocks

When implementing unit tests, it is important to isolate the units we are currently testing. By isolation, we are of course referring to the process of mocking the dependencies of the current subject under test. The mocks can be introduced in various ways, depending on the implementation of the Inversion of Control pattern. If the implementation involves constructor injection, we can mock our dependency interfaces in the first A of our test and pass it on to our target. Otherwise, frameworks such as NSubstitute can replace interfaces as well as concrete classes that are used by the subject.

Taking a look back at our view model and the unit test that was implemented, you might notice that we used the Moq framework to create a mock interface implementation for our `IApiClient` object. Let's now see how to create the unit tests using Moq:

1. Extend the constructor to take the `INavigationService` instance that will be used to navigate to the details view of the selected item; in other words, isolate our `ItemTapped` command implementation:

```
public ListItemViewModel(IApiClient apiClient, INavigationService
navigationService)
{
```

```
_serviceClient = apiClient;
_navigationService = navigationService;
ItemTapped = new Command<ItemViewModel>(async _ => await
NavigateToItem(_));
if (_serviceClient != null)
{
    (Initialized = LoadProducts()).ConfigureAwait(false);
}
}
```

2. Our navigation command will be as follows:

```
internal async Task NavigateToItem(ItemViewModel viewModel)
{
    if (viewModel != null && _navigationService != null)
    {
        if (await
_navigationService.NavigateToViewModel(viewModel))
        {
            // Navigation was successful
            return;
        }
    }

    throw new InvalidOperationException("Target view model or
    navigation service is null");
}
```

> In this example, we are throwing an exception, just for demonstration purposes. In a real-life implementation, it is probably a better choice to track errors internally and/or throw an exception only in debug mode. Moreover, it is not quite SOLID to throw the same type of exception of the scenarios.

3. Let's now implement our unit test:

```
[Trait("Category", "ViewModelTests")]
[Trait("ViewModel", "ListViewModel")]
[Fact(DisplayName = "Verify ListViewModel navigates on
ItemTapped")]
public async Task
ListItemViewModel_ItemTapped_ShouldNavigateToItemViewModel()
{
    #region Arrange

    _navigationServiceMock.Setup(nav => nav.NavigateToViewModel(
            It.IsAny<BaseBindableObject>()))
                .ReturnsAsync(true);
```

```
var listViewModel = new ListItemViewModel(
                        _apiClientMock.Object,
                        _navigationServiceMock.Object);
await listViewModel.Initialized;
var expectedItemViewModel = new ItemViewModel() { Title = "Test
Item" };

#endregion

#region Act

listViewModel.ItemTapped.Execute(expectedItemViewModel);

#endregion

#region Assert

_navigationServiceMock.Verify(
    service =>
service.NavigateToViewModel(It.IsAny<ItemViewModel>()));

#endregion
}
```

4. We have implemented the unit test to check the so-called happy path. We can also take this implementation one step further by checking whether the navigation service was called with `expectedItemViewModel`:

```
Func<ItemViewModel, bool> expectedViewModelCheck = model =>
    model.Title == expectedItemViewModel.Title;

_navigationServiceMock.Verify(
    service => service.NavigateToViewModel(
        It.Is<ItemViewModel>(_ => expectedViewModelCheck(_))));
```

In order to cover the possible outcomes (remember, we are dealing with the view model as if it was a deterministic finite automaton), we will need to implement two more scenarios where the navigation service is `null` and where the command parameter is `null`, both of which will throw `InvalidOperationException`.

5. Let's modify the `Arrange` section of the initial set:

```
var listViewModel = new ListItemViewModel(_apiClientMock.Object,
null);
```

6. In this specific case, the command (that is, ICommand) is constructed from an asynchronous task (that is, NavigateToItem), and simply calling the Execute method on the command will swallow the exception, and we will not be able to verify the exception. Therefore, we will modify our execution to use the actual view model method so that we can assert the exception:

```
#region Act
// Calling the execute method cannot be asserted.
// Action command = () =>
listViewModel.ItemTapped.Execute(expectedItemViewModel);
Func<Task> command = async () => await
listViewModel.NavigateToItem(expectedItemViewModel);
#endregion

#region Assert
await command.Should().ThrowAsync<InvalidOperationException>();
#endregion
```

Notice that, in both test cases, we are still using the the same IApiClient mock without a setup method. The execution of this mock is still possible, since the mock is created with loose mock behavior, which returns an empty collection for collection return types instead of throwing an exception for methods without a proper setup.

7. This brings the tally to ~90% unit test code coverage on ListViewModel, as shown in the following screenshot:

ListItemViewModel	0	0.00%	19	100.00%
ListItemViewModel.<>c	0	0.00%	2	100.00%
ListItemViewModel.<LoadProducts>d...	3	15.79%	16	84.21%
ListItemViewModel.<NavigateToItem>...	2	16.67%	10	83.33%

All of the tests so far have been implemented for the view models. These modules in an application are, by definition, decoupled from the UI and platform runtime. If we were to write unit tests that are targeting a Xamarin.Forms view specifically, or the targeted view model specifically requires a runtime component, the runtime and runtime features would need to be mocked because the application will not actually be executed on a mobile runtime, but rather on .NET Core runtime. The Xamarin.Forms.Mocks package fills this gap by providing a mock runtime that the Xamarin.Forms views can initialize and test.

Fixtures and data-driven tests

As you might have noticed in the previous tests we have implemented, one of the most time-consuming parts of writing a unit test is implementing the arrange portion of the implementation. In this portion, we are essentially setting up the system under test that will be used by the test target. In this setup, our goal is to bring the system to a known state so that the results can be compared with the expected results. This known state is also known as a **fixture**.

In this context, a fixture can be as simple as a mock container that contains the determinate set of components that defines the **System Under Test (SUT)**, or a factory that is driven with a predictable behavioral pattern.

For instance, if we were to create a SUT factory for our `ListItemViewModel` object, we can do so by registering the two dependencies with the fixture. Let's begin:

1. We will start the implementation by initializing our fixture and adding `AutoMoqCustomization`:

   ```
   _fixture = new Fixture();
   _fixture.Customize(new AutoMoqCustomization());
   ```

2. Let's now set up our mocks for the two service interfaces and freeze them (that is, register them to have a singleton life cycle):

   ```
   // Generating 9 random product items
   _expectedProductData = _fixture.CreateMany<Product>(9);

   _apiClientMock = _fixture.Freeze<Mock<IApiClient>>();
   _apiClientMock.Setup(service => service.RetrieveProductsAsync())
       .ReturnsAsync(_expectedProductData);

   _navigationServiceMock =
   _fixture.Freeze<Mock<INavigationService>>();
   _navigationServiceMock.Setup(nav =>
   nav.NavigateToViewModel(It.IsAny<BaseBindableObject>()))
       .ReturnsAsync(true);
   ```

3. After the mocks are set up, let's take a look at the `Arrange` block of our navigation test:

```
#region Arrange

var listViewModel = _fixture.Create<ListItemViewModel>();
var expectedItemViewModel = _fixture.Create<ItemViewModel>();

#endregion
```

4. As we can see, the injection of mock interfaces is already taken care of by `AutoMoqCustomization` and the registered frozen specimens are used for the instances.

   However, what if the data object we were using to execute the test target actually affected outcome in a way that required an additional test case? For instance, the navigation method could have two different routes depending on the data contained in the view-model:

```
if (viewModel.IsReleased)
{
    if (await _navigationService.NavigateToViewModel(viewModel))
    {
        return;
    }
}
else
{
    await _navigationService.ShowMessage("The product has not been
released yet");
    return;
}
```

5. In this case, we will need at least two states of the `ItemViewModel` object (that is, released and not). The easiest way to achieve this is to use inline data rather than the fixture, using the provided the inline data attributes:

```
[Trait("Category", "ViewModelTests")]
[Trait("ViewModel", "ListViewModel")]
[Theory(DisplayName = "Verify ListViewModel navigates on
ItemTapped")]
[InlineData(true, "Navigate")]
[InlineData(false, "Message")]
public async Task
ListItemViewModel_ItemTapped_ShouldNavigateToItemViewModel(
        bool released,
        string expectedAction)
```

6. Using the inline feed of data, we can create a composer that will create the `ItemViewModel` data items using the inline data feed:

```
var expectedItemComposer = _fixture.Build<ItemViewModel>()
    .With(item => item.IsReleased, released);
var expectedItemViewModel = expectedItemComposer.Create();
```

7. Now, we should just make sure that we are verifying the correct `navigationService` method:

```
if (expectedAction == "Navigate")
{
    _navigationServiceMock.Verify(
        service => service.NavigateToViewModel(
            It.IsAny<ItemViewModel>()));
}
else
{
    _navigationServiceMock.Verify(service =>
    service.ShowMessage(It.IsAny<string>()));
}
```

This way, both outcomes of the `ItemTapped` command are, in fact, covered by unit tests.

Maintaining cross-module integrity with integration tests

Most of the time, when we are dealing with a mobile application, there are multiple platforms involved, such as the client app itself, maybe local storage on the client application, and multiple server components. These components may very well be implemented in the most robust fashion and have deep code coverage with unit tests. Nevertheless, if the components cannot work together, then the effort put into individual components will be in vain.

In order to make sure that two or more components work well together, the developers can implement end-to-end or integration tests. While end-to-end scenarios are generally covered by automated UI tests, integration tests are implemented as a pair of permutations of the target system. In other words, we isolate two systems that depend on one another (for example, a mobile application and the web API facade) and prepare a fixture that will prepare the rest of the components to be in a known state. Once the fixture is ready for the integration pair, the implementation of integration tests is no different than the unit tests.

In order to demonstrate the value of integration tests, let's take a look at a couple of examples.

Testing the client-server communication

Let's assume we have a suite of unit tests testing the view models of our client app. We have also implemented unit tests that control the integrity of the `IApiClient` implementation, which is our main line of communication with the service layer. In the first suite, we will be mocking `IApiClient`, and in the latter suite, we will be mocking the HTTP client. In these two suites, we have covered all of the tiers from the core logic implementation down until the request is sent over the transport layer.

At this point, the next order of business is to write integration tests that will use the actual implementation of `IApiClient` to send service requests to the service API facade (also known as the gateway). However, we cannot really use the actual gateway deployment since multiple modules on the server side would be involved in this communication and the system under test would be too unpredictable.

In this scenario, we have two options:

- Create a fixture controller that will maintain the database and other moving parts involved in a known state (for example, a pre-test execution that will clean a sample database and insert the required data to be retrieved from it).
- Create a rigged deployment of the complete gateway, possibly with mocked modules as dependencies, and execute the integration tests on this system.

For the sake of simplicity, let's go with the first option and assume we have a completely empty document collection deployed to run the integration tests. In this case, we can adapt our fixture to register a set of products into a predetermined document collection (that is, the one that the server side is expecting to find), execute our retrieve calls from the application client, and finally, clean up the database.

We will start by implementing our custom fixture:

```
public class DataIntegrationFixture : Fixture
{
    public async Task RegisterProducts(IEnumerable<Product> products)
    {
        var dbRepository = this.Create<IRepository<Product, string>>();
        foreach (var product in products)
        {
            await dbRepository.AddItemAsync(product);
        }
```

```
        this.Register(() => products);
    }

    public async Task Reset()
    {
        var dbRepository = this.Create<IRepository<Product, string>>();
        var items = this.Create<IEnumerable<Product>>();
        foreach (var product in items)
        {
            await dbRepository.DeleteItemAsync(product.Id);
        }
    }
}
```

We have two initial methods, RegisterProducts and Reset:

- RegisterProducts is used for inserting the testing data and additionally registering the product data within the fixture.
- Reset is used to clear the inserted test data. This way, the test execution will yield the same results, at least, at the database level; in other words, the execution of the tests will be idempotent.

Note that the repository creation is done using the Create method so that we can delegate the responsibility of injecting the correct repository client to the test schedule.

Now, let's start working on our tests:

1. We will start by creating the test initialization (that is, the constructor in xUnit) and test tear-down (that is, the Dispose method in xUnit).
2. In the constructor, we will register the repository client implementation that the fixture will be using and register the products using this client:

```
public ClientIntegrationTests()
{
    _fixture.Register<IRepository<Product, string>>(() =>
_repository);
    var products = _fixture.Build<Product>().With(item => item.Id,
string.Empty).CreateMany(9);
    _fixture.RegisterProducts(products).Wait();
}
```

3. And here's the implementation of the IDisposable interface member:

```
public void Dispose()
{
    _fixture.Reset().Wait();
}
```

4. Now that test initialization and tear-down is ready, we can implement our first test:

```
[Fact(DisplayName = "Api Client Should Retrieve All Products")]
[Trait("Category", "Integration")]
public async Task ApiClient_GetProducts_RetrieveAll()
{
    #region Arrange
    var expectedCollection =
_fixture.Create<IEnumerable<Product>>();
    #endregion

    #region Act
    var apiClient = new ApiClient();
    var actualResultSet = await apiClient.RetrieveProductsAsync();
    #endregion

    #region Assert
    actualResultSet.Should().HaveCount(expectedCollection.Count());
    #endregion
}
```

Similar tests can be implemented to test the interaction between the server and the database or other components of the system. The key is to control the modules that are not under test and make sure the tests are executed for the target interaction.

Implementing platform tests

As mentioned earlier, integration tests don't necessarily need to be the assertion of two separate runtimes interacting with each other. They can also be used to test two distinct modules of an application in a controlled environment. For instance, when dealing with mobile applications, certain features are implemented that require interaction with the mobile platform (for example, the local storage API implementation would use the native platform filesystem; even the core SQLite implementation is abstracted to .NET Core).

For integration tests that have to be executed on a specific mobile platform (such as iOS, Android, and UWP), the Devices.xUnit framework can be used. The Devices.xUnit framework is managed by the .NET Foundation. The multi-project template included as part of the SDK creates test harness projects for the target platform and the library projects. Once the execution starts, the tests are executed on the test harness application, providing the real or emulated target platform, hence allowing the developers to execute integration tests on platform-specific features.

Automated UI tests

Arguably one of the most painstaking and costly stages of the development cycle is manual certification testing, also called acceptance testing. In a usual non-automated verification cycle, certification testing can take up to 2-3 times longer than the development of a certain feature. Additionally, if previously implemented features are at risk, regression in those areas would have to be executed. In order to increase the release cadence and decrease the development cycles, it is essential that automated UI (or end-to-end) tests are implemented. This way, the automated pipeline can be verified once and reused to verify the application's UI and integration with other systems, rather than executing manual testing in each release cycle.

App Center allows us to execute these automated tests on a number of real devices and include the automated runs in the development pipeline:

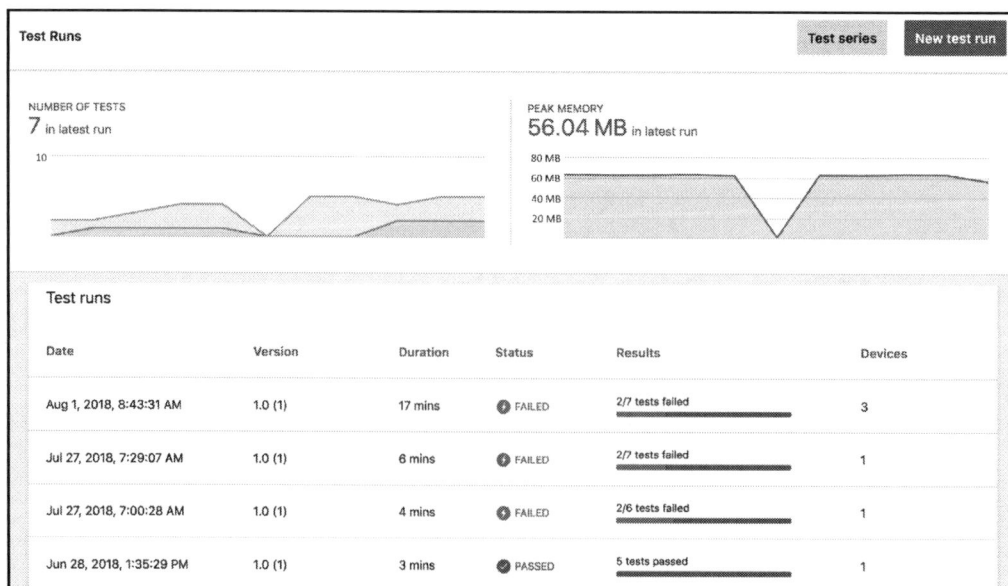

Date	Version	Duration	Status	Results	Devices
Aug 1, 2018, 8:43:31 AM	1.0 (1)	17 mins	FAILED	2/7 tests failed	3
Jul 27, 2018, 7:29:07 AM	1.0 (1)	6 mins	FAILED	2/7 tests failed	1
Jul 27, 2018, 7:00:28 AM	1.0 (1)	4 mins	FAILED	2/6 tests failed	1
Jun 28, 2018, 1:35:29 PM	1.0 (1)	3 mins	PASSED	5 tests passed	1

Xamarin.UITests is one of the supported automation frameworks that can be used to create these automated acceptance tests.

Xamarin.UITests

Xamarin.UITests is an automated UI test framework that integrates tightly with Xamarin target platforms. In addition to applications created using the Xamarin framework, it can also be used to create automated tests for mobile applications created with Java and Objective-C/Swift. NUnit is used together with the automation framework to execute the assertions and create test fixtures.

The framework allows developers to interact with mobile platforms using queries and actions. A query can described as a `select` command that is executed on the current instance of an `IApp` interface, whereas the actions are simulated user interactions with the selected elements (that is, as a result of the query). The `IApp` interface, which makes this interaction possible, provides the required abstraction among the target platforms and facilitates the user interaction with them.

The initialization of an implementation of the `IApp` interface (in other words, the simulated interaction platform) can be done in various ways, depending on the target device and platform.

Here are some examples:

- Initializing the app using an iOS app bundle can be done as follows:

```
IApp app = ConfigureApp .iOS .AppBundle ("/path/to/iosapp.app")
    .StartApp();
```

- Initializing it to be run on an iOS simulator with an already-installed application can be done as follows:

```
IApp app = ConfigureApp.iOS
    .DeviceIdentifier("ABF03EF2-64FF-4206-899E-
FB945ACEA4F2").StartApp();
```

- Initializing it for an Android device that is currently connected to ADB can be done as follows:

```
IApp app = ConfigureApp.Android.ApkFile("/path/to/android.apk")
    .DeviceSerial("03f80ddae07844d3") .StartApp();
```

Once the `IApp` instance is initialized, the simulated user interaction can be executed with the aforementioned queries and actions.

Queries can be written using the various available selectors. The most prominent queries are as follows:

- **Marked**: This refers to the `x:Name` of a Xamarin.Forms element, or an element with the given `AutomationId` object. It's in a similar fashion on native UI implementations, `AccessibilityIdentifies` or `AccessibilityLabel` on iOS and a view's `Id`, `ContentDescription`, and `Text` on Android are used for this query.
- **Class**: This queries the current UI for a specified class name. It's generally used with `nameof(MyClass)`.
- **Id**: This refers to the `Id` of the element that we are trying to locate.
- **Text**: This is any element that contains the given text.

For instance, if we were looking to tap on an element marked with `ProductsView` and select the first child in this list, we'd use this code:

```
app.Tap(c =>
c.Marked("ProductsView").Class("ProductItemCell").Index(0));
```

It is important to note the fluent execution style of the queries, where each query returns an `AppQuery` object, whereas the app actions in fact use a `Func<AppQuery, AppQuery>` delegate.

The easiest way to create structured queries for a certain view is to use the **Read-Eval-Print-Loop (REPL)** provided by the Xamarin.UITests framework. In order to start the REPL, you can make use of the associated `IApp` method:

```
app.Repl();
```

Once the REPL is initialized on the terminal session, the `tree` command can provide the complete view tree. You can additionally execute app queries and actions using the same `IApp` instance:

```
App has been initialized to the 'app' variable.
Exit REPL with ctrl-c or see help for more commands.

>>> tree
[UIWindow > UILayoutContainerView]
    [UINavigationTransitionView > ... > UIView]
        [UITextView] id: "CreditCardTextField"
            [_UITextContainerView]
        [UIButton] id: "ValidateButton"
            [UIButtonLabel] text: "Validate Credit Card"
        [UILabel] id: "ErrorrMessagesTestField"
```

```
[UINavigationBar] id: "Credit Card Validation"
    [_UINavigationBarBackground]
        [_UIBackdropView > _UIBackdropEffectView]
        [UIImageView]
    [UINavigationItemView]
        [UILabel] text: "Credit Card Validation"
>>>
```

The actions differ depending on the selected view element, but the most commonly used actions are as follows:

- `Tap`: This is used for simulating the user tap gesture.
- `EnterText`: This enters the text in the selected view. It is important to note that, on iOS, a soft keyboard is used to enter text, while on Android, data is directly passed onto the target view. This might at times cause issues when interacting with elements that are hidden under or offset by the keyboard.
- `WaitForElement`: This waits for the element defined by the query to appear on the screen. At times, with a shorter timeout period, this method can be used as part of the assertion of the element.
- `Screenshot`: This takes a screenshot with the given title. This represents a step in the App Center execution.

Page Object Pattern

Implementing UI tests within a certain test method can become quite tedious. The automation platform queries and actions would, in fact, become closely coupled and unmaintainable. In order to avoid such scenarios, it is advised to use the **Page Object Pattern (POP)**.

In POP, each view or distinct view element on the screen implements their own page class, which implements the interaction with this specific page as well as the selectors for the view components within this page. These interactions are implemented in a simplified, lexical manner so that the complex automation implementation behind the scenes is not reflected in the actual test implementation. In addition, for interactions and queries, the page object is also responsible for providing a method of navigation to and from another page.

Let's see how to implement our POP structure:

1. Let's start by creating our BasePage object:

```
public abstract class BasePage<TPage> where TPage : BasePage<TPage>
{
    protected abstract PlatformQuery Trait { get; }

    public abstract TPage NavigateToPage();
    internal abstract Dictionary<string, Func<AppQuery, AppQuery>>
    Selectors { get; set;}

    protected BasePage() {}
    // ,..
    // Additional Utility Methods for ease of execution
}
```

The base class dictates that each implementation should implement a Trait object that defines the page itself (in order to verify that the application has navigated to the target view) and a navigation method that will take the user (from the home screen) to the implementing view.

2. Let's now implement a page object for the About view:

```
public class AboutPage : BasePage<AboutPage>
{
    public AboutPage()
    {
        Selectors = new Dictionary<string, Func<AppQuery,
AppQuery>>()

            Selectors.Add("SettingsMenuItem", x =>
            x.Marked("Settings"));
            Selectors.Add("SettingsMenu", x =>
            x.Marked("CategoryView"));
            Selectors.Add("AboutPageMenuItem", x =>
            x.Marked("Information"));
            Selectors.Add("Title", x => x.Marked("Title"));
            Selectors.Add("Version", x => x.Marked("Version"));
            Selectors.Add("PrivacyPolicyLink", x =>
            x.Marked("PrivacyPolicyLink"));
            Selectors.Add("TermsOfUseLink", x =>
            x.Marked("TermsOfUseLink"));
            Selectors.Add("Copyright", x => x.Marked("Copyright");
    }

    internal override Dictionary<string, Func<AppQuery, AppQuery>>
```

```
Selectors { get; set;}

protected override PlatformQuery Trait => new PlatformQuery
{
    Android = x => x.Marked("AboutPage"),
    iOS = x => x.Marked("AboutPage")
};

public override AboutPage NavigateToPage()
{
    // Method implemented in the base page using the App
    OpenMainMenu();
    App.WaitForElement(Selectors["SettingsMenuItem"],
            "Timed out waiting for 'Settings' menu item");

    App.Tap(Selectors["SettingsMenuItem"]);
    App.WaitForElement(Selectors["SettingsMenuItem"],
            "Timed out waiting for 'Settings' menu");
    App.Screenshot("Settings menu appears.");

    App.Tap(Selectors["AboutPageMenuItem"]);

    if(!App.Query(Trait).Any())
    {
        throw new Exception("Navigation Failed");
    }

    App.Screenshot("About page appears.");

    return this;
}


public AboutPage TapOnTermsOfUseLink()
{
    App.WaitForElement(Selectors["TermsOfUseLink"],
            "Timed out waiting for 'Terms Of Use' link");

    App.Tap(Selectors["TermsOfUseLink"]);
    App.Screenshot("Terms of use link tapped");
    return this;
}
}
```

3. So now, using the `AboutPage` implementation and executing actions on `AboutPage` is as easy as initializing the `Page` class and navigating to it:

```
new AboutPage()
    .NavigateToPage()
    .TapOnTermsOfUseLink()
```

The community is divided on including the assertions within the page itself, or simply exposing the selectors so that the assertions are implemented as part of the tests. Either way, it is certain that implementing POP helps developers and QA teams create easily maintainable tests in a short time with ease.

Summary

In this chapter, we have taken a look at various testing strategies for automating the testing and verification process. Creating automated tests helps us to control the technical debt created throughout the development life cycle and keep the source code in check, hence increasing the quality of the code and the pipeline. As you have seen, some of these tests are as simple as unit tests that are implemented at the beginning of the application life cycle and executed almost at every code checkpoint, while some are elaborate, such as the integration and coded UI tests, which are generally written at end of the development stage and executed only at certain checkpoints (that is, nightly builds or pre-release checks). Regardless, the goal should always be to create a certifiable pipeline for code rather than to create code for certification.

In the upcoming chapters, we will discuss the creation of the release pipelines and various tasks that should be included at different stages.

18
Deploying Azure Modules

Azure services are bundled into so-called resource groups for easy management and deployment. Each resource group can be represented with an **Azure Resource Manager (ARM)** template, which, in turn, can be used for multiple configurations and specific environment deployments. In this chapter, we will be configuring the ARM template for Azure-hosted web services, as well as other cloud resources (such as Cosmos DB, Notification Hubs, and others) that we have previously used, so that we can create deployments using the Azure DevOps build-and-release pipeline. Introducing configuration values into the templates and preparing them to create staging environments is our main focus in this chapter.

The following sections will take you through the creation of a parameterized, environment-specific resource group template:

- Creating an ARM template
- ARM template concepts
- Using Azure DevOps for ARM templates
- Deploying .NET Core apps

Creating an ARM template

One of the cornerstones of the modern DevOps approach is the ability to manage and provision the infrastructure for a distributed application with a declarative or even procedural set of definition files that can be versioned and stored together with the application source code. In this **Infrastructure-as-Code (IaC)** approach, these files should be created in such a way that whatever the current state of the infrastructure is, executing these resources should always lead to the same desired state (that is, idempotency).

In the Azure stack, the infrastructure resources created within a subscription are managed by a service called ARM. ARM provides a consistent management tier that allows the developers to interact with it to execute infrastructure configuration tasks using Azure PowerShell, the Azure portal, and the available REST API. Semantically speaking, ARM provides the bridge between numerous resources providers and developer subscriptions.

The resources that we manage using ARM are grouped using resource groups, which are logical sets to identify application affinity groups. ARM templates, which define the set of resources in a resource group, are declarative definitions of this infrastructure as well as configurations that can be used for provisioning the application environment.

If we go back to our application and take a look at the resource group we have been using for this application, you can see the various types of Azure resources that were introduced in previous chapters:

NAME	TYPE	LOCATION
documentdb	API Connection	North Europe
outlook	API Connection	North Central US
HandsOnCrossFunctions	App Service	Central US
NetCoreUserApi-Dev	App Service	Central US
NetCoreWebUsersApi20190330084814Plan	App Service plan	Central US
HandsOn.Azure.Service	Application Insights	East US
XamarinTelemetry	Application Insights	East US
handsoncore	Azure Cache for Redis	Central US
handsoncrossplatform	Azure Cosmos DB account	West Europe
EmailProcessor	Logic app	North Europe
handsoncrossstorage	Storage account	North Europe

While all of these resources could be created using a set of Azure PowerShell or CLI scripts, it would be quite difficult to maintain them without compromising the idempotency of these scripts.

Fortunately, we can export this resource group as an ARM template and manage our infrastructure using the generated JSON manifest and environment-specific configuration parameters. In order to create the initial ARM template, do the following:

1. Navigate to the target resource group and select the **Export template** blade on the Azure portal:

2. Once the resource group template is created, we can head back to Visual Studio and create an Azure Resource Group project and paste the exported template. When prompted to select an **Azure Template**, select the **Blank Template** option to create an empty template.

> It is important to understand that the templates can be created using the provided schema. The export of the template does produce quite a number of redundant parameters and resource attributes that would not occur if the template was created manually, or if we used a base template available on GitHub.

3. When the blank ARM template is created, it has the following schema, which defines the main outline of an ARM template:

```
{                                                              \
  "$schema":
"https://schema.management.azure.com/schemas/2015-01-01/deploymentT
emplate.json#",
  "contentVersion": "1.0.0.0",
  "parameters": {},
  "variables": {},
  "resources": [],
  "outputs": {}
}
```

In this schema, while the parameters and outputs define the input and output parameters of the Azure deployment, the variables define the static data that we will be constructing and referencing throughout the deployment template.

4. Finally, the resources array will be used to define the various Azure resources that will be included in our resource group.

> **TIP**
>
> While dealing with ARM templates, Visual Studio provides the JSON outline view, which can help in navigating through the sections of the template, as well as in adding and removing new resources.

5. Without further ado, let's start by copying our first resource from the exported template into our Visual Studio project. We will start by importing the Redis cache resource:

```
{
    "type": "Microsoft.Cache/Redis",
    "apiVersion": "2017-10-01",
    "name": "[parameters('Redis_handsoncore_name')]",
    "location": "Central US",
    "properties": {
        "sku": {
            "name": "Basic",
            "family": "C",
            "capacity": 0
        },
        "enableNonSslPort": false,
        "redisConfiguration": {
            "maxclients": "256",
            "maxmemory-reserved": "2",
            "maxfragmentationmemory-reserved": "12",
```

```
        "maxmemory-delta": "2"
      }
    }
  }
```

In this resource, the name attribute is referencing a parameter called
the Redis_handsoncore_name parameter, which we will need to add to the
parameters section. Other than the name, some basic resource metadata is
defined, such as apiVersion, type, and location. Additionally, resource-
specific configuration values are defined in the properties attribute.

6. Let's continue with adding the parameter to the parameters section with some
   modifications:

```
"resourceNameCache": {
    "defaultValue": "handsoncore",
    "type": "string",
    "minLength": 5,
    "maxLength": 18,
    "metadata": {
        "description": "Used as the resource name for Redis cache
resource"
    }
}
```

We have added the metadata, which in some sense helps us to document the
resource template so that it is more maintainable. In addition to the metadata, we
have defined the minLength and maxLength attributes so that the parameter
value has some validation (if we decide to use a generated or calculated value).
Additionally, for a string type parameter, we could have defined allowed values.

7. Finally, let's add an output parameter that outputs the Redis resource connection
   string as an output parameter:

```
"redisConnectionString": {
    "type": "string",
    "value": "[concat(parameters('resourceNameCache'),
'.redis.cache.windows.net:6380,abortConnect=false,ssl=true,password
=', listKeys(resourceId('Microsoft.Cache/Redis',
parameters('resourceNameCache')), '2017-10-01').primaryKey)]"
}
```

Here, we are creating the Redis connection string using an input parameter, as well as a reference to an attribute of the given resource (that is, the primary access key of the Redis cache instance). We will take a closer look at ARM functions and references in the next section.

8. Now that our basic template for the resource group (that is, only the partial implementation that includes the Redis cache resource) is ready, we can try deploying the template using Visual Studio.

9. After you create a new deployment profile and designate the target subscription and resource group, you can try to edit the parameters and see how the parameter metadata that we have added reflects on the user interface:

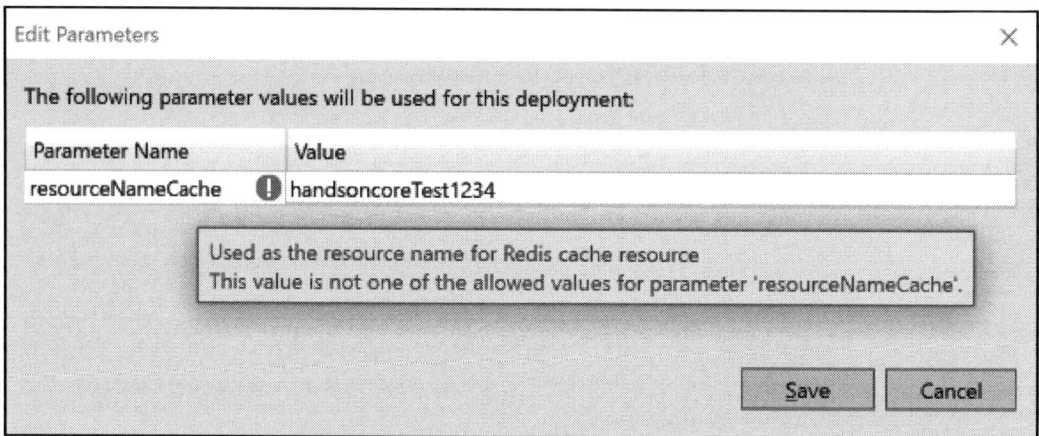

Edit Parameters ☒

The following parameter values will be used for this deployment:

Parameter Name	Value
resourceNameCache ❗	handsoncoreTest1234

Used as the resource name for Redis cache resource
This value is not one of the allowed values for parameter 'resourceNameCache'.

Save   Cancel

10. Once the deployment is complete, you can see the output parameters on the deployment details:

```
09:27:19 - DeploymentName : azuredeploy-0421-0713
09:27:19 - CorrelationId : 4ec72df5-00d3-4194-9181-161a5967235b
09:27:19 - ResourceGroupName : NetCore.Web
09:27:19 - ProvisioningState : Succeeded
09:27:19 - Timestamp : 4/21/2019 7:27:20 AM
09:27:19 - Mode : Incremental
09:27:19 - TemplateLink :
09:27:19 - TemplateLinkString :
09:27:19 - DeploymentDebugLogLevel :
09:27:19 - Parameters : {[resourceNameCache,
09:27:19 -
Microsoft.Azure.Commands.ResourceManager.Cmdlets.SdkModels.Deployme
ntVariable]} 09:27:19 - ParametersString :
09:27:19 - Name                Type                        Value
09:27:19 - ===============     =========================   ==========
```

```
09:27:19 - resourceNameCache    String
handsoncore123
09:27:19 -
09:27:19 - Outputs : {[redisConnectionString,
09:27:19 -
Microsoft.Azure.Commands.ResourceManager.Cmdlets.SdkModels.Deployme
ntVariable]} 09:27:19 - OutputsString :
09:27:19 - Name                    Type                        Value
09:27:19 - ==============          =========================== ==========
09:27:19 - redisConnectionString String
handsoncore123.redis.cache.windows.net:6380
09:27:19 - ,abortConnect=false,ssl=true,password=JN6*******kg=
09:27:19 -
09:27:20 -
09:27:20 - Successfully deployed template 'azuredeploy.json' to
resource group 'NetCore.Web'.
```

In a template deployment, one of the key parameters is the deployment mode. In the preceding example, we were using the default deployment mode, namely `Incremental`. In this type of deployment, Azure resources that exists in the resource group but not in the template will not be removed from the resource group, and only the items in the template will be provisioned or updated, depending on their previous deployment state. The other deployment option available is the so-called complete deployment mode. In this mode, any resource that exists in the resource group but not in the template is removed automatically.

> The resources that exist in the template, but are not deployed because of a condition, would not be removed from the resource group.

ARM template concepts

So, now that we have successfully deployed our first resource, let's continue with expanding our template to other resources:

1. In order to demonstrate the dependencies between the resources, let's introduce next the App Service instance, which will be hosting the users API. Technically, this App Service only has a single dependency—the App Service Plan—that will be hosting the App Service (that is, the `ServerFarm` resource type):

```
{
    "type": "Microsoft.Web/serverfarms",
    "name": "[parameters('resourceNameServicePlan')]",
    "kind": "app",
```

```
                // removed for brevity
        },
        {
            "type": "Microsoft.Web/sites",
            "name": "[parameters('resourceNameUsersApi')]",
            "dependsOn": [
                "[resourceId('Microsoft.Web/serverfarms',
        parameters('resourceNameServicePlan'))]"
            ],
            "kind": "app",
            "properties": {
                "enabled": true,
                "serverFarmId": "[resourceId('Microsoft.Web/serverfarms',
        parameters('resourceNameServicePlan'))]",
                // removed for brevity
            }
        }
```

In this setup, the `sites` resource has a dependency on the `serverfarms` resource type (that is, deployment order will be evaluated depending on these dependencies). Once the `serverfarms` resource is deployed, the `resourceId` of the created resource is going to be used as the `serverFarmId` for the `sites` resource.

Additionally, the users API, from an architectural perspective, has two main dependencies: Redis Cache and Cosmos DB. The resultant resource instances should produce values that should be added to the application configuration of the App Service (that is, connection strings).

2. Since we have already created an output parameter for the Redis Cache instance, let's create a dependency and add the connection string. The connection string can be added as part of the `siteConfig` attribute of a `sites` resource or by creating an additional resource specifically for the site configuration:

```
"properties": {
    "enabled": true,
    // removed for brevity
    "siteConfig": {
        "connectionStrings": [ {
            "name": "AzureRedisCache",
            "type": "custom",
            "connectionString":
"[concat(parameters('resourceNameCache'),
'.redis.cache.windows.net:6380,abortConnect=false,ssl=true,password
=', listKeys(resourceId('Microsoft.Cache/Redis',
parameters('resourceNameCache')), '2017-10-01').primaryKey)]"
```

```
                } ]
            }
        }
```

3. Now when the site is deployed, the Redis cache connection string is automatically added to the site configuration:

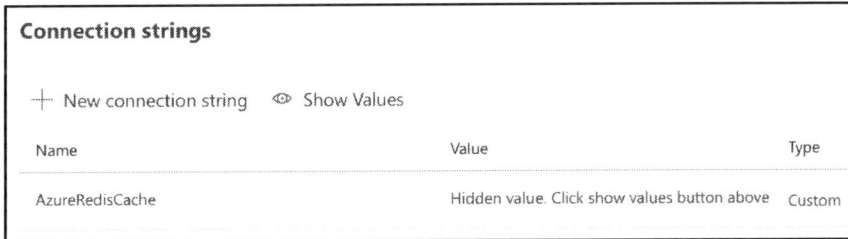

Connection strings

+ New connection string 👁 Show Values		
Name	Value	Type
AzureRedisCache	Hidden value. Click show values button above	Custom

Notice that when preparing the connection string, we are making use of the `concat` function to compose the value, and we use the `listKeys` function to get a list of values from the resource instance that is retrieved according to type and name using the `resourceId` function.

These functions and other user-defined functions can be used throughout the resource template, either while constructing reference values or defining conditions. Some of these functions according to the parameter type are listed here:

String	base64, base64ToJson, base64ToString, concat, contains, dataUri, dataUriToString, empty, endsWith, first, format, guid, indexOf, last, lastIndexOf, length, newGuid, padLeft, replace, skip, split, startsWith, string, substring, take, toLower, toUpper, trim, uniqueString, uri, uriComponent, uriComponentToString, utcNow
Array	array, coalesce, concat, contains, createArray, empty, first, intersection, json, last, length, max, min, range, skip, take, union
Comparison	equals, greater, greaterOrEquals, less, lessOrEquals
Deployment	deployment, parameters, variables
Logical	and, bool, if, not, or
Numeric	add, copyIndex, div, float, int, max, min, mod, mul, sub
Resource	listAccountSas, listKeys, listSecrets, list*, providers, reference, resourceGroup, resourceId, subscription

Using these functions, complex variables can be constructed and reused using parameters and references to deployed resources. These constructs can then be exposed as functions so that they can be reused throughout the template declarations.

As you might have noticed, every time we deploy the ARM template, according to the parameters we define, the `azuredeploy.parameters.json` file is updated. These are the parameters that are provided to the template, and in a multi-stage environment (that is, DEV, QA, UAT, PROD), you would expect to have multiple parameters files assigning environment-specific values to these resources:

```
{
    "$schema":
"https://schema.management.azure.com/schemas/2015-01-01/deploymentP
arameters.json#",
    "contentVersion": "1.0.0.0",
    "parameters": {
        "environment": { "value": "DEV" },
        "resourceNameCache": { "value": "dev-handsoncoreCache" },
        "resourceNameServicePlan": { "value": "dev-handsoncorePlan"
},
        "resourceNameUsersApi": { "value": "dev-handsoncoreusers" }
    }
}
```

With multiple parameters files, unique resource names and addresses can be constructed so that during the deployment, duplicate resource declarations can be avoided. Another method that is generally used to avoid resource name/address clashes is to include the current resource group identifier as part of the resource names, making sure that the resources are specific and unique.

Using Azure DevOps for ARM templates

Once the template is ready and we are sure that all Azure resources that are required by our application are created, we can continue with setting up automated builds and deployments.

In order to be able to use Azure DevOps for cloud deployments, our first action would be to create a service principal that will be used to deploy the resources. A service principal can be described as a service identity that has access to Azure resources within a certain subscription and/or resource group.

So, let's begin:

1. A service principal can be created by adding a new ARM service connection within the Azure DevOps Project settings:

Add an Azure Resource Manager service connection

● Service Principal Authentication   ○ Managed Identity Authentication

Connection name   Principle for HandsOnCrossPlatform

Scope level   Subscription

Subscription   Visual Studio Enterprise – MPN (205bb4f7-2f8a-4904-9f37

Resource Group   HandsOnCrossPlatform

Subscriptions listed are from Azure Cloud

A new Azure service principal will be created and assigned with "Contributor" role, having access to all resources within the subscription. Optionally, you can select the Resource Group to which you want to limit access.

If your subscription is not listed above, or your organization is not backed by Azure Active Directory, or to specify an existing service principal, use the full version of the service connection dialog.

☑ Allow all pipelines to use this connection.

OK   Close

2. Creating the connection will create an application registration for Azure DevOps and assign this service principal the contributor role on the selected subscription.

3. The service creation process can also be executed using the **Authorize** button when creating an **Azure Deployment** task. The **Authorize** button becomes available if an Azure subscription is selected, instead of a service principal:

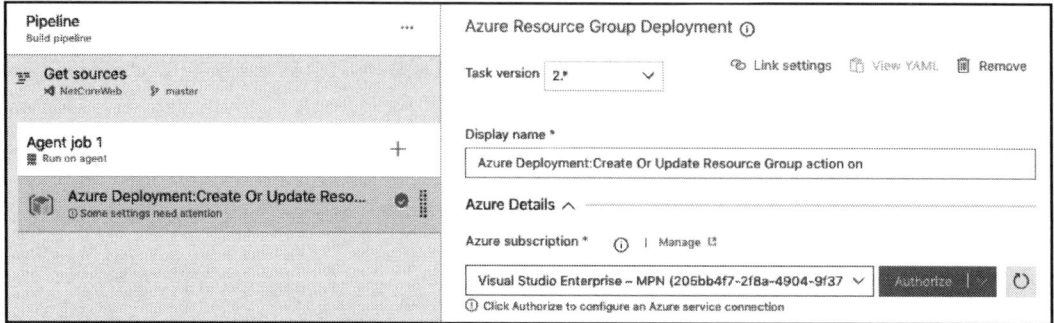

4. Once the service principal is created, we can continue with setting up the deployment for the resource group. Azure DevOps offers multiple deployment options for deploying Azure resource group templates, such as the following:

- **Azure PowerShell**: To execute in-line or referenced Azure PowerShell scripts
- **Azure CLI**: To execute in-line or referenced Azure CLI scripts
- **Azure Resource Group Deployment**: To deploy ARM templates with associated parameters

For this example, we will be using the **Azure Resource Group Deployment** task:

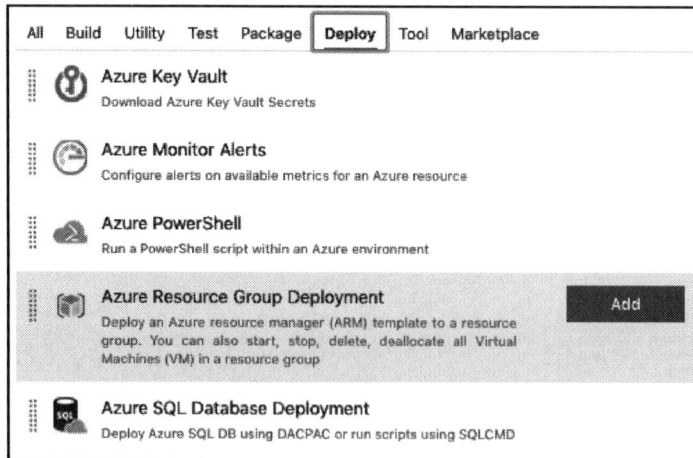

5. For this task, other than the service principal settings, the **Template** section provides the main configuration area. Here, we will be selecting the ARM template and providing the parameters file designated for this environment:

6. Additionally, we can define configuration values such as deployment name and deployment outputs.

> **TIP**
>
> The **Azure Resource Group Deployment** task allows the selection of three deployment modes. On top of the Azure deployment modes (that is, complete, and incremental), the **Validate** option provides the ability only to validate the template. The **Validate** option can be used to validate pull requests and can only execute during continuous integration builds.

7. Now, the ARM deployment can be triggered whenever an update is merged into the master branch that keeps the development (or higher) environment up to date with the ARM template definition:

Deploying .NET Core applications

After the ARM template is deployed and the Azure resources are created, our next step would be to deploy .NET Core applications (that is, microservice applications as well as our functions app).

Azure DevOps provides all the necessary tasks to build and create the deployment package for an app service/web app. The trifecta of creating a .NET Core web deployment package is composed of restore, build, and publish. All these dotnet CLI commands can be executed using the built-in tasks within the build-and-release pipeline. So, let's begin:

1. We will start by restoring the NuGet packages for our users API microservice:

2. The next step is to build the application using a specific build configuration (a pipeline variable can be used for this):

3. After the project is built, we can prepare our web deployment package to be able to push it to the app service resource that was created in the ARM deployment step. In order to prepare the deployment package, we will use the **publish** command:

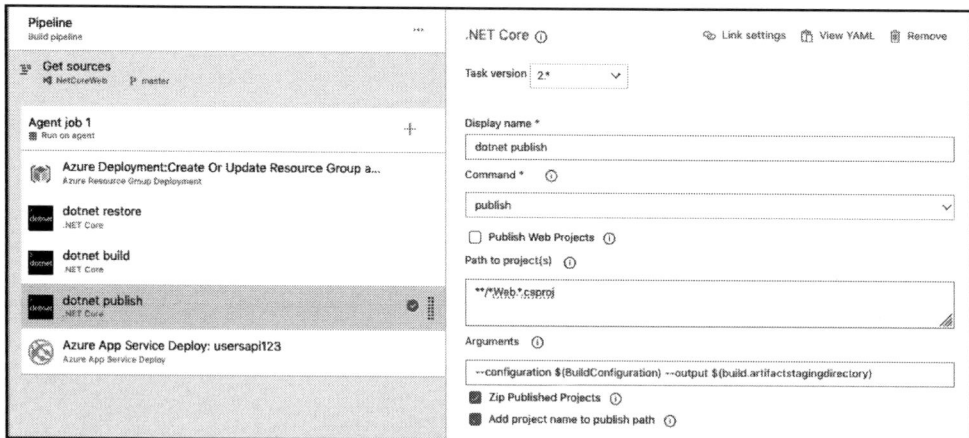

4. Finally, for the deployment, we can make use of the **Azure App Service Deploy** task:

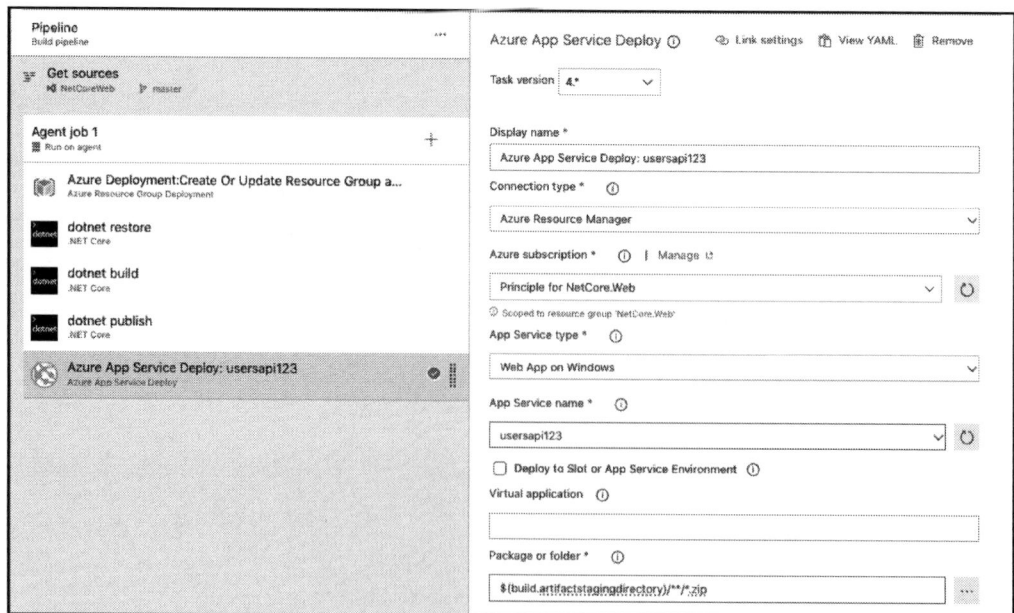

In the deployment step, it is imperative that the package or folder parameter matches the folder that was used as the output directory in the publishing step (that is, $(build.artifactstagingdirectory)).

In the last task, we have used a hardcoded name for the App Service name parameter. This would mean that each time the ARM parameter file is modified, we would need to update the build definition; or, if we were using a YAML file for the build definition, the YAML definition. This could prove to be a maintenance nightmare, since every time a new environment is created, the build and release definitions would need to be updated.

5. One possible solution to integrating the ARM deployment with the actual application deployment would be to output the target application name from the ARM template and use it as the App Service name parameter:

```
"outputs": {
    "redisConnectionString": { //... },
    "userApiAppResource": {
        "type": "string",
        "value": "[parameters('resourceNameUsersApi')]"
    }
}
```

6. Next, we will assign the output from the ARM deployment to a variable named armOutputVariables, using the **Deployment outputs** option from the **Azure Resource Group Deployment** task:

7. Next, we can add a PowerShell script task (that is, not Azure PowerShell) to parse the output into JSON format and assign the required App Service name to a pipeline variable:

```
$outputs = ConvertFrom-Json $($env:armOutputVariables)

Write-Host "##vso[task.setvariable
variable=UsersApiAppService]$($outputs.userApiAppResource.value)"
```

8. At this point, the App Service resource name can be accessed like other pipeline variables (that is, with $(UsersApiAppService)) and can be assigned to the **Azure App Service Deploy** step.

The rest of the build templates can be created in same manner, using the same or similar .NET Core and Azure tasks.

Summary

In this chapter, we have gone through the basic steps of creating an ARM template so that the cloud infrastructure required for our application can be provisioned and managed in line with IaC concepts. Having set up our cloud resources as a declarative JSON manifest, we can easily version and keep track of our environment(s) without environment drift and infrastructure-related deployment issues. The .NET Core build and publish steps that are part of the Azure DevOps services are then used to create the deployment artifacts, which seamlessly integrate with the Azure cloud infrastructure.

We have managed to prepare our build-and-release pipeline for one of the .NET Core services in this chapter. However, what we are actually after is to create the deployment artifacts during the continuous integration build and use a release pipeline to deploy the infrastructure, followed by the deployment of the App Service artifacts. We will create the release pipeline in the next chapter.

19
CI/CD with Azure DevOps

Continuous Integration (**CI**) and **Continuous Delivery** (**CD**) are two concepts that are deeply rooted in the Agile project life cycle definition. In the agile methodology, the DevOps effort is mostly spent on decreasing the CD cycle so that smaller sprints and smaller change sets can be periodically delivered to the users. In return, the smaller the change, the smaller the risks, and the easier adoption will be for the users. In order to minimize the length of delivery cycles, an automated delivery pipeline is vital. Using the toolset provided with Azure DevOps, developers can create fully automated templates for builds, testing, and deployments. In this chapter, we will set up the build and release pipeline for Xamarin in line with the Azure deployment pipeline.

In this chapter, we'll look at the following topics:

- Introducing CI/CD
- CI/CD with GitFlow
- **Quality Assurance** (**QA**) of branches
- Creating and using release templates

Introducing CI/CD

In the previous chapters, we set up various build definitions to create application binaries and packages that can be used as deployment artifacts. While preparing these artifacts, we implemented automated tests that can be included in automated build definitions. This process of automating the build and testing of code every time a team member introduces changes to version control is generally referred to as CI. CI, coupled with a mature version control system and a well-defined branching strategy, is the primary factor in encouraging developers to be bolder and more agile with their commits, contributing a high release cadence.

On the other hand, CD is the (generally) automated process of building, testing, and configuring your application and finally deploying that specific version of your application to a staging environment. Multiple testing or staging environments are generally used, with automated creation of infrastructure and deployment, right up until the production stage. In a healthy CD pipeline, the success of the sequential set of environments is measured by progressively longer-running activities of integration, load, and user acceptance testing. CI starts the CD process and the pipeline stages each successive environment upon successful completion of the previous round of tests.

In CD, a release definition is composed of a collection of environments. An environment is a logical container that represents where you want to deploy your application. Physically, an environment can refer to a cluster of servers, a resource group on cloud infrastructure, or a mobile application distribution ring. Each environment (sometimes referred to as stage) has its purpose, and a subset of the stakeholders from the development pipeline are assigned as owners for that specific environment:

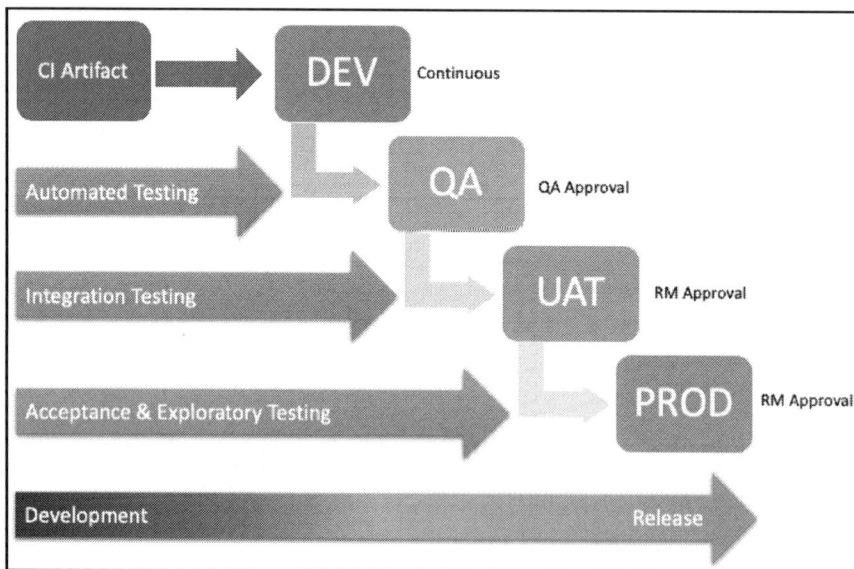

The configuration of the web services and mobile application for each environment could differ depending on the purpose of that specific environment. Nevertheless, it is important to keep in mind that the environments should, at any point in the CI/CD pipeline, host the same application release version and binaries. This way, we can make sure that, once the application is promoted to a higher environment (that is, closer to production), it will function as it did in the previous stage.

As you can see, in DevOps terms, deployment does not always mean release or production. As part of CD, various deployment pipelines are triggered by commits from the development team. If the committed code is verified by unit tests and integration builds, the artifacts from these branches are deployed to the staging environments. In staging environments, smoke and acceptance tests are executed. If the application integrity and new features are verified by these tests on various stages, the new release can be rolled out across the production environment.

In Azure DevOps terms, as you have seen from the previous examples, CI is implemented using Git and Azure DevOps Build templates. CD, on the other hand, is handled by release definitions using the build artifacts prepared with the triggered CI builds.

> CI/CD pipelines can, in fact, be prepared to use TFVC and associated branching strategies; however, Git and its associated branching strategies such as GitFlow provide a more flexible and agile setup.

Using Azure DevOps, the transition between the environments (that is, the promotion process) can be controlled with pre-deployment and post-deployment gates. These gates can be set up to require approval from specific stakeholders in the development pipeline (for example, the environment owner). In addition to manual approval, remote services can be used for gate approvals. For instance, it is possible to synchronize the release of a module for an application with another dependent module, or delay a deployment stage until certain tests are executed.

The examples provided here are only one of the possible designs, and the version control implementation, branching strategy, and the associated release pipeline setup should be designed and executed according to the needs of the development team and business requirements. In our example application, for creating the pipeline, we will be using Git and GitFlow as our version control and branching strategy, respectively. For the release pipeline, we will create a development release that will be automatically deployed with each commit/merge to the development branch (that is, the next version), whereas QA, UAT, and production environments will be deployed from release branches (that is, the current version).

CI/CD with GitFlow

The easiest way to illustrate CI/CD would be walk through the policies and procedures, starting with GitFlow. Here, we are dealing with two separate repositories, namely, web and application, and each of these repositories have their own life cycle.

In other words, while it is not advised, it is possible to have unsynchronized releases and versions of our web application (that is, the service infrastructure) and application (that is, mobile platform releases); hence, it is important to create backward-compatible modules and communicate releases to development team members.

Development

Development of the application or web modules start with the creation of a feature branch (for example, `feature/12345`). The feature branch can be shared between multiple developers or handled by a single developer. If the feature branch is being worked on by multiple developers, user branches can be created following a similar convention along the lines of `user/<user identifier>/<feature id>` (for example, `user/cbilgin/12345`). Once each developer is done with their implementation, a pull request can be executed on the main feature branch.

An important factor for determining the health of a feature branch is the commit differences between the development branch and the feature branch. Ideally, the feature branch should always be ahead of the development branch, regardless of the number of pull requests completed on the development branch while the feature branch is being worked on. In order to achieve this, the feature branch should be periodically rebased on the current development tree, retrieving the latest commits from other features.

The work done on a feature branch can be tested locally by developers with locally running the web application and running the mobile application on the desired emulator/simulator. While iOS and UWP simulators can use the localhost prefix, since the local machine network is shared by the simulator, Android emulators use their own NAT table, where localhost refers the mobile device, not the host machine. In order to access a hosted web service on the host machine, you should use `10.0.2.2` IP interface. The following table shows the different IP address and how they can be used in the context of Android emulator:

Network address	Description
`10.2.2.1`	Router/gateway address
`10.2.2.2`	Special alias to your host loopback interface (that is, `127.0.0.1` on the host machine)
`10.0.2.3`	First DNS server
`10.0.2.4-6`	Optional additional DNS servers
`10.0.2.15`	The emulated device network/Ethernet interface
`127.0.0.1`	The emulated device loopback interface

If the application and web services are handled on separate repositories with their own release cycles, the local web server instance should use the latest commit on development (or master) branch, making sure that the development is done with the latest service infrastructure.

Once the feature is ready to be integrated into the next release, the developer is responsible for creating a pull request with the associated work item that this feature branch represents.

Pull request/merge

In an ideal setup, the only way a feature branch is merged into the development branch, should be through a pull request. Pull requests are also used to execute quick sanity checks and code reviews.

The quality of the code that is delivered by the developers can be verified by the branch policies of the target branch (that is, the development branch). In this example, for the development branch, we will make use of four policies:

- **Work items to be attached to the pull request (feature and/or user story or bug)**: The tasks and user stories are generally attached to the commits within the feature branch. The feature, user story and/or bug work items (these are parent work items to the tasks) have to be attached to the pull request so that, once the release pipeline is created, these work items can be determined from the release build.

- **Review by two team members**: To encourage the peer review process, a minimum of two members are responsible for reviewing the pull request. One of these team members is generally the team lead, who is a mandatory reviewer of any pull request targeting either the development or release branches. Each reviewer is responsible for commenting the code changes, which then are corrected by the owner of the pull request.

- **Review by team lead (included in the minimum count)**: The architect or team lead is generally the person ultimately responsible for the quality of the code introduced into the development or release branches, hence he/she is a mandatory reviewer for the pull requests.

- **Branch evaluation build**: For the mobile project, the branch evaluation build can be a build of the Android project (since Android builds can be executed on a Windows build agent, as opposed to the iOS builds having to be executed on a Mac agent). This build should execute the unit tests and run static code analysis using a platform such as SonarQube or NDepend.

> **TIP**
> In such a setup, it is wise to allow team leads or other accountable stakeholders to have override authority over the policies in cases of emergency, bypassing the evaluation build (rarely) and the review requirements (more common).

Using Azure DevOps, in order to set up policies and enforce developers to create pull requests, the target branch (that is, the development branch) should be configured with the respective branch policies:

Require a minimum number of reviewers
Require approval from a specified number of reviewers on pull requests.

Minimum number of reviewers   2

- [ ] Allow users to approve their own changes.
- [ ] Allow completion even if some reviewers vote "Waiting" or "Reject".
- [ ] Reset code reviewer votes when there are new changes.

Check for linked work items
Encourage traceability by checking for linked work items on pull requests.

Policy requirement

- (•) Required
  Block pull requests from being completed unless they have at least one linked work item.
- ( ) Optional
  Warn if there are no linked work items, but allow pull requests to be completed.

- [ ] **Check for comment resolution**
  Check to see that all comments have been resolved on pull requests.

- [ ] **Limit merge types**
  Control branch history by limiting the available types of merge when pull requests are completed.

Build validation
Validate code by pre-merging and building pull request changes

+ Add build policy

Build pipeline	Requirement	Path filter	Expiration	Trigger	
Validation Tests	Required	No filter	Strict expiration	Automatic	Enabled

Require approval from additional services
Require other services to post successful status to complete pull requests. Learn more

+ Add status policy

Any policy requirement on the target branch should enforce the use of pull requests when updating a branch.

Once the policy validations are satisfactory (that is, all required policies are met), the code is ready to be merged to the development branch. Azure DevOps allows the selection of a merge strategy during completion. This merge strategy should be in line with the branching strategy and design. In our example, we will be using Rebase; however, any of the other three merge options can be used:

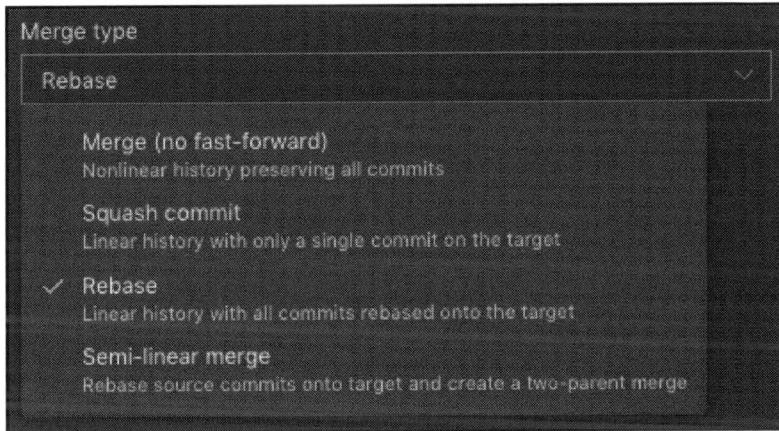

Once the pull request is merged into the development branch, the CI phase can commence.

The CI phase

Any update on a CI-enabled branch triggers the build(s) to build the application and/or web services package. For instance, for the mobile application development pipeline, multiple builds can be triggered for target mobile platforms with development stage configurations (for example, DevDroid and DeviOS configurations). These builds can prepare the application packages as build artifacts and publish it with the next application version and minor revision.

The trigger for the CI build can be set up in the build properties:

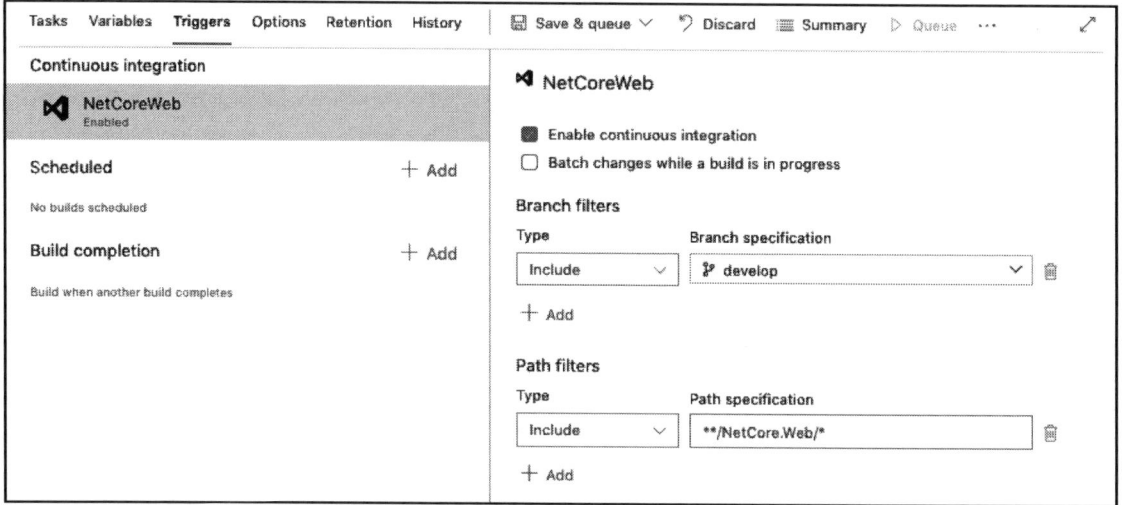

In addition to the triggering branch, path filters can be set so that, depending on the changes introduced to the application code base, different CI builds can be triggered and artifacts prepared.

Additionally, the CI build should execute any available unit tests, along with simple integration tests, that can be run on the build agent, and these results can be published with code coverage to the pipeline. Another round of static code analysis with the current artifact version annotation can be executed and published on the static analysis platform (for example, a SonarQube server). This would help to correlate the source code delta and possible issues occurring with this version of the application.

When the CI build is successfully completed, depending on trigger setup, a release pipeline can be created, deploying the prepared artifact(s) to the target environment (that is, the development environment in this case):

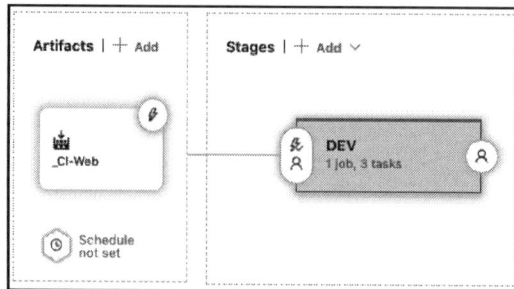

This example deploys the microservice packages prepared by the CI build and deploys them to the development environment. The environment is updated by the **Azure Resource Manager** (**ARM**) deployment to target the resource group. It is common practice to have the development environment release setup without any pre-approval gate so that any code integrated into the development branch is automatically deployed to the development environment.

Azure DevOps offers two tasks to publish the build pipeline artifacts, and created artifacts can be used as the trigger or a secondary artifact for a release pipeline:

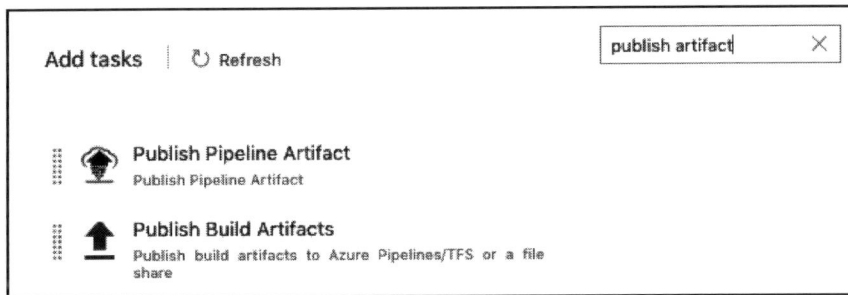

Add tasks   ↻ Refresh   publish artifact   ✕

    Publish Pipeline Artifact
    Publish Pipeline Artifact

    Publish Build Artifacts
    Publish build artifacts to Azure Pipelines/TFS or a file share

The general rule of thumb for artifact publishing is to use the staging directory for the build:

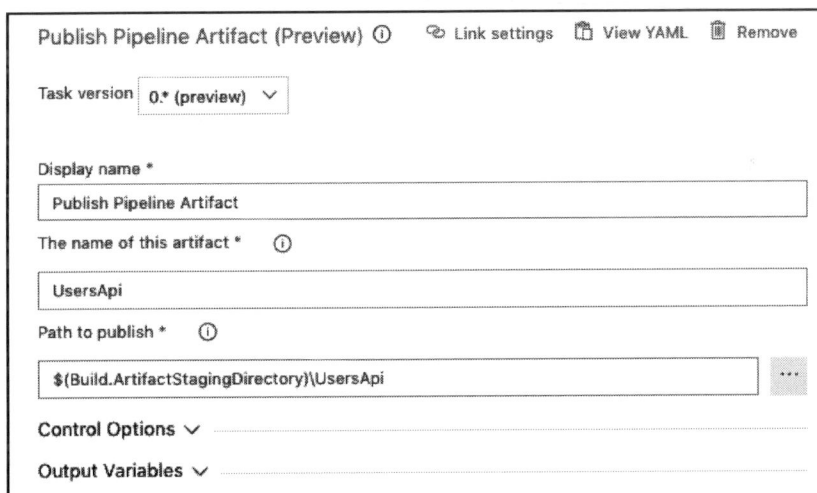

Publish Pipeline Artifact (Preview) ⓘ   ⮞ Link settings   🗋 View YAML   🗑 Remove

Task version   0.* (preview) ⌄

Display name *

    Publish Pipeline Artifact

The name of this artifact *   ⓘ

    UsersApi

Path to publish *   ⓘ

    $(Build.ArtifactStagingDirectory)\UsersApi   ...

Control Options ⌄

Output Variables ⌄

Before the artifact publishing task, for the web application, a .NET publishing task can be used with the same output directory. For the Xamarin packages, a copy task can be created, copying the application package(s) to the staging directory.

Release branch

Once the development branch is verified and the current feature set matches the predetermined release scope, a release branch is created (for example, release/1.8). The creation of the release branch goes hand in hand with the triggering of the release build that will prepare the complete set of artifacts required for the release of a certain environment. The associated release pipeline will, in return, be triggered by the builds on this branch.

We should dedicate a paragraph here for Xamarin application packages because of the environment-specific configuration structure and multiple application artifacts. As previously mentioned, for a native app to support multiple configurations, we would need to create multiple application packages. If you consider a minimal support scenario, such as only supporting the iOS platform, in order to have QA, staging, and production environments in our release pipeline, we would need to have three separate application packages that are created from the same source code version. If we also want to support Android, this would mean for a single environment (for example, QA), we would need to deploy an IPA and an APK to Visual Studio App Center (two separate application rings) and verify these applications using the same web infrastructure. The synchronization of these builds and release stages should be carefully designed and executed. Multi-configuration builds for a single platform can be used in conjunction with multi-agent build templates to create a single artifact with multiple packages, so that the release pipeline is triggered only after all of the required builds are finalized.

Finally, the build template for the release artifacts can also be used to check the quality and process the scope introduced by the release to be created. Release pipelines also support the execution of automated tests so that the validation process can be automated within a release environment. For instance, after a certain service API package is deployed, it would be a good idea to execute functional tests against the deployment URL to verify the success of the deployment. In similar fashion, before the application package is deployed to a certain App Center ring (for example, staging), Xamarin Test Cloud tests can be executed to verify the application features.

Hotfix branches

After the release pipeline is used to deploy the application to either the fast or slow ring (QA or UAT), depending on the testing agreement, the QA team can start testing the application.

At this phase, any bugs created or additional feature requests that are pulled into the current release scope should be introduced by the development team using hotfix branches. Hotfix branches originate from the current release branch and are merged back into the release branch (for example, `release/1.8 -> hotfix/12324 -> release/1.8`) with the user of pull requests. Once the hotfix is merged back into release branch (that is, it has passed the validation), it will need to be merged back to development branch as well to propagate the code changes and avoid regression in the following releases.

The merges and pull requests to the release branch follow a similar (although not identical) methodology as that in the development branch. This way, we can push the hotfix modifications through the same quality validation process.

Production

In the release pipeline, a certain version of the artifact(s) can be promoted from one environment to higher ones until the production release is complete and the new version of the application is delivered to the end users.

According to the pipeline design, production environment can also utilize a staging or phased release strategy with release rings using deployment slots (that is, on Azure App Services), native staging with TestFlight (that is, for iOS applications), or even incremental releases, which both Apple and and the Google Play Store support. This way, application telemetry from the release environments can be collected from beta users and introduced back into the development pipeline.

The QA process

In each phase of a CD process, the quality of the features should be verified preferably with an automated process or, at the very least, with proper code reviews. This is where the pull request creation and validation process becomes even more important. Nevertheless, as mentioned, the QA of an artifact or a branch is not limited to the CI phases of the process, but runs throughout the CI/CD pipeline.

Code review

A healthy development team should be driven by collaboration. In this context, the concept of peer review is extremely important, since it gives the chance for the development team to suggest and advice improvements of a colleague's work. Azure DevOps have two branch policies that directly encourage or even enforce the peer review process. One of these policies is the minimum number of reviewers, and the second one is the automatic code reviewer policy:

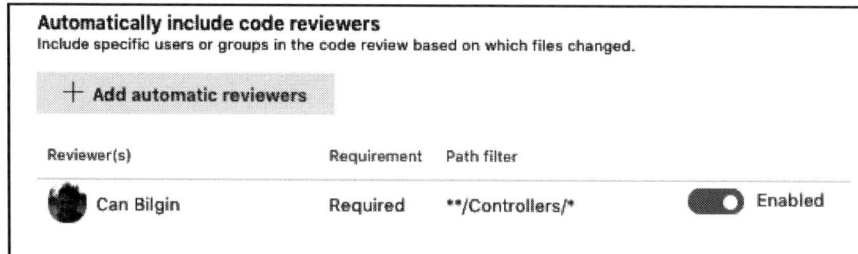

Using the Automatic Code Reviewer policy, multiple optional and/or mandatory reviewers can be automatically added to the pull request review process for different source paths.

This allows the developers to collaborate on the Azure DevOps web portal, creating comments on certain lines, sections, or even files from the commits included in the pull request.

The review process is not only limited to manual developers' feedback, but some partial automation can even be introduced. If included as part of a validation build policy, SonarQube and the Sonar C# plugin can detect that the containing build is executed on a pull request, and the code issues, found as the result of static analysis on the new code, are added to the pull request as comments to be resolved before the pull request is completed.

The review comments added by peers or automated tools can be enforced (that is, they must be resolved before the pull request is complete) using the comment resolution policy:

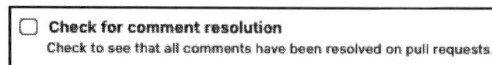

Overall, it is fair to say that code review makes up an important part of code quality maintenance in a CI/CD pipeline.

Testing

As you have seen in `Chapter 17`, *Automated Testing*, various tests can be automated and introduced into the CD pipeline. The tests executed on any CI build can be displayed in the build results summary in a separate section with aggregated report values (taking the previous runs into consideration), giving developers the chance to identify issues before the application artifacts are deployed to the target environment.

Any (failed) test can be used as the starting point to create a work item (e.g. a bug or an issue depending on the process template used) in the product backlog with any associated debugging information, if available. Moreover, automated tests can be associated with actual work items such as features or user stories, allowing the CI process to create meaningful correlations with the project management metadata:

As a result, automated tests can be used not only to identify problems early on, but can also help with the analysis and triage processes. This way, the development team can improve two important DevOps KPIs: **Mean Time To Detect** (**MTTD**) and **Mean Time To Restore** (**MTTR**), creating and maintaining a healthy CD pipeline.

Static code analysis with SonarQube

Because of the compiled nature of C# and .NET Core, static analysis and quality metrics of the source code can help the development team to maintain a healthy development pipeline. Similar to older tools such as StyleCop, and more popular Visual Studio extensions such as ReSharper, SonarQube is an open source static analysis platform providing valuable KPIs and history about the application source. Using SonarQube, certain traits and trends on quality metrics such as complexity, code smells, and duplications can be used to identify issues early on in the development cycle, helping the development team to steer the application in the right direction.

SonarQube supports a number of platforms and languages, including C#, and is deeply integrated with MSBuild and Azure DevOps infrastructure, which makes it an ideal choice for any .NET Core development project. The server component can be hosted on-premise or as part of a cloud setup. On the other hand, SonarCloud is offered as a hosted version of the Java-based platform.

Once the SonarQube server is set up and the required plugins installed (namely, SonarCSharp), the quality profile for a given project can be set up. A quality profile is composed of the quality rules that the source code should abide by, and each rule defines various warning and error levels:

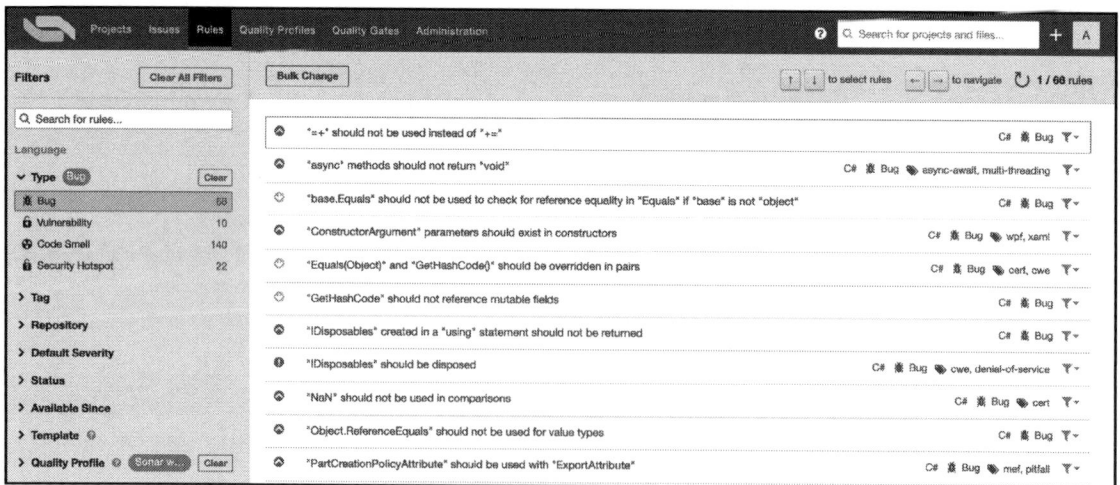

Using the quality profile, a so-called quality gate could be defined, identifying which type of change in source code would trigger gate failure, alerting the development team to possible issues. A quality gate is generally defined on the new code that is introduced into the repository within the leak period (that is, the period in which you calculate the *new* code); however, some aggregate values throughout the project can also be included.

Here, it is important to mention that SonarQube uses a Git extension to access the source code revision history and annotates the code tree so that the code delta and the owner of the commits can be easily identified. A simple quality gate might look similar to the following:

Metric ❓	Operator	Error
Coverage on New Code	is less than	80.0%
Duplicated Lines on New Code (%)	is greater than	3.0%
Maintainability Rating on New Code	is worse than	A
Reliability Rating on New Code	is worse than	A
Security Rating on New Code	is worse than	A

Execution of SonarQube analysis can take place during the CI phase, as well as the CD phase. Whilst the Azure DevOps SonarScanner extension provides a convenient integrated build and analysis experience, SonarLint and the associated Roslyn analyzers provide insight and assistance to the developers within the Visual Studio IDE.

Local analysis with SonarLint

SonarLint is a Visual Studio extension that allows developers to bind a local project to the designated SonarQube server and its associated quality profile. Once the source is associated with a SonarQube project, it downloads the ruleset and these rules are applied to the source using Roslyn analyzers, providing a fully integrated editor experience with highlighting and issue solution options.

Using SonarLint together with SonarQube allows the central management of coding conventions and rules and helps to maintain the code quality within bigger development teams. While the rule definitions are provided by the aforementioned analyzers, the severities defined by the quality profile are included in the project using the ruleset files:

```xml
<?xml version="1.0" encoding="utf-8"?>
<RuleSet Name="SonarQube - App Sonar way" Description="This rule set was
automatically generated from SonarQube.
http://****.northeurope.cloudapp.azure.com:9000/profiles/show?key=cs-sonar-
```

```
way-35075" ToolsVersion="15.0">
    <Rules AnalyzerId="SonarAnalyzer.CSharp"
    RuleNamespace="SonarAnalyzer.CSharp">
        <Rule Id="S100" Action="Warning" />
        <Rule Id="S1006" Action="Warning" />

        <!-- Removed for brevity -->

        <Rule Id="S103" Action="Warning" />
        <Rule Id="S4027" Action="None" />
        <Rule Id="S907" Action="Warning" />
        <Rule Id="S927" Action="Warning" />
    </Rules>
</RuleSet>
```

These rules are periodically synced with the SonarQube server and can be combined with ruleset files that might be defined for other analyzers, such as StyleCop analyzers.

CI analysis

Once the developer commits their changes and creates a pull request, SonarScanner for CSharp can be executed using the Azure DevOps extensions available in the marketplace.

After the extension is installed on your Azure DevOps instance, the setup of the extension is quite straightforward. The initial step is to create an access token on a SonarQube server of your choice (that is, SonarQube or SonarCloud depending on the variant used), and create a service connection on Azure DevOps using this token:

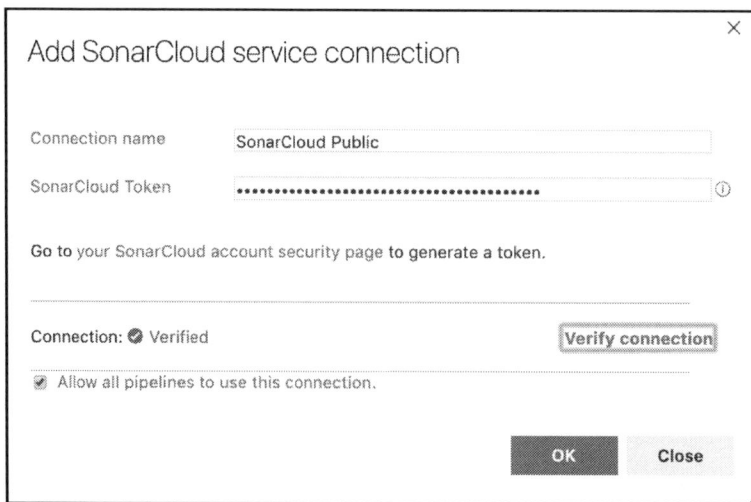

The integrated build tasks will then be included in the desired pull request validation build (or the CI build) as a pair of tasks: Prepare Analysis and Run Analysis.

The Prepare Analysis task downloads the required analysis configuration and prepares the integrated MSBuild execution targets. The Run Analysis task, on the other hand, collects the results that are gathered during the build execution and uploads them to the server. It is important to place the preparation task before any compilation takes place so that the Sonar configuration is ready while the compilation is being executed. A simple build sequence might look like the following:

Finally, the optional Publish Analysis task can wait for the analysis results and then publish them within the current pipeline.

> **TIP**
>
> .NET Core projects do not require a `ProjectGuid` property, unlike the classic .NET projects. However, the Sonar scanner uses `ProjectGuid` to identify the projects and execute analysis on them. In order to make sure the Sonar scanner can be executed successfully, the `ProjectGuid` property should be manually created on each `.csproj` file and set to a random `Guid`.

External Roslyn analyzers

In addition to the built-in set of analysis rules, SonarQube can also consume the warnings and errors that are identified by other Roslyn analyzers, such as the available StyleCop analyzers.

In order to include StyleCop rules into a .NET Core project, it is enough to reference the publicly available NuGet package:

```
<ItemGroup>
    <PackageReference Include="StyleCop.Analyzers" Version="1.1.118"
PrivateAssets="All" />
</ItemGroup>
```

At this point, the coding convention-related issues would be identified and flushed through the build output as warnings. Additionally, the IDE would provide annotations and solutions using the Roslyn infrastructure.

Finally once the project is put through SonarQube server analysis, the issues identified by `StyleCop.Analyzers` would also be stored and included into the quality gate calculations. For instance, the following issues are identified by StyleCop rules but are included in SonarQube:

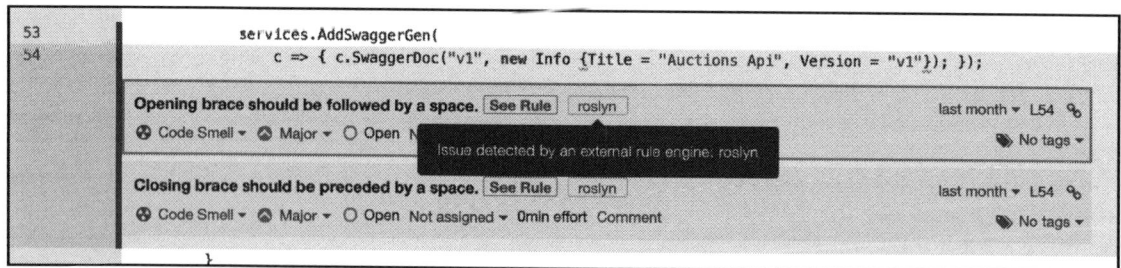

Overall, SonarQube provides a complete code quality management platform that, coupled with Azure DevOps and .NET Core, provides an ideal automated development pipeline and secure the CI process.

Creating and using release templates

As previously discussed, once the CI is complete, published build artifacts should ideally be transferred into the release pipeline, starting the CD phase. Azure DevOps release templates and infrastructure provide a complete release management solution, which, without the need for any additional platform such as Jenkins, Octopus, or TeamCity, can handle the CI/CD pipeline.

Azure DevOps releases

A release definition is made up of two main components: artifacts and stages. Using triggers and gates, the deployment of artifacts to target stages is organized and managed.

Release artifacts

Release artifacts are the elements that provide the components for the release tasks. These artifacts can vary from simple compiled application libraries to source code retrieved directly from the application repositories:

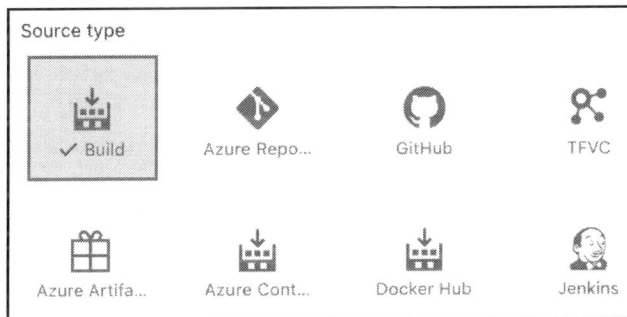

Let's take a closer look at these artifact types:

- **Azure Pipelines**: This is the most commonly used artifact type, which allows the build pipelines to pass the compilation results as packaged components to the release pipeline. Using this artifact type allows the release pipeline to detect the work items that are introduced with the artifacts, creating a direct relationship between project work items and release details. The creation of the new version of an artifact can be used as the trigger for the release.

- **TFVC, Git, and GitHub**: If static content from the source code repository, such as configuration files or media content, or the source code itself, are required for the release pipeline tasks, various repositories can be used as artifacts. Incoming commits to the repository can be used as triggers for the release.
- **Jenkins**: If multiple build and release pipelines are involved in a setup, a service connection can be created for a Jenkins deployment and Jenkins build artifacts can be consumed by Azure DevOps release pipelines.
- **Azure Container Registry, Docker, and Kubernetes**: When dealing with containerized application packages, the images prepared can be pushed to a private container registry and these images can then be retrieved during the release process.
- **Azure Artifacts (NuGet, Maven, and npm)**: Azure Package management artifacts can be retrieved and used to trigger new releases using this source, allowing various packaged components to be included in the release pipeline.
- **External or on-premises TFS**: On-premise TFS infrastructure can also be included in an Azure DevOps release pipeline. In order for this type of integration to work, the on-premise TFS server should be equipped with an on-premises automation agent.

Additional artifacts such as TeamCity can be introduced into release pipelines using the Azure DevOps extensions available on the market place.

In our application pipelines, we would be using the build artifact type that will contain the ARM definition, web API service packages, and the multiple configuration application packages for various environments.

Release stages

In layman terms, **release stages** roughly translates to the environment in which we want the application to be deployed. It is important to emphasize the fact that a stage is only a logical container and does not need to refer to a single server environment. It can refer to various environment infrastructures, as well as a managed distribution ring for mobile applications.

A release stage contains the release jobs that will be executed on release agents. For instance, if we are deploying a build artifact to Azure Stack or a mobile application package to App Center, we can use an agent job on a hosted agent. Nevertheless, if the deployment target is an on-premise server, we would need to use a specific deployment agent or a deployment group.

As mentioned, a release stage can contain multiple release jobs, which can be executed in parallel or sequentially, depending on the dependencies between the components. For instance, in order to deploy the iOS component to the QA distribution ring simultaneously with the Android or UWP packages, we can utilize parallel agent jobs that would select the specific artifacts to download and release. Each job can define which specific artifact it requires. Another example for a multi-job release setup could be a microservice package deployment setup where each service is deployed independently:

In this example, the API deployments could have been configured so that the services are deployed to related to app services only after the main ARM deployment takes place.

Release gates and triggers

In an Azure DevOps release, the transition between stages is controlled by triggers and gates. The release train, as well as manual or external service gates, can be configured using these components.

The main trigger for a release is set up through the artifacts introduced in a release. As mentioned, a build artifact can be set up to trigger a new release with each new version.

The following trigger will be executed every time a build artifact is created from a source branch matching the given wildcard expression (that is, `/release/*`):

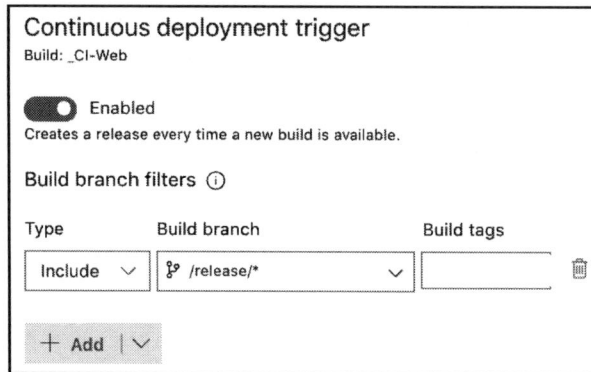

This scenario can be extended to include build tags and additional `exclude` expressions.

On top of the main release trigger, each stage can define a separate trigger. These triggers can refer to the actual release trigger or the completion of another stage. Manual deployment is also included to separate a stage from the main release train. The following trigger defines the QA stage as the trigger for the UAT stage, chaining the two releases:

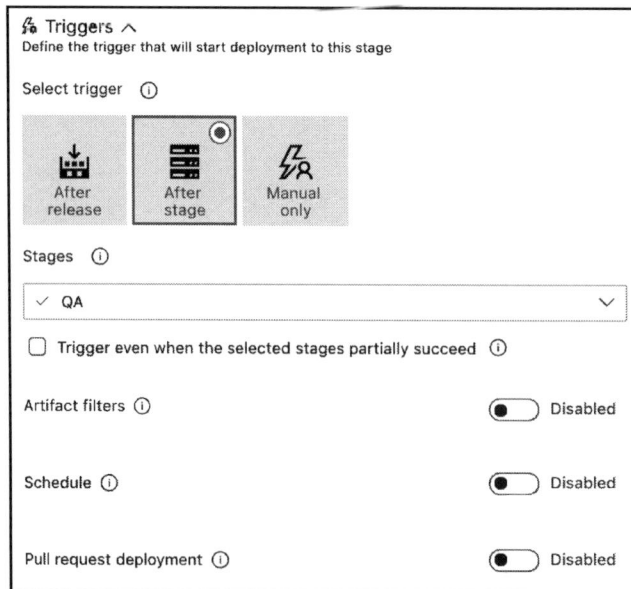

This automatic transition between the stages can be set to expect input from a specific user (manual approval) or an external server. These so-called gates can be defined as pre-deployment or post-deployment gates, with one verifying the availability of an environment to receive the release, and the other verifying the success of the deployment.

Most common application of gates is the manual approval pre-deployment configuration for higher environments, so that on going testing or actual public web application is not jeopardized. Manual approval can be done by any user within the Azure DevOps organization. Approval can be set to expire after a certain time and the selected approvers can delegate to other users.

External gates can be various service endpoints such as Azure Functions, external web services, or even a custom work item query within the same Azure DevOps project:

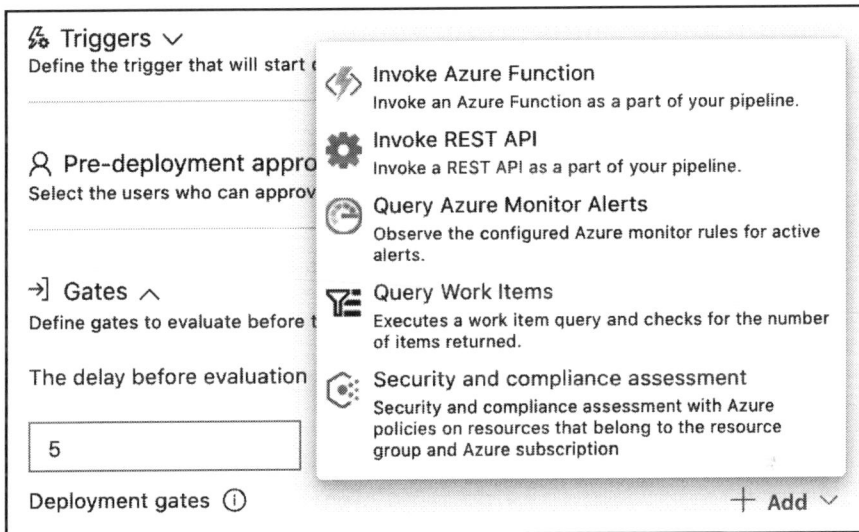

Using the intrinsic trigger and gate capabilities, complex release workflows can be set up and executed on demand or in an automated manner.

Xamarin release template

In the Xamarin release pipeline, we would be receiving multiple application packages for multiple platforms and environments as build artifacts. For instance, consider the following CI build setup for Xamarin Android, in which we would receive three packages for QA, UAT, and PROD, respectively:

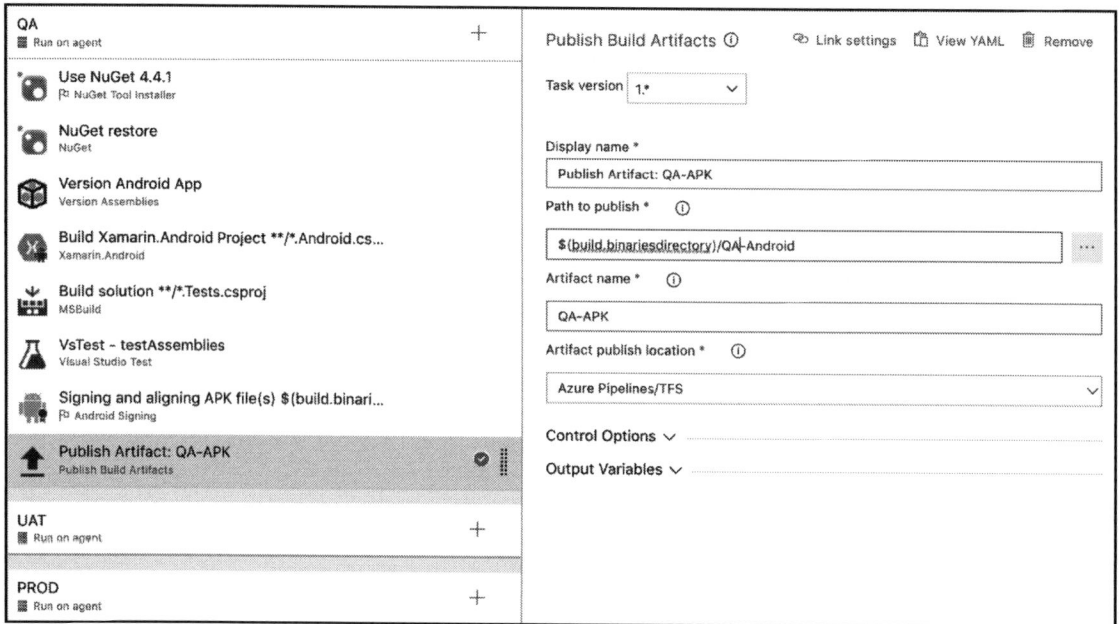

If we were to create a similar multi-agent build for iOS and set these builds to trigger on incoming commits to any release branch, we would be creating a new application for each deployment environment with every new push:

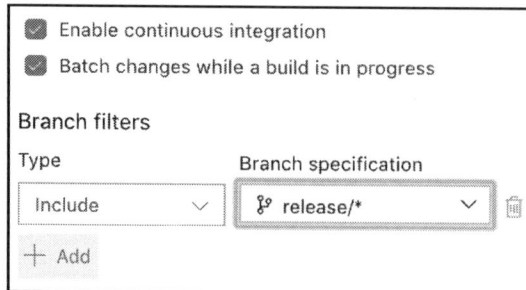

The release pipeline for the App Center releases can now reference these artifacts and deploy them to a specific App Center ring. App Center is capable of pushing the application package to target the App Store, so we can create a production ring on App Center and deploy the package to production from App Center.

The deployment to parallel rings for different mobile platforms can be parallelized as parallel jobs or as parallel stages converging on synchronization stages:

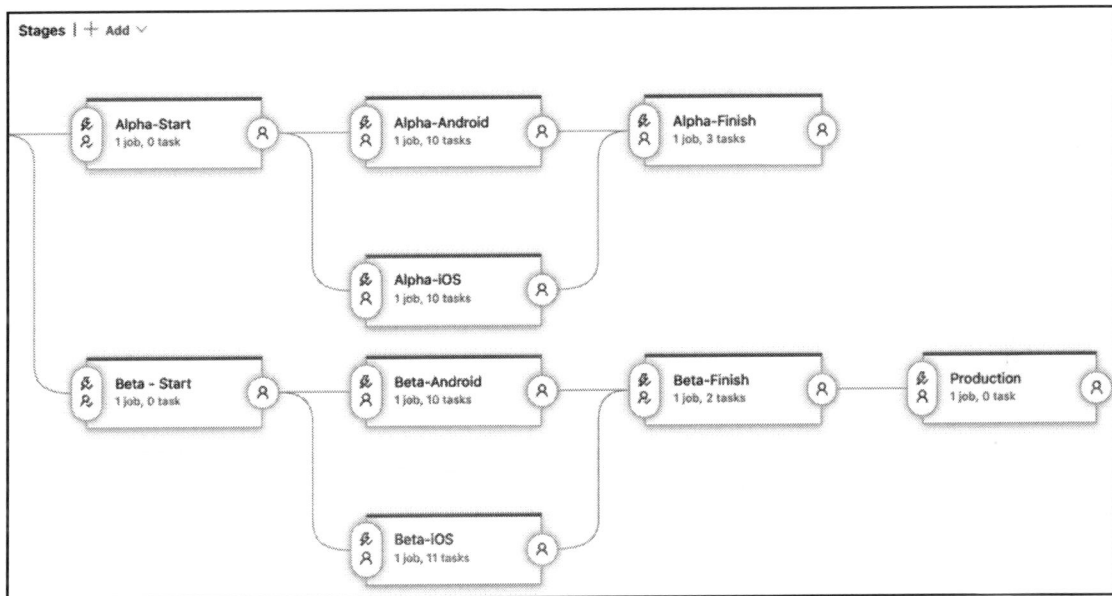

Gates on the synchronization stages (for example, **Beta-Start** and **Beta-Finish**) can be used to control the deployment to certain distribution rings.

Azure web application release

For Azure infrastructure, we would also be receiving multiple packages. The foremost important package of the Azure deployment pipeline is the ARM template and the associated configuration parameters files defining our application configuration. These resources can be retrieved directly from the source repository or they can be packaged during the CI build using the basic utility tasks used to copy files and package them (optionally) in a ZIP container.

> Another useful tool that can be used during the CI build is the validation mode of the Azure Resource Group Deployment task. This way, the CI build can validate the ARM template changes introduced at least against one of the available environments.

The API services, depending on the hosting option selected (that is, containerized, or packaged as a web deployment, and so on), would also be created as deployment artifacts.

The release pipeline would then deploy the ARM template to the target resource group(s). Once the resource manager deployment is completed, the web application packages can be released to the target app service instances or app service slots, depending on the deployment strategy.

Similar to the Xamarin deployment, the release pipeline can be configured to use multiple stages to define an environment or multi-agent stage with multiple deployments. For instance, a sample release pipeline with deployment components separated into stages would look like the following:

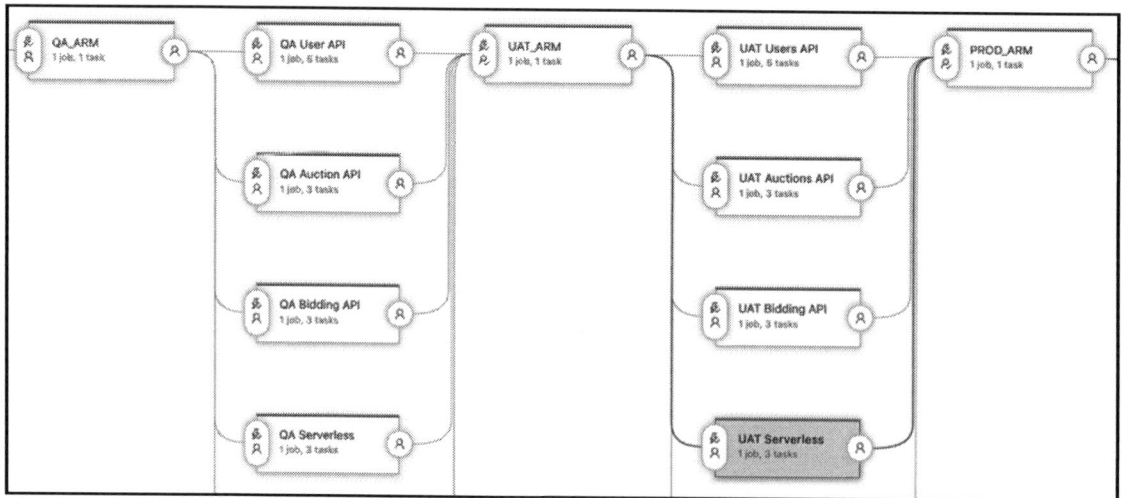

In either scenario, the release management flow should be the same: deploy the infrastructure and configuration, then continue with service package deployments.

Summary

In this chapter, we completed the CI/CD pipelines for both the Xamarin repository and the Azure Web infrastructure. We have seen that the toolset offered by Azure DevOps is perfectly suitable for implementing a GitFlow branching strategy. This toolset is also used for managing the application life cycle by implementing branch policies and setting up CI triggers. Additionally, we have seen how the CI phase should also be used to maintain the code quality and technical debt. Finally, we discussed strategies for implementing release pipelines for both distributed Azure and native mobile applications.

With this final chapter, we have reached the end of the development of our project. In the beginning of the book, after refreshing our knowledge about various .NET concepts, runtimes, frameworks as well as platforms, we moved on to Xamarin development. We have created and customized Xamarin applications using the .NET Standard framework and Xamarin platform runtimes. Hopefully, we have learned where to use which type of customization on Xamarin.Forms framework. Once we have completed the Xamarin project, our focus shifted to the Azure cloud stack and how we can utilize .NET Core on various Azure services that can be used in junction with Xamarin mobile applications. The Azure stack discussion mainly focused on Platform as a Service offerings such as App Services, Serverless components and finally data storage services such as Cosmos DB. In addition, we have learned how we can engage our users in betters ways with the help of external services such as push notifications, Graph API and Cognitive Services. As we have slowly created the mobile application as well as the backend infrastructure, the last section was about how to effectively manage the lifecycle of our projects using Microsoft Azure DevOps and Microsoft Visual Studio App Center. Using Azure DevOps we have tried to realize modern DevOps concepts by creating automated CI/CD pipelines.

While the implementations and practical examples were quite generalized, the hands on concepts we have discussed throughout the book would be good starting point for any cross platform development project you and your team are planning to undertake. For any .NET developer, understanding .NET Core and other implementations of .NET Standard is the key to unlock multiple platforms and create user experiences that span across platforms.

Other Books You May Enjoy

If you enjoyed this book, you may be interested in these other books by Packt:

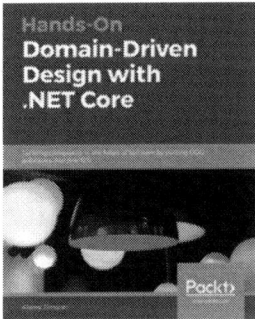

Hands-On Domain-Driven Design with .NET Core
Alexey Zimarev

ISBN: 978-1-78883-409-4

- Discover and resolve domain complexity together with business stakeholders
- Avoid common pitfalls when creating the domain model
- Study the concept of Bounded Context and aggregate
- Design and build temporal models based on behavior and not only data
- Explore benefits and drawbacks of Event Sourcing
- Get acquainted with CQRS and to-the-point read models with projections
- Practice building one-way flow UI with Vue.js
- Understand how a task-based UI conforms to DDD principles

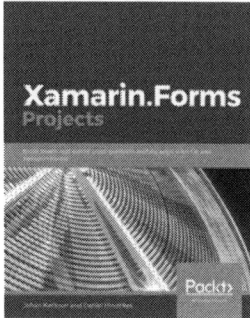

Xamarin.Forms Projects

Johan Karlsson, Daniel Hindrikes

ISBN: 978-1-78953-750-5

- Set up a machine for Xamarin development
- Get to know about MVVM and data bindings in Xamarin.Forms
- Understand how to use custom renderers to gain platform-specific access
- Discover Geolocation services through Xamarin Essentials
- Create an abstraction of ARKit and ARCore to expose as a single API for the game
- Learn how to train a model for image
- classification with Azure Cognitive Services

Leave a review - let other readers know what you think

Please share your thoughts on this book with others by leaving a review on the site that you bought it from. If you purchased the book from Amazon, please leave us an honest review on this book's Amazon page. This is vital so that other potential readers can see and use your unbiased opinion to make purchasing decisions, we can understand what our customers think about our products, and our authors can see your feedback on the title that they have worked with Packt to create. It will only take a few minutes of your time, but is valuable to other potential customers, our authors, and Packt. Thank you!

Index

Printed in Great Britain
by Amazon